Cate Fosl

Struggle for a Better South

STRUGGLE FOR A BETTER SOUTH

The Southern Student Organizing
Committee, 1964–1969

GREGG L. MICHEL

Cate —
Thanks for your support over the years — + for coming to S.A. to get this!

Gregg Michel

palgrave
macmillan

for Rhonda

STRUGGLE FOR A BETTER SOUTH
© Gregg L. Michel 2004

All rights reserved. No part of this book may be used or reproduced in any manner whatsoever without written permission except in the case of brief quotations embodied in critical articles or reviews.

First published 2004 by
PALGRAVE MACMILLAN™
175 Fifth Avenue, New York, N.Y. 10010 and
Houndmills, Basingstoke, Hampshire, England RG21 6XS
Companies and representatives throughout the world.

PALGRAVE MACMILLAN is the global academic imprint of the Palgrave Macmillan division of St. Martin's Press, LLC and of Palgrave Macmillan Ltd. Macmillan® is a registered trademark in the United States, United Kingdom and other countries. Palgrave is a registered trademark in the European Union and other countries.

ISBN 1–4039–6010–0 hardback

Library of Congress Cataloging-in-Publication Data
Michel, Gregg L.
  Struggle for a better South: the southern Student Organizing Committee, 1964–1969 / Gregg L. Michel.
    p. cm.
  Includes bibliographical references and index.
  ISBN 1–4039–6010–0
  1. Southern Student Organizing Committee (Nashville, Tenn.)—History. 2. College students—Southern States—Political activity—History—20th century. 3. Students movements—Southern States—History—20th century. I. Title

LA229.M46 2004
378.1′981′0975—dc22   2004044731

A catalogue record for this book is available from the British Library.

Design by Newgen Imaging Systems (P) Ltd., Chennai, India.

First edition: November 2004

10 9 8 7 6 5 4 3 2 1

Printed in the United States of America.

Contents

Acknowledgments		vii
Abbreviations		xi
Introduction		1
Chapter 1	White Southerners, Civil Rights, and Student Activism, 1961–1964	11
Chapter 2	Building an Organization: The Founding of SSOC	33
Chapter 3	Growing Pains	63
Chapter 4	SSOC and White Student Activists at Mid-Decade: The Agenda Grows	89
Chapter 5	Shifting Ground: From the Campus to the Community	107
Chapter 6	New Message, New Messengers	131
Chapter 7	Short-Term Gains, Long-Term Costs	167
Chapter 8	Falling Apart: The Dissolution of SSOC	189
Appendix	"We'll Take Our Stand"	227
Notes		229
Bibliography		297
Index		315

Acknowledgments

This book grew out of a conversation I had in the fall of 1992 at the University of Virginia with my dissertation advisor, Paul Gaston. I had entered Paul's office one afternoon to ask him if he knew anything about the Southern Student Organizing Committee. I recently had come across a brief reference to this 1960s group of progressive white students and was considering writing about the organization in his upcoming seminar on the twentieth-century South. Little did I know, Paul not only was familiar with the organization, but, as a young professor at Virginia in the 1960s, he had known many of the students involved in the group. As we talked that day, Paul encouraged me to begin preliminary research on the organization, and he suggested that, perhaps, the group would be a good subject for a long-term project.

Long-term, indeed. I find humor in the fact that the Southern Student Organizing Committee was born, lived, and died in less than half the time it has taken me to complete this project. Over these 12 years of researching, writing, and rewriting, I know I have confounded friends and family alike with my seeming obsession with the group. And while many former members of the organization were thrilled that I had undertaken this study, I am sure they eventually wondered what was taking me so long to finish. Now that I have, I am honored to thank the many people and institutions that have supported me throughout this endeavor.

The staffs at all the libraries I visited were generous with their time in helping me locate sources and ever-patient with my unending requests for more boxes, more files, and more photocopies. I especially wish to thank Michael Plunkett, Director of the Special Collections Library at the University of Virginia, for his assistance in establishing the SSOC archive, thus making Virginia the leading repository of extant SSOC materials. For their help in securing photographs, I thank David Pilcher of the Mississippi Department of Archives & History, Dee Anna Grimsrud of the Wisconsin Historical Society, and, especially, former SSOC activists Archie Allen and Tom Gardner, who generously allowed me to use photos from their personal collections.

When I began this study, I was uncertain if the individuals who had participated in the Southern Student Organizing Committee would welcome my efforts to dig around in their past. Happily, though, nearly everyone I contacted supported my efforts. They not only agreed to sit, in some cases, for very lengthy interviews, but they opened their homes to me, helped me locate hard-to-find people, and gave me access to critical and often rare SSOC materials in their possession. This book would not have been possible without their assistance. I particularly thank for their help Dorothy Burlage, Robb Burlage, David Doggett, John Duggan, Gene Guerrero, Ed Hamlett, Roger Hickey, Jack Sullivan, and Sue Thrasher. I also wish to acknowledge those individuals who have donated papers to the burgeoning SSOC archive at Virginia. David and Ronda Kotelchuck, Archie Allen, Bruce Smith, Steve Wise, Tom Gardner, Frank Goldsmith, Harlon Joye, Nelson Blackstock, Ruth Wells, Everett Long, Connie Curry, and Bill and Betsy Jean Towe have given the archive a solid base on which to build.

Several SSOC veterans deserve my special thanks. Two of the group's chairmen, Tom Gardner and Howard Romaine, helped to give my work credibility with others by fully supporting my research from its earliest days and by helping persuade many of their former colleagues to come to Charlottesville for the group's 1994 reunion. By inviting me into her home and sharing with me her recollections about the group, Lyn Wells allowed me to tell the story of the person at the center of many of the group's actions and controversies. Lastly, I am indebted to David Nolan for his sharp memory and his patient, sympathetic replies to my frequent queries about specific events in the organization's history. And his wonderfully rich and detailed tours of St. Augustine, Florida, are among my fondest memories of my research.

The University of Virginia offered a nurturing and supportive environment in which to develop this study. Stipends from the Graduate School of Arts and Sciences allowed me to present my work at several conferences, while the Department of History's Southern History program helped to finance several research trips. My colleagues in the Southern History program, including Scot French, Juliette Landphair, Matt Lassiter, Andy Lewis, Andy Myers, Anne Sarah Rubin, and Phil Troutman critiqued drafts of several chapters and provided me with numerous helpful suggestions.

I completed this book at the University of Texas at San Antonio, where I have found an equally supportive community of scholars and friends. A Faculty Summer Research Grant from the College of Liberal and Fine Arts enabled me to make a final research trip to several important archives in Mississippi. Wing Chung Ng, the History Department chairman, took deep interest in my work and modeled how to balance teaching and administrative responsibilities with one's own research. I always appreciated his frequent reminders to carve out time from the daily grind of the academic semester to work on the book. Stacy Townsend in the History office provided timely assistance with assembling the book's photos, and Patrick Murphey helped compile the index. At a crucial stage in the revision process, my History Department colleagues Kirsten Gardner,

Kolleen Guy, and Patrick Kelly offered compelling and trenchant criticisms of the introduction. Their questions and their insights have made this important section of the book stronger. Any remaining errors of fact or interpretation here or elsewhere are, of course, my responsibility.

I have had the privilege to work with two fine editors at Palgrave. Debbie Gershenowitz was a strong advocate of this work from our first meeting, and her close reading of an early draft of the manuscript gave me a fresh perspective on its strengths and weaknesses. As the book neared completion and my energy level flagged, Brendan O'Malley helped nudge me across the finish line with skill, wit, and grace. His words of encouragement and praise, often delivered precisely when I most had wearied of the book, reenergized me and helped me bring the process to a satisfying conclusion.

My deepest intellectual debt is to the scholars I worked with at Virginia. Their advice and guidance have improved this work in countless ways. Reginald Butler supported this project in its earliest stages and encouraged me to continue working on it even as I embarked on new research during a postdoctoral fellowship at the Carter G. Woodson Institute for Afro-American and African Studies. Julian Bond expressed a keen interest in my work early on as well, and he helped me make contact with other movement veterans. Nelson Lichtenstein was as enthusiastic about this project as I was, and discussions with him always forced me to reconsider my assumptions and to ask new questions. The late Bill Ellwood brought his knowledge of the movement at Virginia to bear on the dissertation, and his perceptive reading of it helped me begin to transform the dissertation into a book.

I owe special thanks to two individuals at Virginia in particular. Ed Ayers has been a mentor and friend since I first met him in 1988 as a new graduate student interested in studying the Civil War. Although my historical interests evolved, Ed remained a staunch ally who offered invaluable insights and criticisms as I completed the dissertation. He expressed faith in my abilities even when I doubted myself, and I valued enormously his support and encouragement. Ed's cheerfulness was the perfect antidote to the trials and tribulations one experiences as a graduate student, and his excitement about history proved contagious. As my advisor, Paul Gaston did more than anyone else to help me grow as a historian. Through his teaching and his scholarship, Paul exposed me to a view of southern history that stressed contradiction, dissent, and subversion. Like so many others who studied with him, I learned from Paul that an honest reckoning with the southern past can help to make the South of today a better place. Since our first conversation about SSOC, Paul has been my sharpest critic and greatest advocate. His thorough engagement of my work has made this book far better than it otherwise would be.

My family has sustained me throughout this project. Though perhaps a bit mystified as to how my Southern California upbringing led to a career studying the American South, my parents, Barbara and Stephen Michel, and my sister, Randi Michel, have supported this endeavor as they have every other I have undertaken. For that, and their unshakable faith in me, I am eternally grateful. My children, Halle Malkah and Jeffrey Abe, have had to put up with me working on this study for their entire lives. In the book's final stages, especially, they showed incredible tolerance as I spent yet another weekend afternoon working on my book instead of attending to the more serious and important business of playing basketball or riding bikes with a five-year-old and a two-year-old. These remarkable children have changed my life and taught me the true meaning of unconditional love. It is to Rhonda, their mother and my wife, that I dedicate this book. Rhonda has lived this project with me from its first day. Countless times she boosted my spirits when they were low and gently reminded me that the book was not yet done on the occasions when I seemed a bit too full of myself. Most of all, Rhonda has reminded me that there is so much more to life than reading and writing about the past. She is, and always will be, the only story I tell.

Abbreviations

AWS	Associated Women Students
COFO	Council of Federated Organizations
CP	Communist Party
ILGWU	International Ladies' Garment Workers Union
MFDP	Mississippi Freedom Democratic Party
NSA	National Student Association
PL	Progressive Labor Party
RYM	Revolutionary Youth Movement
SCEF	Southern Conference Educational Fund
SCHW	Southern Conference for Human Welfare
SDS	Students for a Democratic Society
SGER	Student Group for Equal Rights
SLAM	Southern Labor Action Movement
SNCC	Student Nonviolent Coordinating Committee
SRC	Southern Regional Council
SSA	Students for Social Action
TWUA	Textile Workers Union of America
VCHR	Virginia Council on Human Relations
VSCRC	Virginia Students' Civil Rights Committee
YAF	Young Americans for Freedom

Introduction

> This is an historic moment. Someday historians will write about it.
> —Howard Zinn at first SSOC conference[1]

> See, the history of SSOC is never going to be recaptured.... [T]he papers were burned, and... I think the papers would have documented all this stuff.... I've never liked the fact that there's not an archive of this organization I spent a lot of time working for.
> —Sue Thrasher[2]

Ed Hamlett stood crying at the podium, unsuccessfully trying to contain his emotions. The historic room on the University of Virginia campus fell silent. Some of the middle-aged men and women in the audience, who had gathered on a bright spring morning in 1994 for a reunion of the Southern Student Organizing Committee, began to sniffle, and several dabbed their eyes. "I know what brings the tears. It's whether I can say it or not," confessed Hamlett, one of the group's founders. "So much of our culture has been kept from us," he sobbed. "Part of what held us together was learning about those who'd gone before. And they were not insignificant. They were not small in number.... People worked together," he exclaimed through his tears.[3]

It was a poignant moment. The people of whom he spoke were whites in the South who had supported progressive causes and experienced hostility and isolation for their efforts. Hamlett and the others in the room were among them. In 1964 they created SSOC, a Nashville-based organization of young whites, primarily students, which worked for progressive change. The group initially focused on civil rights. Before long, though, it transformed itself into a multi-issue organization, which, in addition to advocating black equality, organized opposition to the Vietnam War and the draft, encouraged challenges to restrictive in loco parentis policies on southern campuses, supported the women's liberation movement, and drew students into interracial organizing campaigns among southern industrial workers. The students of SSOC were neither the first nor the only whites in the region to speak out on these issues. But in the 1960s their activities made SSOC the most important movement organization of young whites in the region. Hamlett's tears suggested both the sense of kinship that bound progressive whites in the South across generations and the scars that many still bore from their activism.

The 1994 reunion at Virginia, which was a center of SSOC activity, took place 30 years to the week after the organization's founding and nearly 25 years after its collapse in 1969 amid bitter internal discord. The gathering aroused a wide range of conflicting emotions: joy and anger, gratitude and resentment, pride and regret. For many who attended, the reunion was the first opportunity to reconnect with former colleagues since their days working together. Some approached the gathering with trepidation about meeting those to whom they once had been close but had not seen in more than two decades. Others were eager to reestablish relations with old friends and colleagues. A few worried that they would not recognize or remember old comrades or, worse, that old comrades would not recognize or remember them. The reunion was also the first time in decades that many SSOC veterans had paused to reflect on their involvement in the organization. As they reminisced about the group's activities, reminded each other of frightening or funny incidents, and sang the songs that had galvanized them to action, it was obvious that their work with SSOC remained a deeply meaningful and influential part of their lives.

The conversations at the reunion revealed that the former activists passionately believed that SSOC had helped to make white southerners more receptive to social change. The Southern Student Organizing Committee was the only region-wide group in the 1960s that made a sustained attempt to organize white southern support for progressive movements. Throughout its five-year existence, SSOC was active at a broad cross-section of predominantly white southern colleges and universities—public and private, parochial and secular, large research institutions and small liberal arts colleges. It also reached out to white southerners in places as diverse as inner-city Nashville, the Mississippi Gulf Coast, and North Carolina textile mill communities. SSOC activists challenged the notion that the white South was a monolith of opposition to progressive causes. They rejected the suggestion that the group itself was an aberration, an oasis of progressive thinking in an otherwise sterile and reactionary region. In their view, white southerners would become involved in, or at least supportive of, progressive movements if approached by the right people—other whites in the South—and in a way that emphasized moderation, not confrontation. Southern distinctiveness, the idea that the South was a fundamentally unique region, shaped the organization's goals and agenda and attracted whites to the group. In its short life, SSOC's regional orientation and well-developed southern identity enabled it to win more support from students on predominantly white southern campuses than either the black-led Student Nonviolent Coordinating Committee (SNCC) or the northern-dominated Students for a Democratic Society (SDS), the leading activist youth organizations of the decade.

The sense of triumph that permeated reunion conversations about SSOC's work gave way to anguished discussions when the SSOC veterans turned their attention to the personal costs of social activism. Becoming involved in activist causes was not a decision whites in the South made lightly. Stepping outside the "magic circle" and rejecting the dominant beliefs of the white South required great courage. SSOC students faced the sobering reality that their activism could lead to loss of friends, condemnation and rejection by one's family, and expulsion from school. Even in the mid-1990s, the consequences of their activism continued to reverberate in many of the activists' lives; some only recently had reconnected with their families, while a deep rift continued to separate others from siblings and parents.

The willingness of SSOC activists to risk such severe consequences distinguished them from their fellow students in the South. Most college-age whites in the region opposed the causes that inspired SSOC students' activism, and those who shared the beliefs of SSOC members typically chose not to involve themselves in organized movements. Among the young whites in the South who became activists, a few gravitated toward SNCC, since it was the preeminent southern student group and, prior to 1966, welcomed white support. SDS, with its reputation for radicalism and deep ideological discussions, appealed to a minority of others. SSOC, on the other hand, drew support from young whites who sought to take the message of progressive change into white communities and onto white campuses. They were undeterred by the enormity of the challenge they faced and undaunted by the prospect of disrupting their personal lives. Will Campbell, the well-known minister and civil rights activist who was close to many SSOC activists, tried to explain the activists' motivation. "Why don't they join the SNCC or some other militant group?" Campbell rhetorically asked in 1964. "Because they want to give those from whose loins they sprang one more chance. Because they are social reformers and not revolutionists. Or just because they are white. But never because they are scared."[4] Energized by the hope that their work among southern whites could create a South free of poverty, racism, and oppression, the students of SSOC boldly enlisted in the progressive movements of the day.

As the reunion neared its end, the former activists engaged in an animated and spirited debate about the circumstances of the group's demise. No aspect of the group's history was as complicated or as divisive—both at the time of the break-up and in the decades since. SSOC veterans avoided the issue at the reunion for as long as possible. "It took a day-and-a-half to even begin to broach" the subject, Ronda Kotelchuck remembers, "because you knew where the sensitivities were and where the pain was." Not until the final morning, she recalls, did the assembled veterans of the group "edge up on the pain and the anger" about SSOC's collapse.[5] The ensuing discussion revealed not only that they held widely divergent views of what happened to the organization but that many individuals remained resentful and even angry about its fate, blaming one faction or another for its dissolution or accusing certain leaders of wanting to kill it. Significantly, the SSOC activists who were blamed for the demise of the group at the reunion did not attend the event. While schedule conflicts prevented some of these individuals from participating in the reunion, others had no interest in attending an affair at which their rivals of old would be present and at which they thought they would be uncomfortable or not made to feel welcome. For both those who attended the reunion and for those who stayed away, the hurts, wounds, and disappointments of 25 years earlier continued to fester.

Why was it that, as SSOC's Nan Orrock remarked at a more recent gathering of the activists, "people who'd faced death together couldn't even talk together" by 1969, the year the group disbanded—or, for that matter, by 1994?[6] Orrock's comment reveals a key point about SSOC's death: divisions among SSOC activists had become so acute by 1969 that few had the will, desire, or interest to keep the group alive. Although SSOC had maintained a regional orientation and promoted the view that the South was unique, a growing number of the activists, including several key leaders, concluded that the group had come to place too much emphasis on the ways in which the South differed from the rest of the nation. This notion, they argued, obscured similarities in the problems faced by northerners and southerners alike and wrongly suggested that SSOC should remain apart from national New Left organizations, like SDS. The group's reliance on the symbols and rhetoric of the Old South and the Confederacy to suggest the distinctiveness of the South proved especially problematic. Equally debilitating was the growing tension among the activists over whether SSOC needed to develop a clear ideological framework to shape its actions. The group never had a sharp analytical focus. Throughout its existence, a general vision of a more humane, non-racist, and peaceful society undergirded SSOC's activism. By 1969, however, some SSOC activists had begun to argue that the group needed to ground its work in a coherent radical ideology and develop its own critique of American society.

Differences over southern distinctiveness and SSOC's ideological orientation created lasting divisions in the group. It also earned SSOC the enmity of SDS. During SSOC's last year, SDS vigorously and relentlessly attacked the group as insufficiently radical and misguided in its focus on the South. To the more intellectually sophisticated members of SDS, with their position papers and formal ideologies, SSOC's broadly conceived vision of a better society contrasted sharply with their highly developed Marxist analysis of America. And SSOC's emphasis on the uniqueness of the South appalled SDS members who instead viewed the South from a national and international perspective. In the spring of 1969, as SDS experienced its own internal divisions, the competing factions within the northern group united to break fraternal ties with SSOC. Abandoned by SDS and wracked by factionalism, SSOC was unable to mount an effective defense against the SDS assault. In June 1969, the group voted to disband.

This book tells the story of the Southern Student Organizing Committee, from its birth, when it embodied its founders' idealism and hope for racial reform, to its demise amid the factionalism, inter-organizational warfare, and shattered dreams of building a better South. Although the book's focus is primarily SSOC's organizational history, the story I tell is built on the premise that understanding the organization requires an understanding of the people who comprised it, their desires and doubts, ideals and prejudices, personal histories and life aspirations. In this sense, SSOC is a prism through which to explore the world of young white activists in the 1960s South. The book strives to capture the spirit—the sense of promise and possibility, optimism, and idealism—that infused this unique movement group. It examines not only the activists' work within SSOC but their occasionally overwhelming fears of failure, their fervent hope that white southerners would be receptive to their efforts, and their fundamental belief that the individual could be an agent for change.

Although SSOC reached out to both student and non-student whites, it had its greatest successes at the South's predominantly white colleges and universities, where it helped to mobilize student support for a variety of national, regional, and local progressive causes. Especially significant was the group's ability to activate students on issues particular to their campuses, from efforts to safeguard student free speech rights to campaigns to end sexism on campus. Throughout the region, SSOC students formed the activist vanguard on many campuses; they were the first outspoken proponents for these causes and helped to popularize them with other students. SSOC created space for discussion of politically relevant and controversial issues on many southern campuses. By the late 1960s, a wide range of students, including those who previously did not self-identify as activists, and student groups, including student political parties, long the bastion of Greek conformity, moved into these spaces and became the champions of liberal reforms on southern campuses. In short, SSOC helped to make dissent respectable on southern campuses; it contributed to bringing progressive concerns into the mainstream of student life.

Scholars of the 1960s know well that evaluating a movement organization's effectiveness or tangible successes is a task fraught with difficulty. Few, if any, movement groups accomplished their stated goals. What historian Jennifer Frost has called "the mix of frequent defeats and limited achievements" reflects a sense of failure.[7] Certainly, when judged by its own rhetoric, SSOC was a failure; the New South it spoke so frequently about—a South free of the bigotry and poverty that had marred the region's past—seemed just as distant in 1969 as it had in 1964. Yet SSOC's inability to achieve its lofty goals obscures the significance of its survival for more than five years in the harsh and repressive atmosphere of the South. The group's dissolution in the face of internal problems and external pressures highlights the obstacles it had managed to surmount until 1969, making its existence all the more remarkable. That SSOC died broken and divided is less significant than the fact that it lived as long as it did.

Moreover, if SSOC was not able to remake the region, it did help to transform the South's predominantly white colleges and universities. The group did not merely give white students a vehicle by which they could work for change in the wider society. Just as importantly, it provided them with the means, the rationale, and the encouragement to challenge the conformist culture at their schools and the restrictive and confining policies that circumscribed student life and defined their educational experiences. Through its organizing and advocacy work, SSOC played a pivotal early role in the effort to democratize the South's predominantly white campuses, to make them more open, diverse, and modern institutions of higher learning. As activist and historian Howard Zinn has observed, social changes occur "not because of a miracle from on high, but because people have labored patiently for some time." SSOC no longer existed when change occurred on many southern campuses. But its work on these campuses provided the grassroots support that made the broad-based, lasting reforms of later decades possible.[8]

Because it was an organization of young whites working on a wide variety of progressive causes in the 1960s South, SSOC bridges the history of the 1960s, the civil rights movement, and the New Left. Nevertheless, most scholars of this era overlook the group or offer brief and frequently inaccurate descriptions of it. Kirkpatrick Sale and Clayborne Carson, authors of the most thorough histories of SDS and SNCC, respectively, each suggest that SSOC's roots were in the organizations they study. Sale argues that SSOC was "an offshoot of SDS set up in 1964 to carry on SDS-style politics with a slightly gentler Southern accent," while Carson posits that SSOC was founded in order to be "a southern white student counterpart to SNCC." Historians writing in subsequent years have, if they speak of the group at all, tended to echo Sale's or Carson's view. Alice Echols labels SSOC "an SDS offshoot" in a footnote in her account of the radical feminist movement. In his work on the University of North Carolina, William Link ignores the "Southern" in the group's name to explain that SSOC was "a national organization established in 1964 as a white offshoot of the . . . Student Nonviolent Coordinating Committee."[9] To other commentators, SSOC was noteworthy only because it was a novelty. Sociologist John Shelton Reed contends that SSOC was "a group that sixties trivia buffs may recall as a sort of regional affiliate of the Students for a Democratic Society. . . . As I heard it, their SDS comrades from the Northeast and West Coast . . . just couldn't handle it [SSOC's regionalism], and drummed SSOC out of the movement." Not only is Reed's rendering of the group's history shallow, inaccurate, and seemingly based on rumor, but by relegating it to trivia status he suggests that SSOC is unimportant except to those who keep track of the obscure.[10]

Such contradictory views and uninformed representations of SSOC are only partially corrected by the few scholars who have studied the group. These works, though, suffer from their own set of historical problems. Sara Evans incorporates SSOC into her study of the origins of the women's liberation movement of the 1960s. But Evans is concerned with only the women of SSOC—the path they took to social activism, their relationship with men in SSOC, their experience as women in the movement—and thus leaves unexplored much of the group's history. Bryant Simon provides a more thorough treatment of SSOC in his undergraduate honors thesis at the University of North Carolina at Chapel Hill. While the essay is an excellent primer on the organization, it is primarily a narrative account of the group's history.[11] Movement veteran Harlon Joye and historian Christina Greene, in essays published 24 years apart, present brief but thoughtful and informative overviews of SSOC's rise, short life, and collapse. Joye and Greene focus particular attention on the group's difficulties and failings, and thereby flatten and compress SSOC's history. SSOC merits the critical evaluation set forth in their essays, but also is worthy of this book-length treatment that reveals the complexity and richness of the organization's history.[12]

Evans, Simon, Greene, and Joye separate themselves from other scholars by giving serious attention to SSOC. That most historians of the 1960s South and the social movements of the decade ignore SSOC reflects their general disregard for white student activists in the South. By glossing over or bypassing the activism of young whites in the region, their scholarship perpetuates the image of the white South as backward and benighted, unwilling to reform itself and unhesitant to use violence to preserve the "southern way of life." For these scholars, white students in the South are not part of the story of the progressive movements of the era. They appear in their works, if at all, only as the enemies of change, rabble-rousers and thugs who the movements had to overcome. Hence, histories of the antiwar, women's liberation, and university reform movements focus almost exclusively on the activism of northern white students.[13] Likewise, the civil rights movement is told as the story of southern black heroes and northern white missionaries who suffered the scorn and abuse of southern whites in an effort to reform the region's racial order.[14] Of course, northern whites and, especially, southern blacks *were* the agents of change in the South. But by disregarding the contributions of white southern students to the civil rights struggle and lumping all white southerners together as a people unified by their devotion to segregation and, more generally, the status quo, this scholarship simplifies the tale of the movements of the era, denies legitimacy to white dissenters in the South, and leaves the erroneous impression that in the 1960s all whites in the South were on the wrong side of history. Young whites in the region were not monolithically predisposed to oppose progressive movements. The life of the Southern Student Organizing Committee is testimony to this fact.[15]

A Note on Sources

The epigraphs that open this chapter highlight the challenge of writing about the Southern Student Organizing Committee more than 30 years after its demise. SSOC's organizational records do not survive. One year after SSOC disbanded, several former members made a bonfire of the records because they feared that the files, which included rosters of supporters, funding sources, and other sensitive information, would fall prey to government subpoenas or theft by enemy organizations. The loss of the organization's records has helped to render SSOC invisible to scholars and to exacerbate its marginalization in the historical literature of the era.

When the papers were destroyed, many records were lost forever. But an astoundingly large and varied number of sources survive: correspondence from SSOC leaders resides in the collected papers of SNCC, SDS, and other activist groups; budgets and planning documents are part of archival collections of some of SSOC's financial supporters; and the many SSOC publications, from its newsletters to the numerous pamphlets it distributed, survive in alternative newspaper collections, university libraries, and former activists' homes. This last location proved to be particularly important for the research for this book. Numerous individuals have maintained their own personal collections of SSOC materials, ranging from a few items stuck in folders in the back of drawers to reams of papers stored in boxes in basements or closets. Membership information, SSOC publications, meeting minutes, letters from members around the region, carbon-copies of outgoing correspondence, detailed information on individual chapters—all this emerged from individuals' homes. Through the generosity of these former members, these materials now form the basis of the recently established SSOC archive at the University of Virginia. As the central repository for SSOC materials, the archive will ensure their preservation as well as their use by future generations of students and scholars.

The written materials offer an important avenue into the group's past. So too do oral histories. Interviews I conducted with SSOC activists, supporters of the group, and activists in other student organizations provided a wealth of information unavailable in the written sources: details about specific SSOC programs and actions; personal histories of the former activists; information about the birth, life, and death of SSOC chapters and the form and content of student activism on particular campuses; and, most importantly, a sense of the experiences shared by white students involved in social activism in the South in the 1960s.[16] Interviews, of course, are not a trouble-free source; interviewers must be especially mindful of their sources' biases, interest in distorting facts, and faulty memories. For the SSOC interviews, the vagaries of individuals' memories was an especially important issue because the interviews focused on events that had occurred 25 or 30 years in the past and for which people's memories were admittedly hazy. Whenever possible, therefore, I corroborated their recollections with written sources or the recollections of others. Despite the challenges the interviews posed, their value as first-hand accounts outweigh their potential pitfalls.[17] Unlike dry membership lists or tedious grant proposals, the interviews shed light upon the emotions and feelings of the individual activists as they recounted some of the most formative and traumatic events of their lives. Together, the interviews and the contents of the SSOC archive, both new source materials, have created this opportunity to tell SSOC's story.

Chapter 1
White Southerners, Civil Rights, and Student Activism, 1961-1964

The dilemma that is presented here is perhaps the dilemma that faces every Southern white person who becomes deeply committed to the integrationist movement. How do you relate to the white southern moderate or liberal and at the same time relate to a group of people who are as militant and as activist as students in the Student Nonviolent Coordinating Committee?

—Bob Zellner[1]

For too long, we have attempted to bring white people into the freedom movement. I think we must reverse this process and take the movement into the white communities. We must help white people see that the Negro has gained strength by casting off fear and that the white man is still the slave to fear.... all [whites] are enslaved in the South, denied their free speech and their opportunity for a good life.

—Sam Shirah[2]

The ten civil rights marchers were tired, sore, and scared as they approached the state highway patrolmen, national reporters, and 1,500 white onlookers who awaited them at the Alabama state line on May 3, 1963. Although disapproving whites had heckled and harassed the integrated group on the previous two days of the march—in the worst incident, someone had thrown a firecracker at them, narrowly missing five of the marchers—the absence of traffic from U.S. Highway 11 and the presence of Colonel Al Lingo, Alabama's feared director of Public Safety, heightened the group's fear of violence as it prepared to cross from Georgia into Alabama. That such a situation had arisen, however, hardly surprised the marchers. When they assembled in Chattanooga, Tennessee, on May 1 to begin the long, 375-mile walk to Jackson, Mississippi, the civil rights workers knew they were courting trouble. Indeed, a specter of violence hung over the march from the start because the activists were continuing William Moore's Freedom Walk. Moore, a white postal worker from Baltimore who had undertaken the march to dramatize his opposition to segregation and racial intolerance, was gunned down on the highway outside Attalla, Alabama, on April 23. While the march had been Moore's personal crusade, his death prompted these ten activists to resume the journey to Jackson, lest white racists come to think that violence could stop the march.[3]

Sam Shirah, a white 20-year-old worker for the Student Nonviolent Coordinating Committee, the leading student civil rights group of the day, led the group as it moved down the Georgia blacktop toward an uncertain fate. To Shirah, an Alabama native, the tension of the situation was nothing new since he never had been reticent about his support for black rights. As a schoolboy his classmates branded him a "nigger lover" for advocating desegregation, and as a college student he frequently was the only white face in civil rights protests. When the activists set out from Chattanooga, Shirah, wearing the sign Moore had worn when shot— "Eat at Joe's, Both Black and White" on one side and "Equal Rights for All (Mississippi or Bust)" on the other—marched at the head of the line, the same position he held as the group approached Al Lingo and the Alabama state line. The impending confrontation had not dampened his sense of humor, as earlier that day Shirah had lightened the mood among the marchers by wiring George Wallace, Alabama's segregationist governor and Shirah's old Sunday School teacher in Clayton, Alabama, asking that the group be allowed into the state. Wallace made no reply, and when the group reached the border Lingo and his officers blocked the road. Ignoring Lingo's orders to disperse, Shirah led the marchers into Alabama, where they promptly were beaten, shocked with electric cattle prods, and arrested by the troopers. After spending 31 days in jail, where they were housed on death row, Shirah and the others were convicted of breach of the peace and fined $200 each. The freedom march was over.[4]

The beating and arrest of the marchers confirmed Shirah's belief that it was time for white civil rights workers to take their activism in a new direction. While participating in black-dominated protests and waging lonely struggles on the rural highways of the South allowed whites to bear witness to their opposition to segregation, these actions rarely drew other white southerners into the burgeoning civil rights movement. More likely, the physical abuse they suffered kept other sympathetic whites from getting involved. Shirah came to believe that it was the responsibility of whites in the movement to persuade their peers to support civil rights. Activists like himself, Shirah began to counsel, had to begin to organize in the white community and on white-dominated college campuses. As he later explained, "I think we liberal and radical whites have been wrong in our orientation. We've been trying to bring white people into the movement as it now exists. Instead, we should seek to take the Movement into the white communities, make it their movement too." In the spring of

1964, less than a year after the march, Shirah would play an instrumental role in the founding of the Southern Student Organizing Committee.[5]

SSOC's roots were in the black student-led civil rights movement that developed following the February 1960 Greensboro, North Carolina, sit-ins and the subsequent creation of the Student Nonviolent Coordinating Committee. Specifically, SSOC grew out of both SNCC's efforts to reach out to white southern students as well as the involvement of white students in the long-lasting campaign to desegregate public accommodations in Nashville, Tennessee. In the summer of 1961, SNCC, partly in response to Sam Shirah's urgings, created the White Southern Student Project to draw young white southerners into the growing movement for civil rights. Initially, hostile administrators, indifferent students, and the very real threat of physical violence led the project's staff to make only sporadic attempts to organize white college students. Not until 1963 did the project become an active presence on numerous southern campuses owing to the appointment of a more aggressive organizer—Sam Shirah—and the growing receptivity of young whites to calls for racial reform in the wake of events ranging from the violent repression of civil rights demonstrations in Birmingham, Alabama, to the interracial harmony of the March on Washington.

By 1963, growing numbers of white college students in Nashville openly supported the civil rights movement. Dating back to the city's first desegregation campaigns in the spring of 1960, students and a few faculty members from Vanderbilt University, Scarritt College, and Peabody College had walked picket lines, joined sit-ins, and suffered abuse at the hands of white thugs incensed by their activism. While students at these nearby and closely connected institutions maintained contact with the White Southern Student Project, they charted their own course of activism during the early 1960s. In the fall of 1963, as the Nashville student activists worked to integrate a popular neighborhood eatery, they were quick to note the surge in civil rights activism among their peers elsewhere in the South, much of it inspired by Shirah's work with SNCC. Because white student activists typically were a small, isolated minority on their campuses, the Nashville students believed that establishing a communication network among these students would bolster morale and keep far-flung activists apprised of developments at other colleges and universities. As a result, the white students in Nashville, along with the White Southern Student Project's Sam Shirah, initiated a series of meetings in the winter of 1964 to discuss creating an activist organization of white southern students, which soon led to the founding of SSOC.

The White Southern Student Project

The White Southern Student Project represented the first systematic effort to organize white student support for the civil rights movement. The inspiration for the project came from the Southern Conference Educational Fund (SCEF), an interracial civil rights group that in mid-1961 made the first of three annual $5,000 grants to SNCC for the purpose of hiring a white field secretary to recruit white students into the movement. Rooted in an earlier era of civil rights activism, SCEF was the surviving branch of the Southern Conference for Human Welfare (SCHW), an organization of liberal southerners that promoted black civil rights and New Deal economic reforms for the South from 1938 until it disbanded in 1948. Although SCEF survived the dissolution of the SCHW, its staunch opposition to racial segregation and suspicions that it harbored communists made it a pariah throughout the South. SCEF's controversial stands also focused attacks on two of its leaders, Anne and Carl Braden, Louisville-based activists with long personal histories of involvement in social justice movements in the South. In 1954, the Bradens were charged with sedition in connection with their efforts to integrate a white Louisville neighborhood, and Carl Braden was convicted and served eight months of a fifteen year sentence before winning release on appeal. Then in 1958, Carl Braden refused to testify before the House Un American Activities Committee, later serving ten more months in prison on contempt charges. Despite the continuous attacks by SCEF's opponents, Anne Braden managed to serve as editor of the organization's monthly newspaper, the *Southern Patriot*, a publication that played an important role in drawing young white southerners into the civil rights movement since it often was their only source of information on the black struggle.[6]

Because its tax-exempt status limited SCEF to those activities defined as "educational," the group spent much of its time encouraging progressive white activism across the South and then publicizing what activism did occur in the *Southern Patriot*. As Anne Braden remembers, "we had a tendency, if you look through the *Patriot*, if some white person did something halfway decent or stood up . . . to write about it. . . . We were trying to show whites this was their movement; they need to be there in action and not just sitting around talking."[7] Through her work on the *Patriot*, Braden learned about the involvement of white students in numerous civil rights campaigns in the early 1960s. In 1960, for instance, white students from the University of Texas at Austin joined black students from nearby Huston-Tillotson College in holding April sit-ins at lunch counters and restaurants and then in initiating stand-ins at movie theaters near the Texas campus in the fall. And shortly after she encountered a dozen white students at the second SNCC conference in October 1960, several of whom she was surprised to discover attended her alma mater, Randolph-Macon Woman's College in Lynchburg, Virginia, Braden began to develop the idea for the White Southern Student Project, concluding that, as she later put it, "if the white students could hear what was going on and understand it, they would want to become involved."[8]

Braden took the idea for a white student organizer to the SCEF Board of Directors, and by the end of November she had written Ella Baker, the Southern Christian Leadership Conference's Executive Director and the guiding force behind the creation of SNCC, and Marion Barry, SNCC's first chairman, to inform them that SCEF had approved funding to hire a student traveler for white campuses and to inquire if SNCC would be interested in such a project. They were. The final plan called for SCEF to fund the creation of a white campus organizer position on SNCC's staff. It made sense, the SCEF Board argued, for the new organizer to work for SNCC since it was the most prominent black student civil rights group and since SCEF did not have a student wing. Additionally, Braden had convinced SCEF's executive director, James Dombrowski, that it was essential for the organizer not to be on SCEF's staff in order to bolster the student activists' sense of independence from adult groups. Otherwise, Braden wrote Dombrowski, the students would think that "we are just another one of the dog-eat-dog organizations seeking to 'capture' the students for the glorification of our own organization."[9] SCEF considered the project's main purpose to be "education geared toward action," explaining in the final proposal for the project that "wherever white student bodies have had access to information about what is really going on among Negro students, wherever they have been close enough to catch the great moral spirit that permeates the movement, the interest among white students has been great and at least a few have wanted to join actively in the movement." The project's success, SCEF believed, depended on SNCC hiring an organizer with the communication skills and the personal experience in the movement to "be able to bring to other students information and inspiration that will make them want to find their own ways to be active too." Finding the right person for the job, however, proved to be no easy task.[10]

When Anne Braden first conceived of the White Southern Student Project, she believed that Jane Stembridge, a Union Theological Seminary student and Virginia native who became SNCC's first office secretary in June 1960, would be a logical choice for the white organizing position because she was knowledgeable of SNCC's activities and because she was known and trusted by others in the organization. Stembridge, however, did not want to be considered for the position, as she was back in school and more interested in writing than organizing. Braden then turned her attention to other white student activists she knew. But of the students she contacted over the succeeding months, including Sandra Cason, a University of Texas activist, and Rebecca Owen, one of the student activists at Randolph-Macon Woman's College, not one was interested in the position. In May 1961, as SCEF finalized the project proposal, Braden complained to James Dombrowski about the difficulty of finding a student willing to apply for the job. Because of this lack of interest, she explained, "that leaves, at the moment, only Bob Zellner, at

Huntingdon College in Montgomery as an active possibility." Although Clifford and Virginia Durr, respected southern liberals then living in Montgomery and among those who helped found the SCHW, knew and liked Zellner, Braden reported that Virginia Durr "cautions we should wait and give him a little time to prove himself before we go overboard."[11]

Born and raised in Baldwin County in southern Alabama, John Robert Zellner was an idealistic, strong-willed, and naïve senior at Huntingdon in 1961. The son of a Methodist preacher who early in his career had been an organizer for the Ku Klux Klan in Birmingham, Bob Zellner, whom one colleague remembers was a "good old boy . . . BMOC-kind of guy," came to sympathize with the goals and aims of the emerging civil rights movement during his final two years at Huntingdon.[12] As a senior, Zellner ran afoul of teachers, administrators, and even state officials when he and four classmates observed a SNCC-led nonviolent workshop and met Ralph Abernathy and Martin Luther King, Jr., as part of their research for a sociology paper in which they were supposed to propose solutions to "the racial problem." The students then defied school officials' orders and attended a meeting of the Montgomery Improvement Association, and as a result, Zellner recalls, "I was asked to leave school, the Klan burned crosses around our dormitory, and we were called into the office of the Attorney General of the state of Alabama." Unfazed by the reaction they had provoked and apparently unconcerned with his personal safety, Zellner, unlike the other four students, refused to withdraw from school. And when the administration tried to confine him to campus, he brazenly ignored his restrictions in order to attend a meeting at the Highlander Folk School, the integrated labor organizing center in Knoxville, Tennessee, which was a center for movement discussions and planning.[13]

As controversy swirled around Zellner in the spring of 1961, southern liberals, such as the Durrs and the Bradens, took notice of him. And when Ralph Abernathy excitedly told Anne Braden about Zellner and the handful of other students who had taken up a collection for the MIA at Huntingdon during the *Sullivan v. New York Times* libel case, she resolved to interview him for the *Southern Patriot*. After graduating from Huntingdon and then spending a portion of his summer at Highlander, Zellner applied for and was offered the organizer position for the White Southern Student Project. On September 11, 1961, Zellner reported to Atlanta to begin his work for SNCC.[14]

Just 23 days later, Zellner was the only white person to take part in an October 4 civil rights demonstration in McComb, Mississippi. While no white campuses were located anywhere near McComb, SNCC Executive Secretary Jim Forman had suggested that Zellner attend the SNCC staff meeting scheduled to take place in the city in order to get better acquainted with the group and to find out what type of projects were in the works. The protest developed during the meeting, and Zellner suddenly and unexpectedly, found himself in the middle of his first demonstration, a protest at which he was attacked by white onlookers and then jailed with 121 other demonstrators, many of whom were local high school students. As time passed, however, it became clear that Zellner's presence in McComb was not an aberration; throughout his stint as white organizer, he spent most of his time in the black community rather than on white college campuses. Zellner later justified his work in the black community on the grounds that he could not explain to white students "what was going on unless I myself became an integral part of it, and of course my personality make up and psychology also tended to draw me into the area of action." Just as significant, however, was the fact that white-student organizing did not look like promising work. In the fall of 1961, Bob Zellner did not think predominantly white southern colleges and universities would welcome a visit from a son of the South intent on drawing students into a movement for racial reform.[15]

He was right. Indeed, the dearth of white student activism in the South at the start of the 1960s distinguished southern institutions of higher learning from those elsewhere in the nation at a time when other regional differences in higher education were disappearing. By the 1960s, for example, federal research support ceased to be the domain solely of northern colleges and universities. Although northern schools continued to receive the majority of federal research funds, southern colleges and universities won an increasing share of this federal largess. In 1968, for example, 24 southern institutions were among the 100 colleges and universities that received the most federal research and development support, including Duke University (ranked 20th), the University of Florida (39), the University of Texas (40), Vanderbilt University (57), the University of Virginia (65), Georgia Tech University (84), and West Virginia University (100).[16]

Like their counterparts in other regions, southern institutions of higher education witnessed a surge in enrollment throughout the 1950s and 1960s. Between 1960 and 1965 alone, enrollments in the South increased by 75 percent, compared to 72 percent nationally. Public colleges and universities in the South were prime beneficiaries of the growing enrollments, particularly in the first half of the 1960s. At the University of Virginia, for example, the number of undergraduates increased 31 percent from 1960 to 1964, and by 1965 the University of North Carolina, the University of Georgia, and Louisiana State University all enrolled twice as many students as they had in 1958.[17]

In terms of student activism, however, the South departed from national trends, as colleges and universities in other regions began to experience student unrest in the late 1950s. This activism was rooted in students' heightened interest in social and political issues which, in turn, was a consequence of their constituting a new generation of middle-class youths. Although raised at a level of affluence their parents had never known as children of the Depression, many young people were put-off by the materialism and the obsession with wealth that were the hallmarks of the thriving mass-consumption society of the 1950s. Likewise, the looming threat of nuclear annihilation, which civil defense drills and the construction of bomb shelters suggested were real possibilities, raised doubts in their minds about the future of the world. In addition, many could not help but contrast the ease and serenity of their lives with the social injustices afflicting the nation, especially the scourge of racial discrimination. Repulsed by mainstream society, fearful for their own future, and appalled by the nation's social problems, the first of this new generation of students arrived on campuses in the late 1950s ready to act, and they quickly became involved in activist causes.[18]

The overwhelming majority of predominantly white colleges and universities in the South were immune from this student activism in the late 1950s. The University of Texas at Austin, however, was an exception to this southern pattern. A homegrown student movement began to coalesce there in the late 1950s, galvanized by the university administration's 1957 decision to remove Barbara Smith, one of the few black students on campus, from the leading role in the university's spring opera. The move outraged students across campus. In response, leading student organizations, from the Young Democrats and Young Republicans to the most prominent service groups, the Cowboys and the Silver Spurs, organized a successful boycott of the opera. The following year, pro-civil rights students launched the "Student Equal Opportunity Steer Here" drive, a campaign to persuade restaurants near campus to desegregate voluntarily.[19]

In subsequent years, activism centered on the university YWCA and the Christian Faith-and-Life Community, an interdenominational, interracial residence and the only integrated place near campus black students could live. Liberal-leaning, activist-oriented students gravitated to these institutions because they were integrated centers of Christian liberalism and political activism and because they attracted students who, as Dorothy Dawson Burlage, a former Christian Faith-and-Life Community resident, puts it, "were more thoughtful, more serious, more concerned, more substantial than anyone else around."[20] The campus newspaper, the muckraking *Daily Texan*, also played an important role in stirring student unrest. Under the stewardship of such dynamic and politically oriented editors as Willie Morris, later the editor of *Harper's*, and Robb Burlage, a founder of the Students for a Democratic Society, the paper frequently took on the state's most powerful interests and questioned the orthodoxies and pieties of mainstream southern society. Both earned the enmity of university and state leaders for their efforts. While Morris's hard-hitting pieces on oil and gas issues led the Board of Regents to try to censor the paper by restricting the topics it could cover, Burlage's advocacy of desegregation and academic freedom resulted in his removal from the editor's chair. Nonetheless, by 1960 the university, Morris recalled, had "exploded with vitality: protest meetings, sit-ins, stand-ins, debates; . . . and everyone who mattered was stopping there to lecture or to talk with the bright young people engaged in the 'liberal movement.' "[21]

A "liberal movement" of young people existed at few other predominantly white southern colleges and universities when Bob Zellner started his tenure as the White Southern Student Project organizer in the fall of 1961. As Zellner was well aware, students on campuses that, generally, were politically conservative, socially homogenous, and culturally provincial were unlikely to prove responsive to his efforts. Still, he occasionally visited such campuses. In the fall of 1961, for instance, he traveled to Jackson, Mississippi, where he spent time at all-white Millsaps and Bellhaven colleges as well as at historically black Tougaloo College. Over the course of a week he made eight separate trips to Millsaps, a small Methodist school, where he talked with students who were known to support the movement, met several other interested students who wanted to receive the SNCC newsletter, *The Student Voice* (but only, he reported, if it could be sent "in a plain envelope"), and encouraged all of them to attend the nonviolent workshops being held at Tougaloo. But as if to signify his uncertain commitment to white organizing, Zellner departed Jackson before he knew if the students attended the workshops.[22]

Zellner's preference for working along side other SNCC staffers on black campuses and in the black community also made white campus work problematic for him. He much preferred the excitement and camaraderie of the activism taking place in black communities to the lonely and difficult task of trying to win white support for the movement. Thus, he took part in numerous black-led demonstrations, leading to his arrest in places such as Baton Rouge and Albany, Georgia. He soon found that this activism hindered his ability to work with white students. Because he "had become a somewhat active person in the movement and had been arrested several times," he explained, "I felt quite estranged from the Southern students that I was supposed to be talking to." Additionally, the publicity he had garnered as a white person in SNCC limited his effectiveness on white campuses. Since his name was known to segregationists, he remembers, "when I went on college campuses, I couldn't go put a bulletin board notice up, 'Bob Zellner is going to be here such-and-such a night.' I had to go on campus very quietly." And Alabama authorities did their best to demonize Zellner. As he recalls, "every time I ever got beaten up or arrested in Alabama they said 'Bob Zellner, civil rights leader'—they'd always say I was a civil rights leader"— in hopes that by portraying him as a leading radical in SNCC they would reduce his appeal to potentially sympathetic white students. Administrators at Huntingdon and Birmingham-Southern College were particularly effective in this regard, opposing his visit to their campuses and encouraging students to harass him. This was an unnerving experience for him and, he later wrote, it "made me very reluctant to put myself in the situation again where I would feel that everyone thought I was a real subversive or criminal of some sort."[23]

Further complicating Zellner's mission was that most white southern students were afraid to openly support the movement. As Zellner knew from his own collegiate experience, "the few white students who are capable and desirous of participation in the integration movement are burdened with fear, pressured by their families (some of whom are 'moderates' or 'liberals') to 'stay out of this integration mess,' and confronted by the real and sometimes imagined threat of state oppression." The job of the white student organizer, he believed, was "to see that those who might act are encouraged to do so and that, once they do, they are not isolated and quietly destroyed."[24] But this task was made more complicated if the activist maintained ties to SNCC. While, on the one hand, relations with SNCC kept the organizer close to the leading edge of the movement, on the other hand, such ties alienated him from moderate students who considered SNCC a dangerously militant organization. As he explained in 1962,

> Whoever takes the project must recognize the basic dilemma: it is very difficult to relate intimately with the SNCC staff without getting involved in action and in getting involved in action you tend to put some distance between yourself and the white southern moderate. But someway this dilemma must be solved. Someone must reach the white students and interpret to them what is going on.... Someone must try to relate and bridge the gap between them, the white Southern students and the militant people who are fighting on the front lines.

Zellner would not be the person to do this; he resigned the position in 1962. Still, he made an invaluable contribution to SNCC's effort to attract the support of young whites. While he had won little support on the white campuses of the South, the attention he received for his activism in the black community drew a few other young white southerners to SNCC.[25]

Sam Shirah and Ed Hamlett

In 1963, as Bob Zellner prepared to head north to Swarthmore College to pursue graduate studies, Sam Shirah took his place as the SCEF-funded white field organizer in SNCC. Like Zellner, Shirah was a native Alabaman and a Methodist preacher's son. But whereas Zellner felt most comfortable in the midst of the black community, Shirah preferred to work with southern whites, as he was committed to the idea of carrying the civil rights movement into the white community. Loud, aggressive, and charismatic, the guitar-playing Shirah reminded SNCC leader Charles

McDew of "what Woody Guthrie was like, what a *real* Populist was like." Although he tended to float from project to project—as one colleague puts it, "Sam was a person who might be there one day but might not be there the next or might be in a different thing the next"—Shirah was untiring in his effort to organize white support for the movement.[26]

Born in 1943 in Troy, Alabama, the hometown of SNCC's John Lewis, Shirah lived in northern Florida and several different Alabama communities as a youth. By the time his family had settled in Montgomery in 1956, during the famed boycott of the city's buses, his father openly supported desegregation. Shirah, too, became an advocate of black equality, and his participation in civil rights activities resulted in his expulsion from three colleges in one-and-a-half years—Gulf Coast Junior College in Panama City, Florida; Huntingdon College in Montgomery; and Birmingham Southern College. In the first half of 1963, he took part in several high-profile protests, including the desegregation demonstrations in Danville, Virginia, and the resumption of William Moore's Freedom Walk. But as he moved into Zellner's position in SNCC, Shirah had concluded that white activists such as himself needed to work more in the white community and on white campuses instead of simply trying to persuade other whites to join black-led protests. By the middle of 1963 he had decided that "just offering my body in the Negro demonstrations was not enough. I began to feel that something had to be done to reach the great numbers of white people in the South who have felt that this movement is their enemy. It's not their enemy. It might be their salvation."[27]

For too long, Shirah insisted, movement whites had ignored their brethren in the South, partly because the action was in the black community but also because they considered southern whites, especially working-class whites, hopelessly racist. All the more reason, Shirah argued, to organize in the white community. As he asked one white audience, "Where is the Movement that will say to us whites that we are slaves to fear, to hate, to guilt, to an inferiority complex that the Negro has overcome in his movement and therefore we must catch up?" Ultimately, Shirah believed that by organizing white support for the movement, white activists, like the Populists of the previous century and the CIO organizers of the 1930s and 1940s, carried on the tradition of trying to unite whites and blacks in common cause. The earlier generations' efforts failed, however, because "when the chips were down the white Southerner fled back to the false glory of his white skin," Shirah explained. But he was convinced that with the assistance of southern white students such an effort could succeed in the 1960s. "Two years ago," he explained in early 1964, "the frontier for the white student who wanted to be a part of current history was to get on a Negro picketline against a segregated restaurant. That's still needed, and all of us will be doing that too. But the frontier today is elsewhere. It is for the white student to go into the white community and organize."[28]

Besides having a stronger commitment than Zellner to working with southern whites, Shirah also benefited from growing white sympathy for the civil rights movement. As he stepped up his organizing efforts in the fall of 1963, the events of the previous six months made young white southerners increasingly receptive to calls for black equality. In May, the violent repression of the Birmingham movement, which included the use of police dogs and fire hoses against protesters, some as young as six years old, sickened and outraged the nation and prompted some young southerners previously indifferent to the cause to become movement supporters. In June, President Kennedy's statement that segregation was morally reprehensible, made on the same night that the NAACP's Mississippi field secretary, Medgar Evers, was murdered, along with Kennedy's announcement a week a later that he would propose comprehensive civil rights legislation (that eventually became the landmark Civil Rights Act of 1964), gave the goal of black rights national legitimacy. With its rhetoric of love, understanding, and goodwill to all and its vision of a racially harmonious future, the March on Washington at the end of August inspired increased white support for the movement in the South. And in September, the death of four young girls in the bombing of Birmingham's Sixteenth Street Baptist Church shocked people throughout the country and persuaded scores of young white southerners to become active in the civil rights struggle.[29]

With the movement in the news seemingly every day, by the start of the 1963–64 school year white students had joined together to form civil rights groups on numerous southern campuses. At the University of Florida, the Student Group for Equal Rights supported the local Gainesville civil rights movement as well as established a program for tutoring black youths. In Nashville, the Joint University Council on Human Relations provided an organizational base for activist students from Vanderbilt University and Scarritt and Peabody colleges. Local issues typically were the central concern of other groups as well, such as the Georgia Students for Human Rights at the University of Georgia, the Liberal Club at Tulane University, the Students for Social Action at the University of Louisville, the Virginia Council on Human Relations at the University of Virginia, and the Student Nonviolent Action Group at Florida State University. As Sam Shirah remarked in late 1963, "no longer is it one or two white students active; often there are hundreds." Within SNCC, Shirah pressed the group to devote more resources to organizing white students. Likewise, he encouraged SCEF to expand the White Southern Student Project, requesting "that we put more workers in the field—visiting campuses and stimulating white students, and that as we bring them into activity we steer them where possible into community work in white communities."[30]

To take advantage of the increased white support for the civil rights movement, Shirah spent much of the 1963–64 school year on college campuses, encouraging students to organize, leading demonstrations, disrupting normal campus life—in short, doing his best to bring the movement to the students. In the fall, he raced between 20 different campuses in six states, frequently incurring the wrath of administrators and the hostility of segregationist students. In September, officials at Birmingham Southern College, which Shirah had attended the previous fall, banned him from the school's grounds for trying to organize a protest in the wake of the bombing of the Sixteenth Street Baptist Church. In the words of Dean Ralph Jolly, "the policy of the school is that he should leave the campus and never return." Two weeks later, Shirah became the subject of controversy at the University of South Carolina in Columbia, where, working with the campus Methodist group, the Wesley Foundation, he spoke with students about the movement over several days until school officials suddenly had him arrested for trespassing. Although the charges quickly were dropped, a resolution was introduced in the student senate censuring the Wesley Foundation for offering "support and assistance to a member of an integrationist group," identified as one "Samuel Shirer." The resolution led to a torrent of student criticism, and the campus newspaper, the *Gamecock*, published numerous letters opposing it. Most letter writers defended the Wesley Foundation on religious grounds rather than by explicitly advocating civil rights. As one student wrote, "I am shocked that a USC student . . . would stoop so low as to attempt to condemn God's Church for carrying out the teachings of our Master. The Wesley Foundation did nothing contrary to the teachings of The Methodist Church or the principles of Christian Brotherhood." Shirah spent only five days on campus, but the consequences of his visit lingered for several weeks.[31]

In addition to focusing renewed attention on the South's white campuses, Shirah made occasional visits to black colleges and universities. On a swing through Tallahassee, Florida, in November, Shirah visited both historically black Florida A&M and predominantly white Florida State universities, and he was pleased to find that a handful of students from both schools were working together to protest the continued segregation of area restaurants. In an effort to encourage more students to participate in the local movement, Shirah, along with two white Florida State and two black Florida A&M students, tried to integrate the Dobbs House restaurant and subsequently were arrested. One week later, Shirah led an interracial group of 14 students from both schools to the governor's office to complain about the arrest of civil rights protesters in Tallahassee, an event that generated publicity throughout the state.[32]

As 1963 wound down, Shirah continued to urge SNCC's leaders to strengthen the group's commitment to white student organizing. At its end-of-the-year executive committee meeting, he argued that SNCC should hire another staff person in order to help the project grow. The executive committee agreed, and in February 1964 Ed Hamlett, a native Tennesseean who recently had dropped out of graduate school at

Southern Illinois University, became a member of the SNCC staff. Hamlett previously had co-chaired a student civil rights group at the University of Tennessee with Marion Barry, who was there working on a graduate degree in chemistry. In the fall of 1963, Hamlett had expressed to Barry an interest in SNCC's white organizing work, and when SNCC decided to hire another staffer Barry put his name forward. Soon thereafter, Hamlett joined Shirah as a white student organizer for SNCC.[33]

Throughout the winter and early spring of 1964, Shirah and Hamlett drummed up support for the movement at predominantly white colleges and universities across the region. Their mission, Shirah explained, was to "agitate [and] to try to destroy apathy." Hamlett visited more than a dozen schools, including the University of Georgia, where the Georgia Students for Human Rights had become active, and Southwestern at Memphis, where white students, under the tutelage of Reverend James Lawson, who then had a congregation in the city, took part in a months-long campaign to desegregate one of the city's most prestigious religious institutions, the Second Presbyterian Church. Hamlett also attended the Atlanta Area Student Conference on Religion and Race, where he met several students interested in becoming involved in the civil rights movement. For his part, Shirah continued to build student support for the movement wherever he went, including Huntingdon College and the University of Southern Mississippi, where he helped students create an alternative newspaper. By the end of the school year, Shirah had visited 48 colleges and universities—32 of them predominantly white—and Hamlett's efforts pushed their combined total to more than 60. Between the two of them they had contacted more than 1,000 students and faculty members. The project had come along way since Bob Zellner's first tentative steps at white organizing in the fall of 1961.[34]

In May, as he reflected on the project's work over the past year, Shirah noted that in January he had begun to see "the need for a Southwide organization of white students." Shirah believed that such a group could offer sympathetic whites the same type of organizational support that SNCC had given black college students when it first emerged in 1960. He also reaffirmed his desire to organize in the wider white community and not just on campuses, a goal that Bayard Rustin had encouraged at the November 1963 SNCC conference at Howard University. The essence of Rustin's message, Shirah approvingly recounted, was that white activists needed to "go into the white communities and organize the people there to form an alliance with the civil rights movement . . . to the end that these two groups of disinherited people, the Negroes and the downtrodden whites, may work together to achieve a society that will be of benefit to all." As Shirah saw it, an organization of southern white students not only would support white activists across the region but could be the vehicle by which white students enter poor and working-class white communities. While Shirah contemplated how to get such an organization off the ground, white students in Nashville, jubilant over the success of their recent protest actions and hopeful that they could inspire other white students to become active in the movement, set in motion the process that eventually would lead to the creation of a white southern student group—the Southern Student Organizing Committee.[35]

White Activists and the Nashville Movement

During the early 1960s, the civil rights movement in Nashville was one of the most dynamic and successful local movements in all the South, combating segregation wherever it existed in the city and raising to national prominence local leaders James Lawson, Diane Nash, John Lewis, James Bevel, and Lester McKinney. Because several Protestant denominations and publishing houses were based in Nashville—hence the city's moniker as the "buckle of the Bible Belt"—and because the city was home to 13 colleges and universities, including historically black Fisk University, Tennessee State University, American Baptist Theological Seminary, and Meharry Medical College, civic leaders cultivated the image of an enlightened and moderate city that embraced progressive racial policies and considered itself above the brutal forms of discrimination common in the Lower South. Yet although blacks and whites mixed in public schools, on city buses, and in the police department, segregation persisted in restaurants, lunch counters, movie houses, and other public accommodations, particularly those located in downtown Nashville.

In 1959, James Lawson, a regional director of the pacifist Fellowship of Reconciliation and one of the few black students in the Vanderbilt Divinity School, began conducting workshops in nonviolence for black students in preparation for the launching of demonstrations targeting the city's segregated lunch counters. While the students had initiated several brief sit-ins in late 1959, the Nashville movement began in earnest when the Lawson-trained students occupied seats at downtown lunch counters on February 6, 1960, five days after the Greensboro sit-ins had started. So began four years of near constant protest against segregation of public facilities in Nashville. This first wave of sit-ins ultimately succeeded when in April, after two months of beatings and arrests and the bombing of black attorney and movement supporter Z. Alexander Looby's home, Mayor Ben West, confronted by Diane Nash at a protest rally on the steps of City Hall, publicly conceded that segregation was unfair and morally wrong. Shortly thereafter, merchants started to desegregate the city's lunch counters. As this process began, the Nashville students discussed using sit-ins to open other areas of public life, and by the fall the students had unveiled plans to attack a whole range of segregated facilities in the city, from restaurants to movie theaters. These protests continued throughout the 1960–61 school year and made Nashville a center of movement activity. As James Forman, the future executive secretary of

SNCC, remembered, the Nashville he visited for the first time in the summer of 1961 was "an exciting place to be, buzzing with activity and debate in the student community."[36]

By the spring of 1962, though, a wide range of public facilities, other than lunch counters, remained segregated, as the majority of downtown merchants refused to integrate their businesses. Consequently, movement leaders demanded that Nashville become a "complete open city," and on any given day protesters could be found leading sleep-ins in hotel lobbies, conducting stand-ins at restaurants, and organizing picketing of movie houses. White students and faculty members participated in these demonstrations, and in December, David Kotelchuck, a visiting assistant professor of physics at Vanderbilt, was involved in an incident that brought attention to the role of whites in the Nashville campaign. The 26-year-old Kotelchuck, who was in his first semester at Vanderbilt, had come to the school from Cornell University in Ithaca, New York, where he had done his graduate work. While at Cornell he was an active supporter of the civil rights movement, picketing the local Woolworth's in the wake of the 1960 Greensboro sit-ins and helping to organize a 1962 project in which Cornell students participated in a voter registration drive in Fayette County in rural southwestern Tennessee, where white landowners had evicted blacks who had attempted to register to vote. Once Kotelchuck arrived in Nashville, he became the Tennessee contact for the project and gradually began to attend local civil rights demonstrations.[37]

On December 8, 1962, Kotelchuck was the only white among 13 protesters who picketed the segregated Herschel's Tic Toc restaurant in Nashville, a movement target because of its hostility to civil rights demonstrators. During the recent Thanksgiving holiday weekend, for instance, Sam Block, a voter registration worker in Greenwood, Mississippi, who had come to Nashville for a SNCC meeting, was stabbed in the chest with a pen by an angry white person as he and others sought service at the restaurant. As Kotelchuck and the others marched in front of the Tic Toc on that early December day, white onlookers and restaurant employees began to harass, push, and shove the picketers, and someone in an upstairs window dropped a paper bag filled with water on them. Because he was white, Kotelchuck quickly incurred the wrath of the hecklers, and he narrowly avoided injury when he ducked away from an employee's punch. A newspaper photographer captured the moment on film, and the picture of a contorted Kotelchuck twisting away from a large, angry, fist-throwing thug appeared in the newspaper the next day and ignited a heated discussion about the local movement within the Vanderbilt community. While some students and faculty members praised Kotelchuck's actions, the student senate censured him, most letter writers to the school paper, the *Vanderbilt Hustler*, condemned him, and Chancellor Harvey Branscomb, after noting in a letter to dismayed alumni and others that Kotelchuck did not have a permanent appointment, stated that although Kotelchuck did not break the any laws, "what I might say with reference to his judgment, good taste, and sense of corporate responsibility is another matter."[38]

Despite the Vanderbilt community's less than enthusiastic response to the Kotelchuck incident, more whites, especially students, became involved in the Nashville movement in the following months.[39] White students also began to take action on their own campuses. Among these young whites were several students from Scarritt College, the college of the General Conference of the Methodist Church. With a mission to prepare students for "careers in Christian service," Scarritt, which had desegregated in 1952, soon after Tennessee state law allowed for white schools to admit black students, attracted a racially diverse group of students from all over the world.[40] During the 1962–63 academic year, 38 foreigners were among the school's 160 students, and two of them became embroiled in a controversy. First, Belmont Methodist Church, the second largest Methodist church in the city, refused membership to the wife of Abel Muzorewa, the future black Rhodesian political leader who was then a young minister studying at Scarritt. Shortly thereafter, in an incident that foreshadowed a later crisis, the Campus Grill, a nearby greasy spoon, refused to serve Lorine Chan, a dark-skinned Scarritt student from Fiji.[41]

In response to these incidents, several white Scarritt students, including Gerry Bode and two students who would help found SSOC, Sue Thrasher and Archie Allen, formed the Christian Action Fellowship to work against racial discrimination. As their first projects, the group sponsored a symposium on the "Church's Role in Racial Integration" and wrote letters of support to 28 Methodist ministers in Mississippi who recently had spoken out against segregation. Then addressing the Campus Grill situation directly, Allen urged Thrasher, a student council member, to draft a resolution that would put on record the student government's disapproval of the restaurant's actions. But although Thrasher and a colleague on the council presented a modest resolution reaffirming the students' faith in the Fatherhood of God and the Brotherhood of Man—in fact, the resolution was based on passages in the "Discipline of the Methodist Church"—it failed to pass student council. Thrasher was shocked that the council rejected such an innocuous resolution. In retrospect, she believes that the resolution simply was too radical a statement given the context in which it appeared. Indeed, the fact that the resolution had been considered at all, she remembers, "had the impact of immediately isolating us from every other student on campus. . . . [V]ery quickly . . . I was part of the little isolated group that was thought of as being radical." A short time later, at the behest of Scarritt teacher Alice Cobb, Thrasher attended a meeting of the Nashville Christian Leadership Conference, the Nashville branch of Martin Luther King, Jr.'s, Southern Christian Leadership Conference, and was so enthralled that she regularly began attending the group's meetings as well as those of the local SNCC chapter.[42]

In the fall of 1963, the policies of the segregated Campus Grill once again fueled student protest in Nashville. This time, however, a serious crisis resulted as Vanderbilt and Peabody students joined their peers from Scarritt in expressing opposition to the restaurant's policies. The Grill's refusal, once again, to serve Lorine Chan precipitated the crisis. Because the Campus Grill was located directly across the street from the Joint University Library, the facility Scarritt, Vanderbilt, and Peabody students shared, word of the incident quickly spread on all three campuses. Among those to hear what had happened was Vanderbilt graduate psychology student Ron Parker. The incident shocked Parker, for he never before had realized that a restaurant he frequented refused to serve non-whites. He soon started talking about the incident with his friends, one of whom suggested he contact Sue Thrasher, who had graduated from Scarritt in the spring and now worked nearby at the Methodist Publishing House. Parker and Thrasher met at a meeting of the Nashville Christian Leadership Conference, and soon they began to develop a network of people on all three campuses who wanted to work to change the eatery's policies.[43]

In early November 1963, after a letter-writing campaign and discussions with management failed to open the Grill to non-whites, the students initiated a direct action campaign to try to force the establishment to desegregate. On November 6, the students called for a boycott and began to picket the restaurant. The demonstrations went on from 6:00 A.M. until 11:00 P.M. every day for two-and-a-half weeks and involved 124 white students, four black students, and one professor, David Kotelchuck. For most of the white students, this was the first time that they had spoken out against segregation. As Parker remembers, "we decided, . . . as moderate Christian southerners, to do something about it [segregation]. And we hand-lettered our little picket signs and we stood out there between classes. And we protested. And we were scared shitless." White patrons and passers-by frequently heckled and harassed them, such as the time a white man spat in the face of a black protestor. The protestors received a mixed response on their campuses. At Vanderbilt, the *Hustler* editorialized against the campaign and, protestor Chuck Myers remembers, one of the fraternities mandated that its brothers eat at the restaurant as long as picketers marched in front. But the student council passed a resolution calling for the desegregation of all restaurants in the city, and nearly 200 faculty members signed a similar petition co-authored by Kotelchuck. Both the local SNCC chapter and the National Christian Leadership Conference offered assistance, but the students turned them down because, Parker believed, "the white students saw them as 'outsiders.'"[44]

The crisis threatened to escalate on November 22 when, as the protesters observed a moratorium on picketing after President Kennedy's assassination, a second Campus Grill, also segregated, opened its doors to business not far from the first location. In response, the students drew up plans for more radical actions which, Parker remembered, "projected sit-ins and arrests." The possibility that the white students would engage in the same type of activities that black students had used to create havoc in downtown Nashville drew the mayor into the protest, and he, along with white business leaders, negotiated a settlement with the restaurant's owner that called for the desegregation of both Campus Grill locations on January 1, 1964. The protesters were thrilled that they had forced the Campus Grill to desegregate. As David Kotelchuck recalls, it "was really exciting because for the first time we had a movement . . . coming out of the traditionally white university mounting a struggle against discrimination. . . . We were taking some initiative on behalf of justice and we were sustaining it." And not only had the students won their demands, but the demonstrations had created a cadre of student activists at each school who would be receptive to future protest actions.[45]

Emboldened by their success, the Nashville activists founded the Joint University Council on Human Relations to coordinate future activities, and in early 1964 they held a series of meetings and informal discussions to consider the possibility of creating a Southwide white student organization. In addition to local Nashville people such as Thrasher, Allen, Parker, and Kotelchuck, many others participated in these conversations, including the Bradens, Sam Shirah, Ed Hamlett, SNCC staff member Walter Tillow, and Robb and Dorothy Burlage, the University of Texas graduates who had married, participated in the founding of SDS, and now lived in Nashville. The Burlages met with the Nashville activists several times, and while they were enthusiastic about the idea of an independent white student organization in the South, they also encouraged the students to think about joining SDS since it was an established group. Robb Burlage quickly discovered that Shirah would play an important role in the decision-making process in Nashville since the local students paid close attention to what he said. Given his long-standing concern about poor whites, Shirah advocated that the students take their activism into the white community. On one visit to Nashville, the Burlages recounted in a letter to SDS's national office, Shirah "argued very persuasively that Southern whites need to do more than form a 'coordinating group' of predominantly white campus activists." Instead, he wanted the students "to rebel totally against the System" by organizing lower-class whites, to help them understand that "race is just being used as an obfuscation by the System-runners to 'keep the <u>white</u> man (as well as the Negro) down.' "[46]

Although they agreed with Shirah about the importance of reaching out to poor whites, the Nashville activists believed that their first priority was to organize those southerners they knew best, namely, white students. By mid-February 1964, the students had concluded that latent support for the civil rights movement existed on many predominantly white southern campuses but that sympathetic students, thinking they were alone in their feelings and fearing what would happen to them if they spoke out, remained silent. At a crucial meeting at the home of Lester Carr, a Fisk University psychologist who had been involved in the Campus Grill protests, the Nashville group decided that the time was now right to try to build an organization that both would put white southern activists from around the South in contact with one another and provide a vehicle by which liberal students could participate in the civil rights movement. In their vision, such an organization would be decentralized, allowing each campus group to set its own agenda in order to win as much support as possible. According to one of the meeting's participants, these local groups would be "completely autonomous . . . with local actions to be determined by local conditions. . . . No one group has to answer for another's actions except in a general sense." The Southwide group, then, would function more as a coordinating committee than as a membership organization. It would sponsor conferences, send field secretaries throughout the region to encourage action, establish a central office to provide local groups with resources and contacts, and perhaps even develop a newsletter. But it would not dictate actions and programs to local groups.[47]

Shortly after the meeting, the group began inviting white students to come to Nashville on Easter weekend to discuss creating a white southern student organization. They also created the Southern Student Organizing Fund to raise money to bring students to the meeting. Then with Shirah's assistance, the Nashville group coordinated an Atlanta meeting in February to discuss their ideas with other activists—both adults and students, blacks and whites. The meeting's participants, among whom were Highlander founder Myles Horton, Ella Baker, Anne Braden, Jane Stembridge, Todd Gitlin, and Robb Burlage, agreed with the Nashville group that white students needed an organization that would serve their needs and interests. Braden argued that it was important for the group to initiate successful projects right away so that when faced with future hardships the students would not, as white southerners had always done, "fle[e] back to [their] own skin." The fundamental issue the meeting's participants grappled with regarded the proposed group's relationship to SNCC. Horton argued that while it was laudable for these whites to create a white student group, they needed to accept black leadership of the movement. Braden, who would have more to say on this issue in the wake of SSOC's founding, agreed with this sentiment and added that it was essential for young blacks and whites to come together so as not to replicate the divisions in the society they sought to transform. But SNCC's Bob Moses, who attended this part of the meeting, recoiled at the idea of a white group merging with SNCC. "'SNCC and this movement,'" Braden remembers Moses declaring, "'is the first thing that we've ever had that belongs to us.'" Although the meeting did not resolve this issue, most went away believing that the new organization would provide white southern students with an opportunity to bring change to their homeland. Even Sam Shirah, though concerned that the proposed group would be too liberal, believed that it could provide the first step toward radicalizing white students. As Robb Burlage remarked about Shirah's views, "I think he is convinced, as we in SDS should be, that the indigenous organizing of such a South-wide group of activists would be a great thing and a good base for more radical white student activity."[48]

Chapter 2

Building an Organization: The Founding of SSOC

When you decided to come this distance to this conference, why did you do it? Do you know what is needed on your campus, in your community, in the state, and in the South and the nation? And finally and most importantly, are you ready to make sacrifices for significant change? Are you willing to become totally involved? . . . Do we have the courage and the vision to make a revolution?

—Ed Hamlett[1]

It wasn't so much about creating an organization, it was about how you involve more whites in the civil rights movement and pull them to the other side so that they're not against what's going on. And if they're a little bit hesitant, then the organization could provide a way for them to move into more activism and be a support group.

—Sue Thrasher[2]

As April 1964 approached, no one knew how many young white southerners would travel to Nashville for the Easter weekend meeting on white student activism in the South. A large Nashville contingent certainly would be present since activist students at Vanderbilt University, Peabody College, and Scarritt College had called the meeting. For two months, though, Sue Thrasher, Archie Allen, Ron Parker, and the other organizers had worked hard to publicize the conference throughout the region. They sent announcements of the conference to potentially interested students whose names they had culled from the *Southern Patriot* or obtained from Anne Braden. They then appealed to their friends in established student organizations to spread the word about the conference. SNCC staffer Walter Tillow assisted by contacting progressive student groups on more than 20 predominantly white southern campuses, while Sam Shirah and Ed Hamlett promoted the meeting at the colleges and universities they visited in February and March. Connie Curry, the director of the United States National Student Association's (NSA) Southern Project, also extended invitations to the conference, writing to students at 20 southern schools, "There is a feeling that a need exists for white Southern students who are involved in the Civil Rights movement to meet together and exchange ideas about things which are most relevant to their situation."[3]

Students across the region responded favorably to the conference announcement. On campuses large and small, public and private, far removed from movement action and already familiar with protest activity, the call to come together resonated with white student activists. While Harry Boyte, Jr., and Rosemary Ezra came to Nashville from Duke University and the University of North Carolina at Chapel Hill, respectively, Marjorie Henderson arrived representing tiny Maryville College in Maryville, Tennessee. The Students for Integration at Tulane University sent Cathy Cade to the meeting and Marti Turnipseed came to represent the few interested students at Millsaps College in Jackson, Mississippi. Soon after finishing his final exams at the University of Georgia, Nelson Blackstock headed for Nashville, as did Bob Richardson and Bruce Smith of Lynchburg College in Lynchburg, Virginia, both of whom unexpectedly found themselves invited to the conference because, they later learned, their picketing of the local newspaper for redbaiting an antiwar school teacher had brought them to the attention of the conference organizers. Dan Harmeling of the University of Florida was able to travel to the meeting as well since he was one of several students to receive assistance from the Southern Student Organizing Fund, an instrument the conference planners had created to help students finance the cost of travel to Nashville. One African American student even participated in the conference—Marion Barry, Jr., SNCC's first chairman who now was a leader of the integrated Students for Equal Treatment at the University of Tennessee. Finally, several movement organizations were represented at the conference; Connie Curry and Frank Millspaugh of the NSA, Anne and Carl Braden of SCEF, and John Lewis and James Forman of SNCC all attended portions of the meeting.[4]

When the first weekend of April—Easter weekend—finally arrived, 45 young people from 15 predominantly white colleges and universities in 10 southern states had converged on Nashville. As the students listened to speeches calling for increased white involvement in the movements for social justice, participated in workshops on how to draw whites into these movements, and shared their own experiences as activists, they became convinced that an organization of young whites could reach moderate white students who, though sympathetic to liberal causes, previously had steered clear of activist organizations. The meeting's participants firmly believed that other white students would be emboldened to join such an organization once they saw that the white activists were just like them. And the recognition that other whites in the South felt just as they did would create a sense of unity and strength among young white activists in the region, most of whom, as part of a small, isolated group of individuals on their respective campuses, did not feel they were part of a larger movement.[5] By the end of the weekend, the students had created an organization of white southern students to fulfill these goals, naming it the Southern Student Organizing Committee.

At the Nashville gathering and at subsequent organizational meetings over the next month, the students outlined the new group's mission and goals, elected officers, developed a plan of action, began publishing a newsletter, and drafted a manifesto announcing SSOC's creation. As the students worked to build the organization in these initial meetings, they discussed and, in some cases, argued about a wide range of subjects, from who could join the group to how the organization should raise funds. Significantly, many of the issues they considered at the founding meetings remained central topics of debate throughout the group's life. In many instances, of course, this was a healthy development, especially for an organization active during the turbulent 1960s. For example, the fact that SSOC activists continuously reevaluated the group's goals and program of action was a sign of the students' willingness to adjust the group's activities as the context in which it operated changed. But many of the issues that retained their importance throughout SSOC's life did so because they were a continual source of tension and conflict within the organization. On which issues should the organization focus? Should it concentrate solely on civil rights or should it adopt a multi-issue perspective? What would be the organization's structure? Was SSOC a white student organization or would it welcome black students, too? Should SSOC work only with students or should it venture into the wider white community? How explicit should the group's regionalism be? What was SSOC's relationship with other activist organizations? SSOC students never provided a final answer to these and other thorny questions, and, as a result, the conflict over these issues contributed to the group's demise in 1969. SSOC's founders, though, were not troubled that many important matters remained unresolved after the initial meetings. None of them had expected the group to be born fully grown, with clearly defined aims and a set plan of action. Rather, what mattered most to the founders was that the students had managed to come together to create an organization of young white southerners committed to making the South a more peaceful, integrated, and democratic society.[6]

Throughout its history, SSOC attracted young whites who were unafraid of the consequences of joining an activist organization. Since turning against the dominant views of the white South could lead to loss of friends and rejection by one's family, these southerners knew that becoming involved in the movements of the day required an unwavering commitment to the causes for which they fought and the courage to reject the racial mores of the region. Beyond this basic similarity, however, students active in the group at different periods held little in common. Owing to the rapidity with which situations changed, the frequency with which "defining" events occurred, and the general tumult of the era—what historian Terry Anderson has called the "kaleidoscope of activity"—two tightly compressed generations of whites moved through SSOC during its five years of existence.[7] The generations focused their attention on different social and political issues, possessed their own unique organizational styles, and responded differently to the ideological debates of the day. The individuals who attended the spring meetings that founded SSOC, along with those who became involved with the group in its early stages, comprised SSOC's first generation. These students shared certain characteristics and experiences that distinguished them from later SSOC activists. The black student-led freedom movement of the early 1960s had inspired their activism, and, not surprisingly, they made civil rights issues a top priority for the new organization. Additionally, members of the founding generation cared little about the cultural issues that later would dominate the student movement nationwide. In its first months, SSOC was concerned only with working to bring political and social change to the South. Yet the change the first generation sought was mild and reform-oriented when compared to the later generation's talk of revolution and secession. During SSOC's first two years, the first generation shaped its agenda and organizational goals as well as defined SSOC's public image.[8]

Coming Together

Workshops, speeches, and panel discussions crowded the program of the April meeting. With students arriving in Nashville throughout the day on Friday, April 3, no formal sessions were held. Instead, the students gathered for a "get acquainted" party at the home of Sue Thrasher. The formal meeting began on Saturday morning and continued deep into the evening at the Presbyterian Student Center, located across the street from the Scarritt campus. Ed Hamlett opened the gathering by giving the keynote speech. In his address, he told the students that "the civil rights movement, in a sense, has freed the Southern white. It has allowed him to get outside himself, and to think about new ideas." Hamlett challenged students whose activism had been inspired by the civil rights movement "to embrace broader issues to meet the many problems of Southern politics and economics." And echoing Sam Shirah, he encouraged the students to take their activism off campus and into the community, suggesting that they become involved in anti-poverty, open housing, and voter registration campaigns. "If we are concerned about mobilizing more students," he concluded, "I submit that we have almost innumerable alternatives to action."[9]

After Hamlett's address, the students spent the remainder of the morning giving brief status reports on their own campuses, detailing any activism that already had occurred and speculating about the prospects for future action. In the afternoon, Robb Burlage of the Students for a Democratic Society chaired a panel discussion by Anne Braden, Marion Barry, and Sam Shirah on "the History of the movement . . . and where it's going." Barry and Shirah talked about white student involvement in the civil rights movement, while Braden set the current surge of activism in the broader context by speaking about the movement prior to 1960. For the rest of the afternoon, the students considered the different approaches to organizing southern students—black and white—that other movement groups had initiated, as they heard from Connie Curry and Frank Millspaugh on the NSA's efforts in the South, Jim Williams on SDS's programs in Appalachia, and Billie Stafford on SNCC's work on the campus and in the community.[10]

Beginning Saturday evening, the activists turned their attention to the question that had brought them together: Should they create an organization for white southern students interested in supporting and taking part in the black-led civil rights movement? As they talked late into the night, they grappled with a series of important questions, many of which SSOC would struggle to answer throughout its existence: "Who are we trying to reach?"; "How do we reach them?"; "Do we need a coordinating body?"; "What is the relationship of the local group to the southwide group?"; "Is our focal point to be civil rights?" If so then: "What are our relationships to be to other civil rights groups?"; "Should the organization initiate projects on its own?" The students continued the conversation Sunday morning in the conference's final session, "Where do we go from here?" While a consensus did not emerge about how to answer that question, the students, optimistic that they could draw more of their peers into the movement, decided to create a student organization to work toward that goal. Thus, the Southern Student Organizing Committee was born.[11]

Most conference participants were overjoyed by the coming together of the student activists. Sue Thrasher remembers, "there was a real sense of excitement in the room because finally you were meeting people . . . who [were] doing similar things, so that you had some support; so that you were beginning to talk to other people like yourself who had gone through some of the same things white southerners went through when they got involved, which is this immediate isolation from your peers and possibly from your family." Many students left Nashville believing that the creation of SSOC would provide a much needed shot in the arm for local organizing efforts. Clemson's Jerry Gainey, for example, was convinced that with SSOC's assistance, the activist students on his campus could win more white support. "My present state is one of excitement about what can be attempted at Clemson and in S.C. next [school] year," he enthused shortly after the meeting.[12]

Others, though excited to be at the conference and supportive of the new organization, reacted to the meeting with a mixture of confusion and awe. For these students, the conference was an eye-opening experience that served as their introduction to the wider movement. Nelson Blackstock recalls that when he arrived at the meeting from the University of Georgia, he "really didn't know what was going on. It was all new to me." Likewise, Bruce Smith remembers that the students at the conference from Lynchburg College "just sat there 'cause we didn't know what was going on." They were, however, moved by a visit to the Campus Grill, the equivalent of the Greensboro Woolworth's to the whites at the conference. The Lynchburg students also were astonished by "Carl Braden's brashness, the way he talked about standing up to racism, and Anne Braden's organizational thoroughness, and being in the presence of . . . Marion Barry and some of the black leadership of the sit-in movement." Carl Braden, in particular, opened Nelson Blackstock's mind to new ideas. Blackstock vividly remembers that Braden, "this older white-haired man sitting over to the side . . . gets up toward the end and says, 'you know, you've been sitting around here talking about the power structure. This power structure is nothing more than the ruling class.' And I remember I'd never heard anybody say that before, so I was quite impressed to be at a meeting where people were saying things like that." The conference, however, was not a politically transforming event for everyone. For instance, because Dan Harmeling already was part of a vibrant student movement at the University of Florida, questions of how to draw whites into the movement were less significant to him. And for Tulane graduate student Cathy Cade, personal concerns overshadowed the conference's proceedings, as she spent the first day secluded in a Presbyterian Student Center bathroom suffering through a miscarriage as a result of an illegal abortion she had undergone the day before.[13]

Lost in the excitement surrounding the group's founding were several matters the students left unresolved. Some students shared Hamlett's views that civil rights was intertwined with numerous other issues that deserved the students' attention. In the words of one SSOC activist, "while civil rights is the issue in the minds of many, it is by no means the only problem which confronts the South. . . . Economic, political, and social injustices are blatant, and are in a real sense the basis of the civil rights issue. It is the conviction of SSOC that these areas of blight on the South must be faced with the same fervor, dedication, and understanding as the more 'popular' issue of Negro rights."[14] But while the students at the Nashville meeting were open to working on other campus issues, such as academic freedom and in loco parentis, they generally were ambivalent about taking their activism off campus. As temporary residents of the towns and cities where they attended school, most students were strangers to the broader white community, and, given southern whites' well-known opposition to the movement, they were leery about trying to work in the white community. Thus, despite Hamlett's pleas, the students concluded the conference without devising a plan on how to induce young whites to organize among non-student whites.

The new group's racial composition was also left ambiguous. As a result, some students departed Nashville thinking they had just created an all-white organization, while others left just as firmly convinced that the new group would be racially inclusive. Hamlett recalls believing that SSOC was created as a biracial organization to work with "whites and blacks on the predominantly white campuses" of the South. Similarly, Florida's Dan Harmeling was repulsed by the notion that SSOC might consider itself a white organization. "I thought we were an arm of the black civil rights movement," Harmeling later explained. "I couldn't envision any of this white stuff that some people were talking about. It just didn't make any sense to me . . . [t]hat there's a white agenda, a separate white agenda; that we somehow had to represent the aspirations of the white students and what they felt." But others were very comfortable with the idea of a white student group. In their view, only a white organization stood a chance of winning the support of moderate white students in the South. Many agreed with Mary Washington College student Nan Grogan, who recalls that she always saw SSOC "as a thing to penetrate the white campuses in the South, which seemed very laudable and worthy." Sue Thrasher believes that since SSOC had set out to work with whites, it was appropriate, even necessary, for SSOC to be a white group. Additionally, she felt that the effort to organize white students should be carried out apart from SNCC. "My sense was that it was our place to work with other white students and not to be messing around in the black movement and telling people what to do because that's exactly what happened in SNCC. I mean, white people really took leadership roles [in SNCC]. And I never saw that as being my job." These competing visions of SSOC, born and then left unresolved at the Nashville meeting, would be a point of contention within the group throughout its first two years of existence.[15]

"We'll Take Our Stand"

Despite these uncertainties, the students considered the conference a great success, for white activists from across the region had connected with one another and shown an eagerness to coordinate future activities. And while the students had created SSOC only in name, they concluded the conference by taking a series of steps aimed at transforming SSOC from a paper organization into a viable student group. First, they appointed a Continuations Committee to meet in Atlanta in two weeks to devise a specific program of action for the group as well as to meet with the SNCC Executive Committee. Next, they tentatively made plans for a second region-wide gathering in May to discuss the results of the Continuations Committee meeting and to lay the groundwork for summer actions. Finally, and most importantly, the students adopted a statement announcing the founding of SSOC. Entitled "We'll Take Our Stand," the statement captured the spirit in which the organization was founded, expressed the goals toward which it strove, and reflected the values and ideas on which the organization was built.[16]

The principal author of the SSOC statement was Robb Burlage, the former editor of the University of Texas student newspaper who, in the words of one peer, was "full of hilarity and Texas sass." As an undergraduate in the late 1950s, Burlage, along with his future wife, Dorothy Dawson, was involved in the burgeoning student movement on the Texas campus, and they both went on to become founding members of SDS—he, in fact, wrote the economics section of the Port Huron Statement, SDS's founding document. After a brief stint as a graduate student in economics at Harvard, Burlage accepted a position as state planning research director for the state of Tennessee, and in the fall of 1963 the young couple moved to Nashville. The Burlages soon came in contact with the Nashville students, some of whom Robb Burlage had met two years earlier when he spent a summer in the city on a Harvard research grant. As a result, the Burlages participated in the discussions that led to the calling of the April conference, and they eagerly attended SSOC's founding meeting.[17]

Although Robb Burlage never assumed a formal role within the new organization, his emotional attachment to the South, long-standing interest in the region's economic and political history, and connection to the Nashville activists prompted him to write the SSOC manifesto. Recalling that he had been "an admirer of southern regionalism" since he was a child, Burlage spent his first nine years in Chapel Hill, where his father was a professor of pharmacy at the University of North Carolina and Howard Odum, the renowned regional sociologist, was a neighbor. When his father then accepted appointment as dean of the College of Pharmacy at the University of Texas, Burlage moved to Austin and later became an outspoken proponent of desegregation while an undergraduate at the university. At Harvard, the South became the focus of Burlage's scholarly work, as he began doctoral research on southern regional economic development under the guidance of John Kenneth Galbraith.[18]

Within SDS, Burlage quickly emerged as the group's expert on the South. In 1962, he already had begun to express many of the ideas that two years later would become the core of the SSOC manifesto. In the SDS pamphlet "For Dixie With Love and Squalor," he called for the opening of the South to liberal forces, arguing that the South should not "be 'left alone' but that it become a model of democracy and equal opportunity for the entire world." To move toward this idealized future, southerners, whites in particular, had to reject the myths that "clog the Southern consciousness," such as romantic notions of the Old South, a belief in black inferiority, and an unnatural fear of urban, industrial life. Because of their youth, Burlage believed white students needed to lead the movement for change in the South, yet he decried their silence on the important political issues of the day. While he noted that some students were blind to anything other than "the 'collegiate subculture' of football, fraternities, and frolic," he lamented that others feared speaking out would make them pariahs on their campuses.[19]

Burlage tackled southern economic issues in the essay "The South As An Underdeveloped Country." Echoing Franklin Roosevelt's words nearly 25 years earlier, Burlage declared that the South remained the nation's "number one economic problem," burdened by the declining productivity of an inefficient agricultural sector and by a manufacturing base built around northern-controlled extractive industries. Too often, he wrote, southern "'economic development' has come to mean: 'Persuade a yankee plant, any kind, to locate in your town.'" Connecting the South's racial order to the region's economic woes, he argued that racial discrimination kept many southerners mired in poverty while driving others out of the region altogether. Not until the Jim Crow system was eradicated, Burlage contended, would the South begin to prosper. As he explained, "Racism and segregation must be eliminated in all places, all sectors, before the 'New South' has even a chance of being anything more than grafting an industrial complex onto the old social structure with built-in alienation and discrimination. Only . . . [then] can the Southern economy grow to the benefit of all. . . . This is the projection, the goal, the dream."[20]

Assisted by his wife, Dorothy, and several of the Nashville activists, including, Archie Allen, David Kotelchuck, and Sue Thrasher, Burlage spent the days leading up to the April 1964 meeting drafting the manifesto. Working at the kitchen table in their tiny, above-garage apartment in the joint university neighborhood, Burlage crafted a document that called for reforms that nearly all southern whites could embrace, such as improving the quality of schools and ensuring adequate health care for all. The document also championed integration and condemned racism, the most radical positions southern whites could adopt in 1964. Additionally, like SDS's founding statement, "We'll Take Our Stand" is filled with soaring rhetoric and visions of a utopian society in which every individual would "be guaranteed the right to participate in the formulation of the social, economic, and political decisions which directly affect his life." But because the SSOC manifesto specifically addressed southerners, Burlage explicitly grounded the statement in the history and traditions of the South, and, as a result, "We'll Take Our Stand" resonated more deeply with the southern students than the Port Huron Statement ever did.[21]

Nonetheless, "We'll Take Our Stand" frequently echoed the tone and tenor, fears and hopes, expressed in the SDS manifesto. For instance, in its opening section, Burlage wrote,

> We do hereby declare, as Southern students from most of the Southern states, representing different economic, ethnic, and religious backgrounds, growing from birthdays in the Depression years and the War years, that we will here take our stand in determination to build together a New South which brings democracy and justice for all its people.[22]

Drawing on his earlier writings, Burlage emphasized that "Our Southland is still the leading sufferer and battleground of the war against racism, poverty, injustice and autocracy. It is our intention," he proclaimed, "to win that struggle in our Southland in our lifetime—tomorrow is not soon enough." Burlage then went on to outline the new organization's goals. Inspired by the black-led freedom movement, the students in SSOC vowed to return to their colleges and their communities "to create nonviolent political and direct action movements dedicated to the sort of social change throughout the South and nation which is necessary to achieve our stated goals." These goals included the abolition of segregation, "an end to personal poverty and deprivation," and the transformation of the South into an idyllic "place where industries and large cities can blend into farms and natural rural splendor to provide meaningful work and leisure opportunities for all." Bringing the manifesto to its emotional peak, Burlage declared,

> We as young Southerners hereby pledge to take <u>our</u> stand now together here to work for a new order, a new South, a place which embodies our ideals for all the world to emulate, not ridicule. We find our destiny as individuals in the South in our hopes and our work together as brothers.[23]

By titling his work "We'll Take Our Stand," Burlage directly linked the SSOC statement to *I'll Take My Stand*, the 1930 book-length manifesto that condemned the expanding urban-industrial order and extolled the virtues of southern rural life, including segregation. The book was written by 12 southern intellectuals who came to be known as the Nashville Agrarians (because they all either had studied or taught at Vanderbilt), foremost among whom were Allen Tate, Robert Penn Warren, Donald Davidson, and John Crowe Ransom. In *I'll Take My Stand*, these writers emphatically denounced the coming urbanization and the industrialization of the South, believing that these developments would exacerbate racial tensions and make the "good life" impossible; leisure, tradition, religion, and art all would be sacrificed in the frantic, dehumanizing, and single-minded pursuit of material goods and economic gain that characterized urban-industrial society. In their view, the rapidly disappearing southern rural order—a stable agrarian society with a humane, traditional culture—should be preserved at all costs.[24]

Burlage's abiding interest in southern history, combined with the fact that the founding SSOC meeting was to take place in Nashville, inspired him to challenge *I'll Take My Stand* in the SSOC manifesto. Burlage recalls that he saw the Nashville Agrarians "as people to be turned on their head because they were romantically trying, at best, to . . . escape from the modern South and not deal with the racist implications of the old heritage." As he pointedly remarked in "We'll Take Our Stand," the Agrarians were backward-looking southerners who "endorsed the old feudal agrarian aristocratic order of the south and opposed what they saw coming in the new order—widespread industrialization and urbanization with democracy and equality for all people." But unlike many other critics over the years, such as journalist H. L. Mencken, who ridiculed the authors for "spin[ning] lavender fancies under a fig tree," Burlage held a more complex view of the Nashville Agrarians. Burlage had titled his statement "We'll Take Our Stand" both because he saw the opportunity to create a dramatic and progressive statement by playing off their work and because he believed that the Agrarians, in their own way, were radicals who went against the grain by opposing the industrialization of the South. While Burlage and other early SSOC activists did not share the Agrarians' contempt for industrialized society, they did respect them for courageously standing up for their principles and for resisting the dominant trends in southern society.[25]

Burlage's decision to link "We'll Take Our Stand" to the work of the Nashville Agrarians enhanced the statement's appeal to the students at the founding SSOC meeting. Particularly significant was the manifesto's emphasis on southern distinctiveness and southern history, both of which helped to define and shape the organization throughout its existence. To Burlage and the students in SSOC, the South was a distinct region, distinguished from the rest of the nation by its widespread poverty and virulent racism but also by its unique traditions, culture, and history. In particular, SSOC students attached great significance to southern history since they believed it accounted for much of the region's distinctiveness. Yet rather than merely blaming the region's current woes on its past practices—particularly the actions of southern whites—SSOC activists identified whites in the past who worked to subvert the racial and economic order. SSOC students hoped that an interpretation of southern history that highlighted white opposition to slavery and segregation would inspire other whites to join the group by suggesting that there was a long history of white resistance in the South. As SNCC chairman John Lewis remembers, "it was important for them, as individuals and as a group, not to run from their history, from the past—not to try to be something else—because they had to say to the larger society, to the larger white community, that 'we are you.'" Drawing on the southern past, Lewis believes, invested SSOC students with a tremendous "sense of moral authority" because, on the one hand, "they could openly identify themselves as . . . someone [sic] who was very proud of southern history, culture, and the past," and, on the other hand, they could claim, owing to their particular interpretation of that past, "that 'we care about social justice and we want to be part of this growing movement of social justice for social change.'" In short, they sought to make southern history serve their current needs.[26]

The search for a "usable" past led many SSOC students into southern history courses in college. Gene Guerrero, who would become SSOC's first chairman, never found this past at Emory University where he studied under Bell Wiley, a respected older scholar whose tales about Johnny Reb and ruminations about how the climate shaped southern history infuriated Guerrero. "This was at the time when the South was burning and we were talking about whether the climate affected people," he angrily remembers. Others had more positive experiences. At the University of Virginia, many activists were drawn to the southern history courses taught by Paul Gaston, a young historian who was known to be a supporter of liberal causes. Gaston exposed his students to a southern past many had not seen before, introducing them to the writings of such scholars as V. O. Key and C. Vann Woodward and challenging them to see that tension, turmoil, strife, and oppression were enduring features of southern history. For the Virginia students, as well as those on other campuses, Woodward's works, in particular, served as a touchstone because they revealed an interpretation of southern history well suited to the activists' current interests and goals. From his 1958 essay "The Search for Southern Identity," they learned that a history of poverty, defeat, and guilt were the underpinnings of southern distinctiveness. At a time when the activists were preoccupied with building interracial alliances, Woodward's work on the Populists showed them that people could unite across racial lines to pursue common goals. *The Strange Career of Jim Crow* taught them that segregation did not become entrenched in the South until the 1890s, thus suggesting that they could destroy this relatively recent creation. Not surprisingly, when the Virginia history department sponsored a talk by Woodward in the mid-1960s, nearly every activist on campus attended.[27] And when SSOC published a "Bibliography on Southern History," six of Woodward's works appeared on the list.[28]

Drawing on their reading of southern history, SSOC students embraced an interpretation of the southern past in which dissidence and radicalism, though hardly commonplace, were defining characteristics of southern history, particularly the white South's history. They believed that shining light on the often overlooked radicals in the southern past could place the group in a tradition of indigenous southern radicalism, win it support from other whites, and, ultimately, enhance its ability to bring progressive change to the region. As one SSOC activist later remembered, "We in SSOC were constantly studying Southern history, looking for antecedents that would satisfy a need for rootedness. . . . We grasped at the Grimké sisters . . . the early, populist Tom Watson, the agrarian reformers . . . even slaveholder Thomas Jefferson in search of our ancestors in radicalism." Anne Braden, whose activism in the South pre-dated the birth of many SSOC activists, readily saw the students in SSOC as following in the footsteps of earlier generations of white radicals, from the Populists of the nineteenth century to the Communist Party of the 1930s to the League of Young Southerners of the 1940s. The students who formed the group, she believed, "were simply one of the succeeding generations of young whites in the South who have been propelled into activity for a better world by the African American struggle for freedom."[29] Besides helping to further their cause, SSOC students believed that identifying with the white radicals of previous decades would enable them, in the words of one member, "to feel okay about being southerners and being active anti-racists; that we didn't have to throw out being southern with the whole thing."[30]

SSOC and SNCC

The students left Nashville believing that they had taken the first steps toward creating a viable white southern student organization. Most of the conference participants considered it highly significant that the meeting had taken place at all. As Sue Thrasher enthused shortly afterward, "I had the feeling many times that weekend as we sat around tables and 'talked' that we were taking our stand." But Thrasher knew that the work of organizing the group had only just begun. Thus, after the meeting, she and the other conference organizers immediately began planning for the meeting of the newly created Continuations Committee in Atlanta on the weekend of April 18–19—just two weeks away.[31]

When the committee convened, its first task was to adopt a program of action, since the students at the Nashville gathering had not outlined specific plans and goals for the new organization. After a brief discussion, it adopted a six-point program that a subcommittee had drafted in the days between the Nashville and Atlanta meetings. The program committed SSOC to developing "campus service and educational programs" designed to awaken white southern students to problems on campus and in the community and to draw these students into the movement for social change. To do this, the program called on SSOC to initiate informational campaigns on such issues as university reform, poverty, segregation, and "Negro deprivation," to begin circulating a newsletter on student activism, and to show white students that through indirect action, such as taking part in letter-writing efforts and political campaigns, "they may perform service for the Movement in ways supplementary to militant direct action or fulltime staff work." The program also advocated that the students take their activism off campus, suggesting that they join "community organizing projects at the coalition level not only in the Negro community but in disinherited predominantly white communities as well—around the issues of unemployment, low wages, union organization, poverty, lack of community facilities, etc."[32] The committee then decided that with the end of the academic year approaching, SSOC should refrain from initiating the program until classes resumed in the fall. In the meantime, committee members agreed that SSOC should issue the first edition of the newsletter envisioned in the program and begin to consider projects the group could co-sponsor with other activist student organizations during the intervening summer months.[33]

The committee next turned its attention to defining the new organization's relationship with SNCC, the subject that dominated the remainder of the weekend. Despite the fact that several SSOC founders either had worked for SNCC or were well acquainted with its leaders, many white activists were ambivalent about establishing close working relations with the black-led group. Some SSOC activists doubted that SNCC would make white organizing a priority. Cathy Cade expressed this concern when she later remarked, "there was a real question as to . . . whether SNCC would be interested in putting its resources into white community organization." Although most of these activists felt that SSOC should maintain informal ties to SNCC, they strongly believed that the new group should develop independently in order to attract sympathetic, moderate whites frightened by SNCC's radicalism. Adopting such a position was problematic, however, for organizing apart from SNCC implied a rejection of the integrationist principles to which the white students were committed. To avoid creating such a perception, SSOC activists suggested that at some point in the future the southern whites organized by SSOC would join forces with the southern blacks radicalized by SNCC to create an interracial movement for social change in the South. While SSOC leaders had only the vaguest ideas about when such an alliance would emerge, how the two movements would be fused, and what its consequences would be, the building of a biracial coalition of southerners was the goal toward which these activists strove.[34]

For the students wary of working too closely with SNCC, the fact that SNCC had begun to reorient itself away from the urban black campuses where it had started and toward the poor, rural, black communities of the Deep South was decisive. While SNCC was not averse to working with college students in these communities, Bob Moses, who in 1961 had led SNCC's charge into the rural hamlets of Mississippi, later explained, "You couldn't really get a base on the campuses in the Deep South, black or white. So the focus of SNCC's work shifted to the community." Additionally, the group cast aside its original position that nonviolent direct action by students would lead to the enforcement of civil rights laws in favor of a more militant program that advocated mobilizing local blacks to organize and demonstrate for their own rights. Although SNCC would not complete this transformation until after the tumultuous events of the summer of 1964, by April most white activists recognized that SNCC was heading in a new direction. To Ed Hamlett, who, as a staff member in both organizations saw his role as keeping the two groups closely allied, it was obvious that SNCC was not "making any effort to do things with students; they were [trying] to get students off the campus to come and work in the community." As a result, many white activists believed that SSOC's focus on the campus required the group to organize independently.[35]

Sentiment opposing the complete organizational separation of the two groups also existed within SSOC, as some activists argued that developing within SNCC would enable SSOC to draw on the veteran organization's vast experiences and resources. Others who shared this view believed that unless SSOC joined with SNCC the two groups soon would find themselves beset by distracting organizational problems, ranging from overlapping programs to leadership disputes. Additionally, these students believed that, given SNCC's declining interest in campus organizing, great prestige would accrue to SSOC if it were to become SNCC's student arm. But even these students were well aware that such a close alliance with SNCC could cause moderate students to shy away from SSOC, fearful that it was as radical and as forbidding as its parent organization. Just as Bob Zellner had suggested two years earlier in his annual report on the White Southern Student Project, the students realized that any SNCC-related organization that hoped to attract moderate white students would have to walk a tightrope. As Georgia Tech student Roy Money wrote to a colleague, "I think the idea of SOC [*sic*] being a student wing of SNCC sounds terrific. I just hope there can be some way of approaching the college students without them knowing that SOC is a student wing of SNCC, at least until the general public understands more about who and what SNCC really is."[36]

Of all the people who favored organizational unity between the two groups, the most vocal was Anne Braden. In conversations with SSOC's leaders and in a long letter to the Continuations Committee, Braden strenuously argued that SSOC should become an autonomous affiliate of SNCC, developing its own program and electing its own leaders but retaining formal organizational ties to the black-led group. Uniting the two groups, Braden believed, would reduce tensions between black and white activists and show the world that an interracial movement could succeed. Dismissing the view that SSOC's attachment to SNCC would frighten away many whites, Braden wrote SSOC's leaders, "your job is to <u>challenge</u> the moderate white student. History is camping on his doorstep and you must help him see it. I do not think you help him by slipping up on him with an organization that looks as if it won't demand too much of him. . . . I think you help him . . . by being just what most of you are: fully committed young people who yourselves understand and identify with the militancy of SNCC."[37]

Braden also believed it was hypocritical and even condescending for the white activists to assume that ties with SNCC would make the group a pariah on predominantly white southern campuses. If contact with the black movement had changed these young whites' lives, she later reasoned, "why wouldn't it work for many, many more? Why not take that message out to the hinterlands?" Even if association with SNCC *did* taint SSOC in the eyes of moderate whites, Braden reflected, "that just says to me that this is a battle that it's essential to wage—because no 'white' movement that is not rooted in the struggles of people of color in this country (and therefore by necessity assaulting racism) is going anywhere." Ultimately, Braden feared SSOC would undermine its own efforts if it separated from SNCC, warning the students, "you might really harm the moderate white student if you . . . merely provide him with an excuse to postpone the day of true confrontation with his conscience and with the problems of the South."[38]

The relationship between the two groups was the focal point of a Sunday meeting with the SNCC Executive Committee to solicit its support for the new organization. After the Continuations Committee formally presented the SNCC leadership with "We'll Take Our Stand" and the six-point program, SNCC–SSOC relations moved to center stage. In the friendly but at times heated discussion that ensued, several Executive Committee members expressed support for the students' efforts to create an independent white organization. Ella Baker, the sexagenarian who was the guiding force behind the creation of SNCC, believed that SSOC should be allowed to develop naturally, even if that meant it would move away from SNCC. Baker counseled the students not to get bogged down in these sorts of organizational issues. SSOC should not "get too structural too early," she told the Continuations Committee. "Don't sit and talk about structure—let the thing grow together like Nashville." Hamlett and Shirah, both of whom continued on SNCC's staff as the White Southern Student Project organizers, encouraged the organization to support an independent SSOC. At one point, Shirah, in an attempt to force the Executive Committee to think of SSOC as an independent entity, asserted that the key "issue [is] not whether SSOC can service SNCC but vice versa."[39] Many of the black student leaders also readily supported the idea of an independent SSOC. Thrasher was surprised to discover, she noted shortly after the meeting, that these officials seemed "more enthusiastic about SSOC than we were ourselves." Several of them, including Courtland Cox and Stokely Carmichael, later became ardent proponents of black separatist thought within SNCC. As Thrasher recalls about the discussion, "the people who really first began to talk about Black Power liked this idea of white people working with white people."[40]

But not everyone on the SNCC Executive Committee favored the creation of an independent SSOC, and some were hostile to the idea of supporting any type of white southern student group. Stanley Wise, SNCC's southern campus organizer, remembers that some of his colleagues felt "that we were wasting time in the white community, that to participate with whites and encourage that kind of group . . . was just unrealistic. . . . [T]hese were not the people in the white community. . . . that is not *the* white community." More common was the view that SSOC should develop within SNCC's organizational structure because it was the ideal vehicle for carrying SNCC's message to moderate whites. According to SNCC Chairman John Lewis, "we felt they could have a major impact on organizing and mobilizing white students and creating a cadre of supporters in the larger white community." These activists also believed that keeping both groups under one organizational roof would demonstrate SNCC's commitment to interracialism and prove that the group was not recreating the divisions it was trying to break down in the wider society. Just as those who later advocated Black Power tended to support an independent SSOC, those who most identified with SNCC's beloved community heritage, including Lewis and Marion Barry, opposed separation, fearful that, as Thrasher later realized, "we were setting up a black movement and a white movement."[41] Although they did not want SSOC to develop as an independent group, these SNCC leaders were not going to expend too much energy on this issue. As Anne Braden recalls, SNCC executive secretary James Forman told her, " 'I don't think they should set up a separate thing, but I'm not going to beg them to stay.' "[42]

At meeting's end, a consensus had not emerged on inter-organizational relations, although it was clear to all that bonds of friendship and respect would ensure close ties between the groups. The meeting did, though, yield some tangible results. The SNCC Executive Committee agreed to give SSOC $300 to help finance the group's first newsletter and a meeting it had called for May 9–10 in Atlanta. The groups also agreed to create a committee comprised of three people from each organization to continue to discuss future SNCC–SSOC relations as well as to consider possible programs on which the groups could work together, one of which was the upcoming Freedom Summer project in Mississippi. Expressing the view of many Continuations Committee members after the meeting, Sue Thrasher remarked with relief that she was "cheered by the fact" that those on the Executive Committee saw "SSOC as something that should not be crushed, at least in its beginning stages."[43]

Officers, Newsletter, and Fraternal Ties

Three weeks after the Continuations Committee met, 40 students, most of whom had attended the initial SSOC meeting in Nashville, gathered in a YMCA building in Atlanta to continue the process of building the organization. The critical task facing the students at this meeting was to develop the group's organizational structure. Rather than creating a tightly knit, hierarchical organization in which the power to set policies and make key decisions resided with the top leadership, the students established SSOC as a loosely structured staff—rather than membership—group in which local affiliates defined their own specific goals and determined the issues on which they would act. SSOC, then, would exist as a decentralized organization in which ultimate authority belonged to the local groups, whether they were SSOC chapters or independent local formations. The role of the group's officers and staff would be to serve these local organizations, furnish them with resources, supplies, advice, and encouragement, and provide a communication network through which the campus-based groups could remain in contact.[44]

Despite SSOC's local orientation, its elected officials would oversee the implementation of the group's program and represent SSOC to the public and to other activist organizations. Not surprisingly given the instrumental role Nashville students played in the founding of SSOC, two of the three officers elected at the meeting came from the city—Ron Parker became treasurer and Sue Thrasher was elected executive secretary. For chairman of the organization the students chose Gene Guerrero, a student at Emory University in Atlanta. Although Guerrero's desire to remain in school prevented him, at least initially, from relocating to Nashville, where SSOC set up its office, he won election as chairman precisely because he was not from the city.[45]

The students approved several projects at the meeting. In an effort to attract other students to the new organization, the activists launched plans to sponsor a Southwide conference in the fall. They also began to lay the groundwork for a campus traveler program that would enable staff members to visit colleges and universities throughout the region to drum up interest in the movement and provide activist students with guidance and support. Jim Williams, a Continuations Committee member from the University of Louisville, then won support for a proposal to create a SSOC newsletter. "If we unite our forces to put out a newspaper," Williams argued, "that work will prepare and bring forward, not only the most competent propagandists, but also the most skilled organizers and the most talented leaders." The others agreed with Williams, and before the end of May SSOC had published the first issue of its newsletter, the *New Rebel*.[46]

Just as at the first SSOC gathering in Nashville, the proceedings of the May meeting revealed SSOC activists' self-conscious effort to highlight the group's southern orientation and roots. The students believed that evoking southern distinctiveness not only displayed their pride in their heritage but would enable SSOC to reach whites who normally ignored activist organizations. To this end, the students worked to ensure that SSOC would become known for its regional orientation and its traditionalism. Provocatively calling the newsletter the *New Rebel* was one manifestation of this effort. More controversial was the symbol, created by Claude Weaver, a black Harvard student who went South to work for SNCC, which the students adopted as the logo for the group: a drawing of clasping black and white hands superimposed over the battleflag of the Confederacy. While a significant minority of students opposed linking SSOC to the Confederacy, no matter how indirectly or creatively, most of these young rebels believed that appropriating the preeminent symbol of last century's rebellion could aid their efforts in trying to initiate a very different type of uprising in the 1960s.[47]

The students also discussed the merits of allying with the Students for a Democratic Society, the northern-based activist organization that had begun to make a name for itself with its anti-poverty work and radical critique of contemporary American society. Ironically, while SSOC had yet to formalize its relationship with SNCC, the organization with which most of the activists identified, it quickly and quietly became a fraternal organization of SDS. But whereas SSOC leaders participated in a searching discussion with their counterparts in SNCC about relations between the groups, they did not engage in a similar dialogue with the leaders of SDS. As a result, differences between the two groups over such issues as SSOC's regionalism, SDS's radicalism, and both groups' desire to organize the South, went unnoticed or were glossed over. And, most significantly, these differences, though seemingly minor in the spring of 1964, eventually ruptured the relationship between the two organizations and hastened SSOC's demise.

In a letter to the students gathered at the May meeting, SDS president Todd Gitlin, citing the two group's common goals, invited SSOC to become a fraternal organization of SDS, thereby allowing SSOC representatives to cast votes at SDS meetings and permitting the group to distribute SDS literature. Extending such an offer revealed SDS leadership's growing enthusiasm for SSOC, an enthusiasm based on the view that SDS could use SSOC to draw southern students to the northern group. Two months earlier, Gitlin had been much less optimistic. "I really can't ingest the reality" of SSOC, he wrote the Burlages, who had been keeping him informed of the ongoing effort in Nashville to create a white southern student group. "[I] am bothered by Tom's [Hayden] comment . . . that the thing sounded 'gossamer.' One of his favorite words, to be sure, but a kernel of truth in it. . . . Inform as to progressive [credentials]." In his May letter, however, Gitlin was decidedly more upbeat, telling the students that SDS and SSOC " 'each have a great deal to offer each other.' " He was, though, much clearer on what SSOC could do for SDS, namely, deliver progressive southern students to it. Key to this process would be the fact that SSOC, unlike SDS, was a staff rather than a membership organization. Thus, as SDS envisioned the relationship between the groups, SSOC would radicalize southern students, who then would join SDS.[48]

While the focus of Gitlin's letter was SSOC–SDS relations, several of his comments raised issues that later divided the organizations. For instance, he took exception to SSOC's emphasis on southern distinctiveness. Gitlin believed that the South was not much different from the North, arguing that "we have all waited too long to realize that the problems of the South are also the problems of the nation." "This is not to say that the South does not have its special problems," he gently explained, but whether or not they were unrelated to northern problems "is a debate we can and should carry on elsewhere." Gitlin also raised the specter of a battle for organizational ascendancy in the South when he noted that SDS was planning to appoint a southern field secretary in the near future. As a fraternal group, SSOC could look forward to working with this campus traveler: "While this field secretary will be trying to organize SDS chapters on Southern campuses . . . [he] would also offer his services to SSOC." Although Gitlin meant this as a gesture of goodwill, SDS organizing in the South would emerge as a major source of conflict between the groups.[49]

But such concerns were far from SSOC leaders' minds in May 1964. Instead, they were pleased that SDS had taken an interest in their two-month-old organization, and they happily accepted Gitlin's offer of fraternal ties. Shortly thereafter, several of them drafted a resolution on SDS–SSOC relations to present at the upcoming SDS national convention. Echoing much of what Gitlin wrote, the resolution stated that "SSOC will serve as the nerve center of the Southern movement, a sort of Ann Arbor of the South. Its role is to make contact with all students in the South who are interested in liberal-radical activity." But because SSOC was only a staff organization, once it reached these students it would have to steer them toward membership-based groups. "This is where SDS enters the picture," the resolution continued. "SDS chapters are needed as <u>membership</u> organizations. . . . SSOC is primarily a <u>coordinating</u> organization. It is necessary that SSOC be complemented by a broad based <u>membership</u> organization." Thus in SSOC's view, the two organizations could work together to organize the white South.[50]

The establishment of fraternal bonds, however, could not mask important differences between the groups. Most notably, SDS attracted students who were more radical politically and more sophisticated intellectually than the founders of SSOC. Sue Thrasher was startled by this discovery, recalling, "when I met SDS people, I thought they were from another planet. They were intellectuals. None of the people I ran around with were intellectuals. . . . They didn't analyze things. You acted on gut instinct, you acted on intuition, and you acted on what you believe." SSOC, with its commitment to action and its lack of ideological moorings, stood in stark contrast to SDS, which had devised a well-developed ideology of social democracy. Ed Hamlett remarked that southern activists were "rooted in time and place in a way that, say, Tom Hayden isn't. . . . Down here in the South, our radicalism is less ideological than it is soul, heart and gut stuff."[51] The differences between the groups were such that Hamlett questioned whether SSOC legitimately could be regarded as a New Left organization. "Most of us do not know enough about the 'Old Left' to reject it," he wrote in 1965.[52]

Despite these differences, the southern students chose to unite with SDS because such an alliance would prove advantageous to SSOC. Joining with SDS would give SSOC instant credibility. Additionally, many in SSOC believed fraternization would permit the fledgling organization to tap into SDS's considerable—at least from SSOC's perspective—resources. As Charles Smith, a University of Texas student, wrote about SSOC's resolution on relations between the groups, "Frankly, the purpose . . . is to force the national student movement to spend some money on us." Thus, the resolution not only detailed the southern activists' vision of SDS–SSOC relations, but requested that SDS provide SSOC with literature to distribute, fund two campus travelers, and hold one of its conferences in the South. Consequently, SSOC students were just as ambivalent about establishing close ties with SDS as they were with SNCC. But there was an important difference: while SSOC and SNCC always maintained cordial and respectful, if not always close, ties, SSOC and SDS would drift apart and develop serious reservations about each other's goals and tactics. Thus, even as SSOC and SDS cultivated fraternal ties, the seeds of their future disagreements already were planted.[53]

The Founding Generation

For the students who founded SSOC and participated in its first meetings, becoming involved with an activist organization was not a decision they made lightly. These young southerners understood that participation in SSOC could alter their lives irreversibly, branding them "race traitors" in the eyes of other whites, leading to rejection by their friends and their families, and closing career opportunities to them. What drove SSOC's founding members to speak out for progressive social change in the South and thereby risk making themselves pariahs in their communities? What impelled them to reject the southern racial order, to step "outside the magic circle" of friends and family to support racial equality?[54]

For SSOC's first generation, shared backgrounds, values, and experiences propelled them into action and distinguished them from their northern contemporaries as well as from later SSOC activists. SSOC's founders were not secularized, privileged youth who came from affluent, highly educated, politically liberal families, as some scholars have suggested about the early SDS.[55] Instead, most had middle-class upbringings, and they typically received little overt familial support for their progressive views. For some, deeply held religious beliefs sensitized them to the sufferings of others and inspired their activism, while for others specific experiences during their collegiate careers motivated them to take action. By rejecting the dominant views of the white South and refusing to remain silent, SSOC's first generation created the vehicle by which later SSOC activists challenged the status quo and worked to bring progressive change to the region.[56]

For many of SSOC's first members, religious faith and their parents' moderate racial views encouraged them to support civil rights.[57] The Southern Baptist Church was a dominant influence in Ed Hamlett's life growing up in the small town of Jackson, Tennessee. In 1959, as a freshman at Union University in Jackson, he was elected an officer in the Baptist Student Union and became licensed to preach. By his second year in college, however, Hamlett "had begun to question a number of things which I had been taught as a southerner and as a Southern Baptist. . . . It seemed to me that the New Testament Christianity which was constantly stressed did not allow for color bars." His parents' moderate views on the race issue also shaped Hamlett's outlook on civil rights. Not only were they not "rabid segregationists," but his father's belief that everyone should have the right to vote awakened him to the injustices of the Jim Crow system.[58]

Born in Atlanta and raised in Gastonia, North Carolina, Anne Cooke grew up in the Presbyterian Church. A deeply religious person, she fell in love with the music of the church and had fond memories of going to tent meetings as a high schooler. Intent on becoming a missionary, Romaine matriculated at Queens College, a small Presbyterian women's school in Charlotte. Her parents, especially her father, held progressive views on civil rights issues. A North Carolina state senator during the 1950s, he had worked on liberal U.S. Senator Frank Porter Graham's failed 1950 reelection campaign—a campaign in which the anti-segregationist Graham was branded a race traitor by his opponent, Willis Smith—and he openly had supported the *Brown* decision.[59]

Sue Thrasher grew up in and around Savannah, Tennessee, in the rural southwestern part of the state. Her parents were devout Methodists, though they rarely attended church. She, however, was very active in the local congregation, becoming president of the Methodist Youth Group and teaching Bible School. Thrasher considered her parents "paternalistic racists": though they referred to blacks as "niggers," they supported desegregation and they treated blacks, in Thrasher's words, "as human beings who were less fortunate than they were but nevertheless were human beings." For the time and place, these were liberal views, and Thrasher attributed much of her progressivism on race to her parents' moderate stance.[60]

A native of Texas who moved to Atlanta during his high school years, Gene Guerrero was a Southern Baptist who took the church's teachings very seriously. His father, who was from Mexico, greatly influenced young Guerrero's views on race relations. Though his father never spoke about civil rights, he set an example by his actions. For instance, Guerrero remembers as a boy "being in the car with the family and pulling up to a gas station and having a black man come up to wait on the car and my father saying 'yes, sir' to him, which struck me at the time as being a little strange."[61] Likewise, Steve Wise's parents' progressive racial views influenced their son's perception of racial matters. Wise, who came from an Episcopal family in Newport News, Virginia, recalls that his father frequently wrote letters to the city's newspaper protesting Jim Crow ordinances on the grounds that "segregation was unchristian." Partly because of his father's views, Wise was always sensitive to civil rights, and as a youth he remembers reading books from the "Boy's Life" series on Booker T. Washington and George Washington Carver.[62]

Bruce Smith, who came to SSOC from Lynchburg College, was drawn to civil rights both by his experience in the Methodist Church and the influence of his racially progressive mother. Smith and his four siblings were raised in suburban Northern Virginia by his widowed mother, an Ohio native who supported civil rights, was active in the local Republican Party, and taught her children that racist Virginia Democrats were to blame for racial discrimination in the state. Reflecting his mother's partisan preferences, Smith remembers growing up "in a household where we talked about the 'damn Byrd machine' and the one party South and the travesties of not having a two party system." Smith's participation in the teen discussion group at his church also encouraged him to think critically about civil rights. Although he never particularly liked going to church, the discussion group at Dulin Methodist Church, with its focus on civil rights and other current events, excited Smith and helped to shape his liberal views on racial matters. That his stand on civil rights had diverged from most of the other congregants became clear to Smith when the church's minister was forced to step down shortly after meeting with several local black clergymen. While Smith was shocked that several powerful church members called for the minister's resignation, he was more appalled that other adults did not defend the minister. As a result, Smith vowed never to return to the church.[63]

Nearly all early SSOC activists could identify a particular event or experience that motivated them to openly support the civil rights movement. Only such a jarring and transformative incident—a conversion experience of sorts—could bring the students, even those raised in racially moderate families, to stand apart from other whites on civil rights, to break free from what one commentator has called the "deep local whiteness" that colored their lives.[64] The turning point for the students frequently occurred early in their collegiate careers. For Ed Hamlett, the change came in 1959 when he moved to Washington, D.C., to attend George Washington University. Although he left Tennessee "a segregationist of a very moderate sort," Hamlett's three years in the nation's capital altered his view on civil rights. First, a friendship he developed with a black woman forced him to reconsider his racial views. Then, he was embarrassed to learn that black sharecroppers in Haywood and Fayette Counties, Tennessee, near his hometown of Jackson, had been evicted from their homes for trying to register to vote. As a result, Hamlett decided he had to act, and he quickly made contact with the Washington NAACP chapter. Shortly after he returned to Tennessee in 1962 to enroll at the already desegregated University of Tennessee at Knoxville, he attended his first SNCC conference and joined with another student, Marion Barry, to become co-chairman of Students for Equal Treatment, an interracial group that successfully desegregated several off-campus facilities. Hamlett then returned to Union University, where he had started his college education, to earn his Bachelor's degree in 1963, and he had a brief stint in graduate school before going to work for SNCC in early 1964.[65]

A series of events early in Gene Guerrero's career at Emory University motivated him to take an active part in the civil rights struggle. Moving away from the Southern Baptist Church in which he was raised, Guerrero became involved with a Unitarian student organization through which he met black students from the Atlanta University Center, the first time he had encountered black youths as equals. Guerrero also had to confront the race issue when he learned the fraternity he had joined, Sigma Nu, had a racial exclusion clause. By the fall of 1963, Guerrero had committed himself to civil rights as he resigned from the fraternity and attended a student civil rights conference in Washington, D.C. He became active in the movement later that fall when he was arrested during a sit-in at a Krystal's restaurant in Atlanta. Ironically, he ended up in jail not for taking part in the sit-in—which he did not do—but for inadvertently crossing the police line to get a better view of the action. Angry and scared, Guerrero was placed in the "drunk tank" where he stayed up all night talking with one of the demonstrators, Sam Shirah. After this incident, Guerrero began to participate in civil rights protests, though his actions were limited: while black students conducted sit-ins, he and other whites, too frightened to participate themselves, would stand across the street holding signs declaring their support for the protesters.[66]

The transformative event in Cathy Cade's life occurred even before she arrived at Carleton College in Minnesota. Born in Honolulu, Hawaii, to midwestern parents, Cade moved frequently throughout her childhood until her family settled in Memphis. Raised as a Unitarian, Cade's views on race were shaped by her mother, an activist liberal who was involved in local church affairs. Although she attended a segregated high school, Cade became active in Liberal Religious Youth, her church's youth group, where she met other young whites who shared her disdain for segregation. When she became president of the group, she helped to arrange meetings with the youth group of a local black Baptist church. The meetings were her first personal encounter with black teenagers, and they proved to be a moment of awareness for her, as she realized, "all of a sudden we were, you know, teenagers. It's not that there weren't differences, but personhood was right there." As a junior at Carleton, she was an exchange student at Spelman College in Atlanta, where she became active in the local movement and began to spend much of her time in the SNCC office. This experience confirmed her desire to work in the movement, and, after graduating Carleton in 1963, she took part in the Albany, Georgia, movement as well as the Freedom Summer project in Mississippi in 1964.[67]

Anne Cooke's life began to change when she worked as a Presbyterian missionary in Mexico after finishing her second year at Queen's College in 1962. She quickly became disillusioned with her work: although the people she dealt with were dirt poor, the missionaries seemed to care only about converting them from Catholicism to Presbyterianism. After three months, she returned to the United States and immediately entered a semester-in-Washington, D.C., program at American University. Just as for Ed Hamlett, her time in Washington was transformative, as she met new people and began to take a serious interest in civil rights. Upon returning to Queens, she recalled, she was "a different person; I'd lost my religion . . . and shifted it over to civil rights." After graduating, she entered the University of Virginia as a graduate student in history and became involved with a local activist group.[68]

On her first day at Virginia, she met her future husband, Howard Romaine, who, with his long hair and jeans, stood out, Cooke fondly remembered, as "this bright spirit of excitement and determination and commitment to something beyond UVa." For Howard Romaine, a native of New Iberia, Louisiana, the turning point in his life also came during his second year in college. The son of a devout Presbyterian mother and a Catholic father— "they were always fighting the sixteenth-century wars of religion between themselves," he only half-jokingly recalls—Romaine attended Southwestern at Memphis. James Meredith's entry into Ole Miss in 1962 had first made him aware of civil rights and prompted him and several of his peers to question the administration as to why so few minority students attended Southwestern. He became actively involved in the civil rights movement the next year when he began participating in Reverend James Lawson's workshops on non-violence at his Memphis church. Inspired by Lawson, Romaine and nine others faced a mob and eventually were arrested in an unsuccessful attempt to integrate the Memphis Tea Room, a restaurant across the street from the Memphis State University campus. Then as a member of the local NAACP Youth Group in 1964, he took part in demonstrations at the segregated Second Presbyterian Church, the largest Presbyterian church in the city. During these demonstrations Romaine met Ed Hamlett, who, in his capacity as white student organizer for SNCC, had traveled to Memphis to try to find recruits for the upcoming Freedom Summer project in Mississippi. Romaine was among the few southern white students who participated in the summer project, working in the Jackson area before going off to graduate school at the University of Virginia, where he would continue his protest activity.[69]

For University of Georgia student Nelson Blackstock, the hard-core segregationist view espoused by many of his peers, along with a chance encounter with a group of peace activists, prompted his involvement in the civil rights movement. When he arrived on the Athens campus in the fall of 1962, Blackstock already had concluded that segregation was wrong. The murder of Emmett Till—a photo of whose bloated, badly deformed body Blackstock had stumbled across in *Look* magazine—the widely televised Little Rock school desegregation crisis, and Harry

Golden's book of humorous but biting critiques of the Jim Crow system, *Only in America*, all turned Blackstock against segregation. While this was a lonely position to take in the working-class neighborhood just south of Atlanta where he grew up, Blackstock was hopeful that he would meet people at college who also opposed segregation. He was disappointed, however, to find that many students were, in his words, "more backwards than the ones I grew up with, arguing a defense of slavery some of them." Though committed to desegregation, Blackstock did not feel compelled to speak out or act on his convictions until one day during his second fall in Athens when he came across an impromptu rally by a peace activist who was part of the Committee for Nonviolent Action's Quebec-to-Guantanamo Peace March. Impressed by the marcher's commitment both to pacifism and civil rights—he wore a SNCC button and a peace button—and incensed that campus authorities had forced him to stop speaking, Blackstock and a friend went to the dean's office to protest the silencing of the activist. Shortly thereafter, Blackstock began to frequent Westminster House, the Presbyterian Student Center, which had hosted the peace marchers in Athens. There, he met other students who shared his opposition to segregation, and soon he was among a group of students that traveled to Atlanta on weekends to take part in sit-ins and picketing at downtown restaurants.[70]

Sue Thrasher and Archie Allen, classmates at Scarritt College, were moved to action both by specific experiences at Scarritt and by the violence that accompanied the movement throughout the South. The fatal bombing of the Sixteenth Street Baptist Church in Birmingham especially affected Thrasher. The bombing, she believed, stained all white southerners and forced her to realize that unless she spoke out, she would be represented by the words and actions of those she despised. After learning of the attack, she later wrote, "I made a silent vow to make my own voice heard."[71] Archie Allen, a Methodist minister's son from rural southwest Virginia, went to Scarritt with the intention of becoming a missionary. He immediately was won over to the cause of civil rights after hearing a guest classroom lecture on "Techniques of Christian Nonviolence" by John Lewis, then a leader of the local Nashville movement. Soon, Allen was playing an active role in the Nashville campaign, participating in prayer vigils against segregation at the YMCA, and helping to organize the Campus Grill demonstrations.[72]

Nan Grogan's commitment to activism, like Ed Hamlett's and Anne Cooke's, was fueled by her experience in Washington, D.C. Grogan, a student at Mary Washington College in Fredericksburg, Virginia, spent the summer of 1963 doing secretarial work in a government office. Neither in Fredericksburg nor in her hometown of Staunton, Virginia, had Grogan ever known blacks as equals. In Washington, however, many of her co-workers were African American, including the head secretary in her office. The March on Washington later that summer was the pivotal event in her Washington experience, revealing to her for the first time that black Americans throughout the nation—including in her native Virginia—were engaged in a broad-based struggle for equal rights. Energized by this discovery, she returned to school and immediately joined the YWCA race relations committee on campus. Over the coming months, Grogan became a leading activist on campus, and later worked in the SNCC office in Atlanta and participated in civil rights projects in Mississippi.[73]

The March on Washington also was a transformative event for Bruce Smith, who, like Grogan, worked a government job in Washington in the summer of 1963. As a Transportation Department employee, Smith was designated an emergency ambulance driver on the day of the march in case it degenerated into a race riot, as many whites feared. Smith reported with his Transportation Department truck to Southeast Washington on the day of the march, but the only action, he happily noted, was the unending procession of cars filled with blacks streaming toward the march. Soon after, Smith learned that several of his high school buddies actually had attended the march. "That just made a massive impact," Smith recalls, because "those were my jockey friends and my drinking-party-regular-old-good-ol' guys." Consequently, Smith thought, "if there's that many people in this country that are committed to standing up, I wanted to [too]. So I went back to Lynchburg College to find others. I said 'there's got to be some people there.'"[74]

For virtually all of the students who joined SSOC in its early days, their protest activities caused trouble at home. No matter how progressive their families were on the race issue, the students' parents disapproved of or, at the very least, were uncomfortable with their children's activism. From the perspective of SSOC's founders, the distance between their racial ideals and their parents' views—regardless of how moderate they were—widened as their activism increased. While living at home during the summer of 1964, Sue Thrasher recalls watching the Republican Convention nominate Barry Goldwater and feeling "light years away from my family . . . although they were not Republican." To her, the Republican Convention "was threatening to what I believed in and wanted to work for. And I just felt—I was antsy to get out of there [home] again and go somewhere. Go to Nashville or go somewhere else."[75]

Similar tensions developed among northern activists' families, too, especially over antiwar activism and cultural issues—drugs, sex, clothing, music, lifestyle. Infrequently, though, did civil rights activism lead to family break-up. In the South, this was an all too common occurrence. In 1963, Joan Browning, a white civil rights activist from Georgia who was arrested during the Freedom Rides in Albany, Georgia, wrote to a northern friend of the "agonizing loneliness and isolation of the 'new' Southerner who is freed of racial or ethnic bigotry and is alone. We are denounced by friends and family as traitors, distrusted by Negroes, and mis-understood by nonSoutherners." Dorothy Burlage concurs, recalling that social activism radically altered the lives of young southerners like herself: "As you got deeper into the movement you, in fact, did burn your bridges. . . . I don't think northern white kids burned their bridges in the same way; that we kept getting further and further from anything we knew about how to live a life. . . . And northern kids usually kept their ties to academic institutions, to other liberal friends, to people who had ties to the system. We lost our ties to the system." As a result, most southern activists were keenly aware of the sacrifices social activism entailed for them. Writing after Freedom Summer in the SCEF newspaper, Nan Grogan noted that while northern students returned home to a hero's welcome, southern students went home to "ostracism and pain." Given this fact of life, most white southern students in the movement were extremely committed to their work. Thus, Grogan explained, "you think a long time [before deciding to participate] and when you make up your mind you really believe."[76]

Still, commitment to the cause did not make home life any easier, and perhaps even led to more open conflict between students and parents. Concern for their children's safety and well-being prompted many parents to oppose their activism. Gene Guerrero's parents, for example, fearful that he might be targeted for abuse, unsuccessfully tried to dissuade him from going to Mississippi for Freedom Summer. Nan Grogan remembers her father's exasperation when he saw that she was living in inner-city Atlanta while she worked for SNCC in the summer of 1964. Dismayed that his daughter would choose to live in a poor neighborhood and work for very little money, he uncomprehendingly told her, "I worked all my life to get out of poverty. How could you return to this?"[77] Howard Romaine and Ed Hamlett met sterner resistance at home. Romaine's parents disapproved of his participation in the campaign to integrate Memphis's Second Presbyterian Church. Fueling their anxiety were the pictures of the interracial group of protesters that the church's deacons sent to the students' parents. Romaine recalls having telephone conversations with his parents in which, while his mother sobbed in the background, his father would ask accusingly, "'Boy, you still riding those niggers around in your car?'"[78] Ed Hamlett initially concealed his protest activities from his parents. Once they learned of his activities, however, Hamlett and his parents exchanged a series of emotional and tension-filled letters. In one, his father tried to convince Hamlett that he was wasting his time:

> As I have told you before, I have been around negroes all my life, have played and worked with them, and I think I know a little more than you of their way of life. I have talked with them in regard to race situations. They want to be left alone, to live their lives as they have thru the years, to carry on as their fore-fathers taught them, just like we have.

In another letter, his mother implored, "I have cryed [sic] so much lately.... Why oh why do you have to do these things.... I know you want to help the colored but if you could do it in trying to win them to Christ through our Baptist work people would understand you better." Hamlett was unmoved by these pleas, and, in an angry reply typical of the many letters he penned during these years, wrote:

> When you write letters like that you only strengthen me in my aims. Can't you begin to see that something is wrong with you if you are upset because I want people, yes friends—close friends of mine to be free. I cannot be free until they are.... I'm practically torn apart that you react to my goals in this manner, but I cannot turn back.... p.s. The majority of Negroes are as good church members as the white folks I know. They don't need the god of most Southern Baptists.[79]

Sue Thrasher's parents, unlike those of her peers, were always very supportive of her activism. Though at times worried about her safety, they never tried to suppress her growing involvement in the movement. Even though her racial views separated her from her parents and occasionally created uncomfortable moments, their support was extremely important to Thrasher. " 'If I had been cut off by my family it would have been a great trauma. But they made it very clear to me that the family relationship would continue. They did warn me from time to time to be careful, but that was because they loved me.' " Her parents showed this mixture of support and concern at her graduation from Scarritt. When two of her friends from SNCC, Lester McKinney and John Lewis, arrived at the ceremony, she openly hugged them, bringing stares from other whites, including her father. " 'For a southern man to see his daughter hugging the necks of two black guys must have been quite an experience. But he didn't reprimand me. He just said I should be careful.' "[80]

Sharing similar backgrounds, values, and principles, the white students who filled SSOC's ranks during its initial months of existence comprised a unique group. While subsequent SSOC students, driven by fresh concerns and united by different values, would reorient the group, SSOC's first generation bequeathed to later activists the organizational structure through which they could work to effect their goals. Beginning as an intriguing idea among a handful of white Nashville activists, SSOC had become a full-blown organization by May 1964, complete with officers, a manifesto, a program of action, and a newsletter. But as the creators of SSOC well knew, founding an organization was not the same as building an organization. Making SSOC a stable, well-grounded organization required time, hard work, and an unwavering commitment to social change. Reflecting on the creation of SSOC in the wake of the May meeting, Thrasher realized that much work lay ahead. "It is my feeling that we have 'taken our stand' and that 'together we will build a new South.'

However, our work is just beginning. The rhetoric of our statement will be meaningless unless we begin right now to work and build for a new social order. SSOC is designed to help us do that. But as an organizational structure it can do very little; as dedicated individuals it can accomplish very much." As the summer of 1964 approached, these dedicated individuals set about to initiate their program and to change the South.[81]

Chapter 3
Growing Pains

> My sense was that, you know, it was the worst possible time to go to Mississippi to try to work with white people because Mississippi was just totally polarized. We tried, we actually went out and tried to engage with people. But it was very hard to do.
> —Sue Thrasher[1]

In a speech at SSOC's first Southwide conference in Atlanta in November 1964, J. Metz Rollins, a black minister who had been involved in the Nashville movement, counseled the students not to "talk yourselves to death. You don't have to have all the ultimate answers before you act." SSOC's early history suggests that the activists fully agreed with Rollins. In the weeks and months following SSOC's creation, the group did not allow discussions about its structure and program to prevent it from actively participating in the civil rights struggle. In fact, more than at any other time during the group's existence, civil rights issues dominated SSOC's agenda in 1964. As a result, SSOC quickly gained legitimacy among activists on predominantly white southern campuses, particularly those where student anti-segregation work pre-dated SSOC's founding. Heartened by the activism of their fellow white collegians, the students in the new group believed that SSOC, as an organization devoted to coordinating local activities, was poised to lead a progressive movement of young white southerners.

But SSOC's development during its first months of existence was neither a smooth nor an easy process. As the founding activists worked to build the organization, the group experienced growing pains that complicated the students' efforts. While SSOC had implemented its campus traveler program, started fundraising, and opened an office in Nashville's joint university neighborhood by the start of the 1964 fall semester, serious and potentially debilitating conflicts emerged within the group in the latter half of the year. These disputes focused on important organizational and structural issues, such as the efficacy and appropriateness of the group's hands-over-flag symbol and the view of some leaders that SSOC should dissociate itself from individuals and organizations perceived to be the slightest bit pink, lest SSOC be tainted as an un-American or communist-sympathizing group. SSOC activists also had to contend with setbacks in the group's program. In the summer of 1964, for example, SSOC struggled, with little success, to make inroads into several white communities in Mississippi during the Freedom Summer project. Additionally, the evolution of the group's programs created a sense of uncertainty both inside and outside SSOC regarding its goals and direction. SSOC's attempt to reach southern whites beyond the confines of southern campuses during Freedom Summer reflected a significant change in direction, though the difficulties the group encountered in Mississippi persuaded most of the activists to restrict their work thereafter to the region's colleges and universities. Another important shift occurred near the end of the year when the students decided that SSOC should not develop as an all-white organization but instead should become a biracial group dedicated to working for progressive change on all southern campuses, both black or white. Thus, as 1964 wound down, SSOC was a growing organization struggling to initiate successful programs and to resolve important organizational issues. Although the students had established SSOC as a dynamic new force on southern campuses, internal disputes, failed actions, and a lack of direction hampered the group's efforts and hinted at the difficulties that later would beset the organization.

The White Folks Project

Impressed by the enthusiasm of the students who attended the founding SSOC meetings in April and the May, the organization's leaders spent the first heady weeks after the founding working to spread word of SSOC's creation and to gain access to the resources that would ensure the group's survival through the summer months. Of primary importance was rushing into print the first issue of the SSOC newsletter, the *New Rebel*, before classes adjourned. In a mere two-and-a-half weeks the group's leaders compiled a four-page newsletter that provided pertinent information about the new organization, offered articles on white student involvement in civil rights actions in Gainesville and Nashville, and included an essay by Robb and Dorothy Burlage that argued that the key to organizing a progressive movement in the South was to persuade whites to focus on the economic concerns that transcended racial lines. Mailed to individuals who attended the founding meetings, sympathetic organizations, and people on the original mailing list, the newsletter was SSOC's first effort to make itself known around the South. While the newsletter later became a polished, glossy, and sophisticated publication, the first issue, with its handwritten title, awkward layout, and photocopied pages, had the look and feel of an alternative paper that had been cobbled together hurriedly on the front lines of the movement.[2]

Organizationally, SSOC was still in flux. Although the group quickly had obtained a post office box, Sue Thrasher's home functioned as SSOC's administrative center; the address on SSOC's first letterhead, in fact, was her residence. The Nashville apartment of Ron Parker, the Vanderbilt graduate student elected treasurer, became the focal point of the group's early fundraising activities. SSOC was in particularly dire financial straits in the late spring of 1964. Until then, the group had operated on a shoestring budget: its only revenue had come from grants of $100 from SCEF to hold April's founding meeting and $300 from SNCC for the May meeting in Atlanta, along with a handful of small individual donations. Recognizing that the lack of financial resources posed a serious threat to the group's long-term viability and short-term survival, SSOC leaders issued a plea for donations in the first issue of the *New Rebel* and carried on conversations with SNCC and SCEF about continued financial support. The students also believed that securing the support of liberal, northern-based, philanthropic organizations would be crucial to SSOC's financial health. To that end, Parker and Archie Allen, who would become SSOC's first campus traveler, took a fundraising trip to New York City in the interval between the April and May founding meetings. Parker and Allen considered the trip a success, as they introduced SSOC to several organizations and established contacts within the liberal foundation world. They especially were pleased about their meeting with a representative of the Anti-Defamation League, they later explained, for the "ADL man said he'd give us entrée to New York groups interested in SSOC."[3]

At the same time that the SSOC activists were trying to secure the group's financial base and establish a stable organizational structure, they also were finalizing plans for the group's first program of action, the White Community Project in the summer of 1964. Conceived by Sam Shirah and led by Ed Hamlett, the White Folks Project, as it came to be known, started as an attempt to win support for the goals and aims of the civil rights movement among Mississippi whites and evolved into a drive to organize working-class whites along Mississippi's Gulf Coast around economic issues relevant to their communities. The White Folks Project developed as part of Freedom Summer, the project that brought hundreds of white students into the Magnolia state in an effort to galvanize the movement in Mississippi. Although the black Mississippians who lived, died, struggled, and eventually triumphed were the true heroes of the Mississippi movement, and although the presence of large numbers of young white northerners gave the story of Freedom Summer its dramatic edge and national spotlight, the small number of white southern students who went to Mississippi to work in the White Folks Project were the only volunteers to reach out to local whites. Unlike the northern volunteers, those from the South believed that transforming Mississippi required working with whites in the state—the opponents of racial reform—and not just with Mississippi's black populace. That the students met with little success, however, clearly indicated that organizing the white community would be a difficult task. Other than a very few individuals, the overwhelming majority of white Mississippians who came in contact with the project turned a deaf ear to the students' pleas for black equality and racial reconciliation. The experience of the White Folks Project volunteers profoundly shaped SSOC's future, as after the summer the SSOC activists quickly retreated to the familiar confines of the southern campus.[4]

Freedom Summer grew out of the November 1963 Freedom Vote organized by the Council of Federated Organizations (COFO), the umbrella organization of Mississippi activist groups dominated by SNCC and headed by Bob Moses. Planned by Moses and Allard Lowenstein, a white liberal Democratic Party activist who had become involved in the Mississippi movement, Freedom Vote was designed to demonstrate that African Americans would vote if given the chance. On election day, as Mississippi whites went to the polls, more than 80,000 black Mississippians cast ballots in a mock election for statewide offices. The high turn-out gratified Moses, as did the national media attention the vote received, a fact due, in part, to the presence of approximately 100 white Yale and Stanford students recruited by Lowenstein to work in Mississippi during the two weeks prior to the election. Hoping to build on the success of the Freedom Vote, Moses and Lowenstein persuaded COFO to bring hundreds of young whites to Mississippi during the summer months to take part in a much larger project focused on voting rights.[5]

The students who founded SSOC took an immediate interest in finding a role for the new organization in Freedom Summer. Because many SSOC leaders maintained close ties to SNCC, they were well aware of the developing plans for the summer project. Ed Hamlett and Sam Shirah, for instance, attended SNCC staff meetings, which, as the summer neared, increasingly focused on the details of the project. These details grew more complicated over time, as COFO planned to open freedom schools and community centers in addition to coordinating a massive voter registration campaign. With the expansion of the program beyond voting rights, SSOC leaders saw an opportunity to initiate a side project working with Mississippi whites. When SNCC and SSOC leaders met for the first time at the April 19 SNCC Executive Committee meeting in Atlanta, the SSOC students asked for and received SNCC's support for incorporating a white organizing project into the summer program.[6]

Starting on June 13, the white volunteers for Freedom Summer gathered at Western College for Women in Oxford, Ohio, to begin training for the project. Among those in attendance were the 25 whites, more than three-fourths of whom were southerners, who had volunteered for the White Folks Project. The non-southern volunteers came from across the nation: Soren Sorenson, for example, arrived in Oxford from Washington state, where he was a student, while Grenville Whitman and John A. Strickland had traveled from Harvard University. Among the southerners were several SSOC founders, including Sam Shirah, Ed Hamlett, Sue Thrasher, Gene Guerrero, and Nelson Blackstock, and a contingent of students from the SDS chapter at the University of Texas at Austin—Charles Smith, Bruce Maxwell, Judy Schiffer, and Robert Pardun.[7] While COFO ran two one-week sessions for the white students who were set to join the larger project, the white community volunteers participated in their own two-week training program. Initially, the White Folks Project volunteers participated in many of the same sessions as the other volunteers. They sang Freedom Songs, practiced the techniques of nonviolent self-defense, and learned from movement veterans about the white hostility that awaited them in Mississippi. That latter issue seemed both terrifying and absurd. Doug Tiberiis, a White Folks Project volunteer from the University of Arkansas, recalled being "amazed" by the "seemingly ridiculous notion that we were to love these people [Mississippi whites]. We received a terrific scolding," he wrote shortly after the summer of a well-known incident, "for our laughing at the remark of a white citizens' councilman in one of our training films. We need to help these people? This advice was necessary if we were to work effectively with these people."[8]

The first week of the Oxford training did not proceed without controversy. First, Anne and Carl Braden precipitated a crisis. The Bradens, who intended to stay out of Mississippi during the summer months so that segregationists could not use their radical political views to taint the project, traveled to Oxford to lend support and encouragement to the volunteers who had taken up the challenge of working with Mississippi whites. Upon their arrival in Oxford, however, the National Council of Churches, the liberal clerical organization that provided substantial funding for the training sessions, prohibited the Bradens from meeting with the students on campus, thus necessitating that the meeting take place at a professor's house nearby. The Bradens's brief appearance in Oxford did, in fact, attract attention from Mississippi segregationists, as an informer for the Mississippi State Sovereignty Commission who attended the Oxford training dutifully noted their presence in a report on the sessions. Then on June 21, shortly after the Braden controversy, Bob Moses called all the volunteers together to announce that three Mississippi civil rights workers had disappeared while investigating a church burning in Neshoba County. The announcement worsened an already chaotic atmosphere, as it came just as the first group of volunteers were preparing to depart for Mississippi and the second group was arriving for its training session. With tension increasing in Oxford and with space on the campus at a premium, the White Folks volunteers agreed to decamp to the Highlander Center for the second week of their program.[9]

In the calm and quiet atmosphere of Highlander, the goals of the White Folks Project came into focus. The students agreed that their primary objective was to begin the process of building support for the civil rights movement among white Mississippians. Few of the activists considered this problematic because SSOC had expressed an interest in taking their activism beyond the campus since its inception. In "We'll Take Our Stand," for instance, the students committed themselves "to work in all communities across the South to create nonviolent political and direct action movements." But tension did develop among the volunteers regarding the specific goal of white community work in Mississippi, as some wanted to organize working-class whites around their own grievances and not merely try to solicit white support for black equality. Several of the activists previously had articulated such a vision of white organizing. Nearly a year before SSOC's founding, Shirah had argued that SNCC should encourage white students to go into the white community "to organize the unemployed and the downtrodden, to develop programs against poverty, slums, poor schools, etc." More recently, the six-point program developed at the founding SSOC conference declared that the students were committed to fighting unemployment, poverty, and low wages in white communities around the South. But a majority of the volunteers were not enthusiastic about white organizing, preferring instead to concentrate on winning general support for the movement from Mississippi whites. Their view carried the day. While the project would not ignore working-class whites, the activists intended to focus primarily on middle-class whites because most of the organizers believed those were the whites most likely to respond positively to their message of racial reconciliation and black equality.[10]

Ed Hamlett, who coordinated the project, had the difficult task of determining in which communities the volunteers should work. Because he and Sam Shirah originally planned for the project to focus on all parts of the state, his attention was drawn to the Delta town of Greenville, the Gulf Coast communities of Biloxi and Gulfport, and the state capital of Jackson. Hamlett ultimately decided to scale back plans for the Delta. Rather than focusing directly on Greenville, seven of the twenty-five volunteers would be based in Jackson, from where they would make occasional forays to Greenville as well as Vicksburg and Meridian. But these volunteers would spend most of their time in Jackson because, Hamlett believed, the presence of "a few liberals" in the city's white community inspired optimism that some whites would be receptive to the volunteers.[11] The project assigned the 18 other students to Biloxi and Gulfport since these cities had large white populations.[12]

Additionally, the activists' work was complicated by the fact that city officials in Biloxi and Gulfport spent the summer devising plans to comply with the court-ordered desegregation of their public schools the following fall. While the planned-for school desegregation might redound to the volunteers' advantage since local whites might become less resistant to black civil rights given that one part of the edifice of segregation was set for removal, it just as easily might inspire a deeper white commitment to the decaying institution and greater hostility toward the advocates of desegregation. Thus, unsure of what to expect but hopeful that they could reach out to local whites, the students left Highlander at the end of June for Mississippi, the state that Bob Moses eerily had called "the middle of the iceberg."[13]

Throughout the summer, the students in Jackson trained their attention on the middle-class white community. To reach these whites, the volunteers frequently chose not to identify themselves as being associated with COFO but instead as working with the local Human Relations Council, a more "respectable" group. The volunteers primary goal was to educate middle-class whites about the civil rights movement, and they worked to explain and interpret for them the meaning and purpose of the freedom schools, voter registration drives, and other aspects of COFO's Freedom Summer program. In particular, they worked to educate the white community about the recently created Mississippi Freedom Democratic Party (MFDP), a black-led, interracial political organization that was planning a challenge to the lily-white state party at the Democratic National Convention in August.[14]

The vast majority of whites who the Jackson-based volunteers approached proved uninterested in or hostile to the civil rights movement. Outside of Jackson, the situation was particularly bleak, as the students made little headway in establishing a beachhead in Meridian, Vicksburg, or Greenville. Partially, this was because they did not devote much attention to these communities. In Greenville, for instance, Ed Hamlett and University of Texas student Charles Smith spent the evening at the home of Hodding Carter, the editor of the *Delta Democrat-Times* and an outspoken opponent of segregation. While this was a memorable evening for them, it was the only time anyone from the project spent in the Delta town. But even when the group did make an effort to build support for the movement outside Jackson, the results were discouraging. In Vicksburg, volunteer Bob Bailey left the town after eight days of trying to make contacts in the white community. "There has been no response and there is no hope of any," he complained about Vicksburg's whites. "They want nothing to do with us, want to keep us isolated so that we will have to go away." For Bailey and the other Jackson volunteers, small-town Mississippi's well-deserved reputation for opposition to the civil rights movement convinced them to concentrate their efforts on the capital city.[15]

The students, though, had few successes in Jackson either. Nearly all middle-class whites the students approached either ignored them or refused to do or say anything that demonstrated support for the movement. Although, the students could report a few positive developments, such as the growth of the Mississippi Council on Human Relations, which volunteer Sue Thrasher believed was the most significant accomplishment of the Jackson campaign, the majority of whites they met scorned them while the few moderates they encountered believed it was futile and potentially dangerous for them to publicly support of the movement. And they were right, Thrasher came to believe. "The tension level in Mississippi was going out the ceiling," she remembers. "It was the worst of all possible times to think that you could talk rationally to white people because it had just so divided people and people were scared. They didn't want to have anything to do with this group of people." The students, of course, knew that they had taken on a difficult task, and they recognized that they were unlikely to win much support. But many of the Jackson-based volunteers, Thrasher recalls, believed that they could "cool people out. Just try to find a few people who would be supportive of the student project and not . . . make any great conversions." Even this modest goal proved largely unattainable given white moderates' reluctance to break with their neighbors and identify themselves as supporters of black rights. By summer's end, the racial climate in Jackson had not yet begun to thaw.[16]

Like their counterparts in Jackson, the White Folks Project volunteers working along Mississippi's resort-lined Gulf Coast strove to build support and understanding for the movement among middle-class whites by working through union locals, civic groups, churches, and businesses. In a hint of things to come, the volunteers ran into difficulties the first day they arrived on the coast when their accommodation plan collapsed: the owners of the house and apartment the men and women, respectively, intended to rent reneged on the agreement, apparently because they had discovered the students' identity. The volunteers spent the first night divided between two cheap hotels. After spending the next two days unsuccessfully trying to find suitable accommodations for the group, the volunteers united in one of the hotels, the Riviera Hotel.[17]

Once they began their work, the Gulf Coast volunteers did not fare much better than those in Jackson. Sue Thrasher, who worked in both Jackson and on the Gulf Coast, recalls visiting several of the local ministers who publicly had expressed support, or at least tolerance, for the movement and the summer project. She hoped that they could be persuaded to speak out further. But she soon discovered that "they couldn't really do more than they were doing. They were hounded and isolated and some of them were moving out of the state, too." For Gene Guerrero, the effort to reach out to Mississippi whites frequently took him to the small Unitarian Church in Gulfport, most of whose congregants worked at nearby Keesler Air Force Base. Guerrero quickly learned that these whites were uninterested in getting involved in the movement. As he puts it, "they didn't know what they could do and were too scared to do it even if they knew what they could do." While the local churches frequently expressed a passing concern with social justice issues, the Biloxi-Gulfport business community made no effort to mask its opposition to Freedom Summer and the White Folks Project. None of the area's business leaders responded favorably to the students' overtures, and most actively opposed the presence of the volunteers in the area. On July 19, the owner of Barney's Restaurant in

Biloxi went so far as to summon police to arrest one of his employees for trespassing when he learned that the employee, Bruce Maxwell of the University of Texas, was one of the summer volunteers. Reflecting on the attitude of the business community, Charles Smith bitterly concluded, "seems that the scaredest white men in the state are the business and professional people. They've just enough of nothing to think that they have something."[18]

Internal divisions added to the problems the Gulf Coast volunteers faced. As the summer progressed, interpersonal differences emerged and relations deteriorated among the volunteers. Undoubtedly, some of the strain and tension was a product of the project's inability to attract supporters, as day after day the students experienced the rejection and absorbed the contempt and hostility of Gulf Coast whites. But problems also resulted from the fact that the project's organizers were too optimistic about Gulf Coast whites' receptivity to the volunteers and thus had overestimated the number of volunteers needed in Biloxi and Gulfport. At the end of July, volunteer Soren Sorenson concluded that 18 workers were several too many "in a community which already expects all manner of chicanery from the 'invaders,' and watches every move so closely that little could be done." That all 18 lived together in one Biloxi hotel exacerbated the problems, as conflict arose over such issues as dress and personal behavior in addition to such work-related matters as organizational style.[19]

Though seemingly trivial and mundane concerns, these issues reflected a deeper split within the group. Led by Sam Shirah, some of the volunteers sought to begin working among lower-class whites. From the beginning, Shirah had wanted the project to focus on white workers. Just as SNCC had turned its attention from middle-class campuses to working-class communities, Shirah believed white activists needed to focus more on working whites and less on white students. But Shirah and his supporters had acquiesced to the middle-class orientation of the project since theirs was the minority view. The project's inability to gain support from middle-class whites, however, prompted these volunteers to turn their attention to the working-class. Hence, clothing became a point of conflict within the group because the style of dress adopted by those seeking to work with blue-collar whites was markedly different from that of the middle-class workers. As Sorenson complained at the end of July, the jeans and workshirts of Shirah and his allies tainted all the volunteers as "a bunch of beatniks. . . . [T]hose of us involved in the middle class were always after the others to improve their appearance for the sake of our success."[20]

By the end of July, it was clear to all that the project was in trouble. The middle-class community work had yielded little fruit, and a better harvest in August seemed unlikely. Relations among the staff remained tense, and before July ended six volunteers had quit and left Biloxi. In an effort to revitalize the project, the remaining activists headed to the

Highlander Center to consult with veteran organizer Myles Horton. Then, under Horton's direction, they convened a weekend retreat in Gulfport. As a result of these gatherings, the volunteers decided to shift their focus to working-class whites, setting the twin goals of developing "a political base" among them and organizing them around important issues in their community. They agreed that for three weeks in August—the final three weeks of the project—six of the volunteers would move into a house in Point Cadet, a working-class neighborhood of Biloxi where many shrimpers and oyster boat workers lived. At the same time, four other volunteers would remain in the Riviera Hotel and continue to try to reach out to middle-class whites.[21]

The reorientation toward working-class whites did not revivify the project. While most blue-collar people they encountered were cordial, few showed an interest in the project. One group of whites that initially seemed interested in the students were the men who earlier had founded an independent union of shrimpers only to see it crushed by a price-fixing suit. Believing that the busting of the union was an issue around which to organize, the students made contact with the fishermen and offered to join the union. But although the shrimpers would have welcomed the support, fear of a white backlash in the community led them to spurn the volunteers' offer. In a meeting of the volunteers, Sam Shirah argued that "these types of problems are inevitable. Talking to people on the wharves . . . is all part of an effort to be part of the community. We've achieved something in that everyone knows who we are now—our frustration is that no one knows what to do."[22]

Other initiatives the volunteers undertook also foundered on white suspicions and fears. For instance, rumors circulated in Point Cadet that the volunteers' "real" goal was the integration of the schools, leading many whites to turn against them. And Point Cadet residents overwhelmingly rejected the volunteers' efforts at political organizing. White suspicion combined with the activists' inexperience to doom their efforts to draw whites into the MFDP. When the volunteers put up SNCC and COFO posters in the storefront office they rented in Biloxi and then called an MFDP organizing meeting, local whites concluded that, in actuality, they were running an employment bureau for blacks. Fearful that the organizers were trying to find jobs for blacks in their neighborhood, whites pressured the landlord to evict the volunteers, which he did on August 12. Recalling the incident several months later, Ed Hamlett matter-of-factly wrote, "An office was rented. A meeting was set. But rumors flew that 'those nigger lovers want to turn our jobs over to niggers.' The office was lost as pressure was applied to the owner." Dispirited and frustrated by their ineffectiveness, the volunteers left the Gulf Coast within ten days.[23]

Most of the volunteers believed that the project was a failure, convinced that it had served no useful purpose and accomplished little. Gene Guerrero thinks it was "ill conceived" and a "waste of time," while Nelson

Blackstock remembers that he and several others spent more time traveling around the state visiting other projects than they did organizing in Biloxi and Gulfport. The situation was no better in Jackson, where the project lacked a clear focus and purpose. For several of the students, the most meaningful and rewarding part of the summer occurred near the end of August when they spent some of their last days in Mississippi in freedom houses, community centers, and COFO offices in black communities around the state, where they filled in for staff who were attending the Democratic National Convention in Atlantic City with the MFDP. The experience was exhilarating for the students; they reveled in the opportunity to work in the black community and to be in some of the state's hot spots. Guerrero worked for ten days at the freedom house in McComb, which occasionally was shot at by local whites. Sue Thrasher spent a week in Hattiesburg, where she worked at a community center and stayed at the home of Victoria Gray, the MFDP's Senate candidate. Compared to her experience in Jackson and Biloxi, Hattiesburg "was like a different world because . . . you were in this beloved community, literally. You felt safe. And it was the outside world that was hostile to you." Working at the community center and seeing how a freedom school operated gave her "the flavor of what the real Mississippi Summer project was really about." On the Gulf Coast, she later explained, "we spent a lot of time sitting around talking about what we should do . . . but by and large we didn't have any idea of what solid role we could play. And we weren't in a community. We couldn't operate a freedom school. . . . So our role necessarily became then a[s] sort of white liberals . . . just trying to talk to people."[24]

The White Folks Project on the Gulf Coast, however, was not completely devoid of success. As the white students later realized, the project afforded them a better understanding of the needs of working-class whites, such as improved schools and better job opportunities. There were more tangible signs of success, too. On August 6, an interracial audience of 150 people gathered in the Biloxi Community Center to watch "In White America," a performance by the Free Southern Theater group. Additionally, a few whites responded favorably to the volunteers' organizing on behalf of the MFDP. During August, 20 white Gulf Coast residents signed Freedom Registration forms, openly giving their support to the MFDP. And one of them, fisherman Robert Williams, an ex-convict who had moved into the volunteers' hotel in early July, initially as part of a plan with his friends in the local Klan to undermine the project, was won over to the volunteers' cause and was among the 68 MFDP delegates who traveled to Atlantic City for the Democratic Convention.[25]

Despite these accomplishments, participation in the White Folks Project was a humbling experience for the volunteers. Anne Braden, herself a long-time advocate of white organizing, concluded that the volunteers' work "was slow going: they learned more about how not to proceed than they accomplished." For the SSOC activists who took part in the project, the setbacks they suffered in Jackson, Biloxi, and Gulfport served to refocus their attention on white southern students and away, at least for the time being, from the broader white community. While the failure in Mississippi weighed heavily on Sam Shirah and others who were deeply committed to working with poor whites, the majority of SSOC activists considered students to be their prime constituency and thus did not view the project's collapse as catastrophic for the organization. To these students, neither SSOC's identity nor its long-term prospects were bound up with the fate of the White Folks Project. In Thrasher's view, the White Folks Project was only "a summer project.... I never thought of us [SSOC] as a community-oriented organization or that that was our job, to go into a community and organize. And I would have fought against that because... if you're going to do that then that's what you focus on and make a long-term commitment to." Ed Hamlett felt differently, perhaps because he had served as the project's director. Shortly after the summer, Hamlett noted that SSOC's "most important thrust must be toward the lower-class white community." He could sympathize with volunteer Bruce Maxwell who at summer's end pleaded for expansion of the program. "The most important work to be done is with poor folks and not with moderates and liberals," declared Maxwell. "It is the responsibility of the freedom movement, before the threat of the movement to the white poor further increases, to include them in our efforts." But in the wake of the young activists' experience in the White Folks Project, nearly two years would pass before SSOC again tried to reach out to working-class whites.[26]

Navigating the Politics of the Movement

Soon after the conclusion of Freedom Summer, SSOC activists worked to give the group a higher profile in Nashville and around the South. First, the group opened a temporary office in the Tennessee Council on Human Relations' building. Then in October, it opened its own office on the second floor of a house at 915 Eighteenth Avenue, South, in the middle of the joint university neighborhood. Sue Thrasher, with help from Ron Parker, Ed Hamlett, and others, kept busy with mundane but crucial administrative tasks—sending out mailings, laying plans for a fall conference, corresponding with interested students around the region, and helping to coordinate the activities of Archie Allen, SSOC's first campus traveler.[27]

The campus travelers were key individuals in SSOC. Their mission was to visit campuses throughout the South to encourage local action and to provide activist students with advice and guidance. To the southern students they encountered, the campus travelers *were* SSOC. While interested students in such far-flung places as Farmville, Virginia; Athens, Tennessee; and

Gainesville, Florida, might receive mailing and correspondence from the Nashville office, the campus traveler usually was the first, and sometimes the only, SSOC person they came to know. As a result, SSOC's campus travelers played a critical role in building support for the organization throughout the South.

SSOC leaders first began discussing a campus traveler program at the May meeting in Atlanta. The activists readily agreed that SSOC needed people traveling throughout the region on behalf of the organization, but they were unsure if they could attain the funds to pay for the program. As the meeting progressed, however, one possible source of funds revealed itself—SDS. In his letter to the Atlanta gathering inviting SSOC to become an SDS fraternal organization, its president, Todd Gitlin, remarked that the northern group planned to appoint a southern traveler who, though working to build SDS in the South, could also assist SSOC. Nothing initially arose from this offer; SDS did not immediately hire a southern organizer. But shortly after Freedom Summer, the two groups agreed jointly to fund a southern campus traveler, and in early September they hired Archie Allen for the position.[28]

Given Allen's ties to SSOC and SDS's unfamiliarity with the South, SDS leaders did not expect him to accomplish much for their organization. They only hoped, as one official put it, that Allen "would write us an occasional letter and mention SDS to folks when it is appropriate."[29] From SDS's perspective, the South was uncharted territory, and the group's leaders assumed that SDS's growing reputation as a militant student organization would limit its appeal among the supposedly uniformly conservative white students of the region. Allen, in fact, concentrated his efforts on building interest in SSOC among southern students, though less because he felt students would not respond to SDS than because he sought to attract supporters to the new group. In his first two months on the job, Allen visited twenty-nine schools in six states, ranging from nationally prominent universities such as Duke University and the University of Virginia to such regional institutions as Hiwassee and Maryville colleges in east Tennessee. At each stop he sought out interested students and introduced them to SSOC, informing them about the new organization's purpose and goals and encouraging them to attend SSOC's first Southwide conference, scheduled for early November. In his reports back to the Nashville office, Allen happily noted that he not only had found numerous students excited to learn about SSOC but that many of them already had been involved in campus activism. As he later enthused, "<u>Every</u> campus I visited, I found persons who were interested in SSOC's program, persons who were or had been active in a movement of students." Allen remained the group's sole campus traveler until late December when, newly married and struggling to survive on the traveler's paltry salary, Allen resigned and was replaced by Kathy Barrett of Loyola University in New Orleans and SSOC chairman Gene Guerrero of Emory, both of whom dropped out of school to work full-time as campus travelers beginning in January 1965.[30]

In addition to leading SSOC's charge on southern campuses, Allen joined several other of the group's leaders in the fall in trying to win financial backing for the group, as the only funding it had received since its founding was a small travel grant from the National Student Association and several personal donations, including $1,000 from David Kotelchuck. In early October, SSOC had just $500 to its name and was in dire need of funds to put on its first Southwide conference in November. To raise funds, the group appealed to labor unions for support. On a trip to New York in December, Archie Allen met with a number of labor leaders from progressive-leaning unions, and though he did not receive any contributions, he was heartened that the unionists had kind words for SSOC. Thrasher, on the other hand, was not encouraged about a meeting she participated in with the head of the AFL-CIO's civil rights division. Much to her surprise, the union official declined to give money to SSOC because, she recalled him saying, "we like to put our money on the front lines of the civil rights movement." His comment made her realize that despite the work she and the other founders had put into building SSOC, the group had a long way to go before it would be considered a major player in the southern movement.[31]

Like other activist groups, SSOC also turned for financial support to northern philanthropic organizations, including the Taconic Foundation, the Stern Family Fund, the Field Foundation, the New World Foundation, the Aaron E. Norman Fund, and the New York Foundation. Throughout the latter half of 1964, Allen, Guerrero, Parker, and Kotelchuck courted these organizations, writing letters asking for financial assistance and making fundraising trips to New York City, where many of the organizations were based. While the foundations expressed an interest in SSOC, they hesitated to support it until the group met two conditions. First, they required SSOC to initiate the process of becoming a tax-exempt organization, thereby signaling that the group would not engage in partisan political campaigns or give money to for-profit organizations. To that end, in September, SSOC retained the legal services of Phillip Hirschkop, a Washington, D.C., attorney who had come to SSOC's attention through his work helping the Law Students Civil Rights Research Committee quickly obtain tax-exempt status. Second, given the foundations' unfamiliarity with the new organization, they would not consider assisting it unless it was endorsed by respected southern liberals. The philanthropists had many questions about the new group. Was SSOC a legitimate organization? Did it promote goals and tactics that meshed with the foundations' agendas? What company did the group keep—was it closely aligned with organizations deemed too radical by the philanthropists' standards? This last question was the most troubling to the foundations given SSOC's close ties with SCEF, which most foundations considered unacceptably radical. For the foundations, the imprimatur of southern liberals would resolve these issues and assuage their concerns about the group. Archie Allen was quick to realize that SSOC would have to negotiate a narrow path in order to secure the support of northern philanthropic groups. "The foundations offer us the greatest hope financially," he wrote near the end of 1964. "But it will require keeping in touch with all prospective foundations, obtaining the tax exemption, and then carefully deciding which foundations are most likely to make a grant, knowing what type of project or activity they will support, and making personal and written appeals for funds."[32]

Tax exemption and southern liberal approval figured prominently in SSOC's first foundation grant, a provisional $2,000 grant made by the Norman Fund in September as a result of a fundraising trip Allen had made to New York earlier in the summer. The grant would be administered by the Southern Regional Council (SRC) and was contingent upon the SRC's endorsement of SSOC. The need to win the SRC's approval created consternation among SSOC leaders, who feared that their ties to SCEF would offend the SRC, the model of white southern liberalism and an organization that had had its differences with SCEF.[33] As Sue Thrasher remembers, "SRC was *the* southern liberal organization. And if you wanted to work in the South you didn't go very far unless you had their blessing.... Unless we got ... [it] on our side we weren't going to get any foundation money from New York."[34] Ron Parker put it more succinctly: "If they are not giving us an endorsement, we are in trouble." SSOC leaders' concern grew when they learned that SRC chief Leslie Dunbar, with whom they were scheduled to meet in mid-October, planned to ask the students, Thrasher nervously noted, "that we stay within the law on tax exemption and that we define our relationship with the Bradens." At the meeting, Thrasher and the others pleaded their case to Dunbar, stressing that despite their ties to the Bradens SSOC was not a communist group. Dunbar, though, was more concerned that the students comply with tax-exempt regulations and not openly support Lyndon Johnson at its November Southwide conference. To the SSOC leaders' great relief, Dunbar expressed his support for the group. With his approval, then, the Norman Fund grant was made official, providing SSOC with a much needed infusion of funds.[35]

The Norman Fund, however, was the only foundation willing to support SSOC in 1964, as others remained leery about the group's orientation and direction. As Thrasher put it near the end of the year, "Will SSOC be radical, or will SSOC be moderate? This is the question being asked by the foundations." SSOC's closeness to SNCC caused concern in the foundation world owing to the black-led organization's increasingly strident rhetoric and its growing impatience with its liberal supporters. Not long after SSOC's founding, in fact, Bayard Rustin, who met with Allen and Guerrero along with Socialist Party leader Norman Thomas and

Tom Kahn of the League for Industrial Democracy, SDS's parent group, pointedly warned SSOC that its it had to "choose sides," that its ties with SNCC would alienate the new group from liberal and moderate movement supporters.[36]

Moreover, SSOC's relationship with SCEF was problematic in the world of liberal foundations. SCEF's supposed communist links, along with the Bradens' alleged ties to the Party and their avowed contempt for the nation's internal security apparatus—exemplified by Carl Braden's 1958 refusal to testify before the House Un-American Activities Committee—made the group an anathema to liberal foundations, most of which were rigidly anti-communist. SSOC's relationship with SCEF also created tension within the organization itself. Many of the students sought to preserve close ties between the organizations. To these activists, distancing themselves from SCEF and the Bradens in order to enhance SSOC's appeal to foundations both would compromise the group's integrity and be a display of wanton disrespect to those who had supported SSOC since its founding. In the months leading up to SSOC's creation as well as throughout the organization's young life, SCEF had provided SSOC with essential organizational and financial support while the Bradens had become informal mentors to the students. As a result, many of the activists refused to turn away from SCEF. In the late fall, for instance, Sue Thrasher, the SSOC leader who had grown closest to the Bradens, accepted an invitation to become a member of SCEF's Board of Directors.[37]

While SCEF's supporters in SSOC were not unaware of the Bradens' reputation, they were optimistic that others would evaluate SSOC on its own merits as an independent organization and not on its ties to SCEF. The sentiment among a number of SSOC's founders, Sue Thrasher remembers, was "that red-baiting was old hat. It had nothing to do with the movement at the time, and that people who did that were just—you know, had their foot in the Dark Ages. . . . [It was] just sort of 'a-a-ah, those people don't know, it's all fifties stuff. You know, this is a new movement.'" As Ron Parker commented, "The situation is so simple with these students that they feel if they tell the honest truth that they are democratic Americans and not agents of the Communist conspiracy this should be enough, the matter should be dropped, and everyone should . . . get on to the more pressing job of transforming the South."[38]

But Parker and several others, including Archie Allen and David Kotelchuck, viewed the situation differently, and, as a result, SSOC's association with the Bradens and SCEF had a polarizing effect on the organization. Although these activists appreciated the support and assistance SCEF had given SSOC, they believed that its relationship with the group would force SSOC to engage in a debate over its right to associate with SCEF, thereby diverting time and attention from its work to bring progressive change to the South. Many of the SSOC activists, Parker explained, "see civil liberties and freedom of association and speech as a present peripheral concern in light of the tremendous central needs (e.g., gaining voting privileges in Mississippi)."[39]

More importantly, these activists feared that ties to SCEF would create the impression, particularly among liberal foundations, that SSOC was a radical organization with links to suspected communists, thereby tarnishing its image and allowing for the group to be red-baited. Kotelchuck urged SSOC to steer clear of radical organizations because as a 28-year-old, he was one of the few who had personal recollections of the problems the McCarthyite campaigns of the 1950s had caused for progressives. Parker went further, disparaging the Bradens as "extreme civil libertarians" and arguing that most in SSOC did "not want the respectable southern and national groups such as the Southern Regional Council or Anti-Defamation League to doubt their composition . . . or intentions." Although the stridency of his rhetoric suggested a unanimity that did not exist on the subject, Parker, as SSOC's treasurer, recognized that the group relied heavily on support from SCEF during its first months. This created a serious dilemma for the organization. As he wrote to Hayes Mizell of the National Student Association,

> given a choice, I hope that we can choose to work with and be supported by someone other than SCEF. This is not red-baiting or negating the Bradens et. al (you know how strongly I feel about their guts and contribution to the movement when us moderates don't get off our asses) but rather coming intellectually to the decision that one would prefer not to be burdened with self-imposed limitations by SCEF association. . . . [I]t logically follows that this other "choice" must be available—i.e. someone else must be willing to help us get started. I want to communicate to the "powers that be" (SRC, et. al) this position and enlist their support; myths must be dispelled early lest SSOC die in the bud.

Archie Allen concurred with Parker, though he clearly was anguished by the situation. As long as SSOC remained close to the Bradens, he reasoned, the group could expect to be red-baited. Writing to the group's leadership, Allen insisted that, "from the position of fundraising, image creation on a broad level, etc. . . . it would be exceedingly harmful to include the Bradens in a prominent position [in the group]. My conscience and morality scream 'To Hell With What People Say' but the cold facts are deadly ones." Uncertain how to finesse the situation, Allen closed by tentatively suggesting that "Perhaps expediency is best at present and work out the problem of association after SSOC is established??"[40]

The majority of the group's leaders shared Allen's sentiment. In October, in an attempt to reaffirm SSOC's independence, the Executive Committee passed a tentatively worded resolution stating that although

SSOC was not officially associated with an adult group, it welcomed the support of any adult organization that shared its goals. Then in December, the Executive Committee, prompted, ironically, by a positive essay Anne Braden had written about the group's November Southwide conference, took more forceful action to separate SSOC from SCEF. While SSOC's leaders appreciated Braden's praise, some were distressed that the essay had appeared in *The National Guardian*, a socialist publication. In response, the committee passed a new resolution, this time explicitly declaring that it opposed communist participation in SSOC activities. Like the resolution passed two months earlier, however, the December statement was ambiguous and its impact was uncertain. Parker believed it merely formalized SSOC's well-known antipathy toward communists, but Gene Guerrero felt it was a cosmetic action that was intended only to appease potential liberal allies.[41]

In the short term, however, passage of the resolution had one clear consequence: It precipitated a debate within the organization that led to Thrasher's resignation from the SCEF Board. Given her warm relationship with the Bradens and her strong desire to maintain close ties with SCEF, this was a painful and deeply troubling episode for Thrasher. "The reason for my resignation from the SCEF board, when I'm honest with myself, is that I'm too cowardly to fight it from within the group," she wrote the Bradens. "So, rather than give you all the rational and realistic reasons that will be given to others, let me be honest with you and say that I know I'm compromising, and that I'm compromising on some things that are very important to me." Although she preferred not to relinquish her seat, Thrasher agreed to step down out of deference to the wishes of the majority of Executive Committee members. By way of explanation, she wrote the Bradens,

> I don't really think that we have enough faith in this idea of what we are trying to do. . . . But you see, Anne and Carl, this is my vision and my faith in an idea. I can't give it to anyone else. And Ron Parker thinks that it is a blind faith. He is a realist and he believes that SSOC will be judged for its responsibility and respectability—and not the idea. And Ron Parker is just as much a part of SSOC as I am. He, too, has shared in the forming of this group and he, too, has a vision.

It was her duty, she believed, to abide by the majority's decision. "I'm tired of having to defend SCEF," she sadly told the Bradens. "I have tried for eight months and yet have watched SSOC steadily giving in on this issue. And now, I'm giving up. I just don't have the energy to fight it anymore. And all I can say to you is I'm sorry." Though they understood the dilemma Thrasher faced, the Bradens were disappointed that she had decided to leave the board, telling her several months later, "Someday I think you are really going to have to turn and search your own souls—both individually and organizationally—on this whole question." James Dombrowski, SCEF's executive director, more directly conveyed his disappointment with Thrasher's resignation. Shortly after Thrasher had notified the Bradens of her decision, she received a one-sentence letter from Dombrowski: "Dear Sue, your letter makes me very sad."[42]

Thrasher's resignation and the turmoil caused by SSOC's relationship with SCEF proved to be an important learning experience for the young southerners. In the wake of her leaving the SCEF board, Thrasher vowed never again to sacrifice her principles on the altar of organizational unity. In the future, she pledged herself, she would maintain her principles, "the collective be damned." For the rest of the activists, the struggle to reshape SSOC's relationship with people for whom they had great respect and deep personal admiration awakened them to the broader politics of the movement and forever disabused SSOC's first generation of the notion that the group could exist as an independent progressive organization in the South, free from the burdens and legacies of past efforts to build a progressive movement of southerners. In Thrasher's estimation, this was the "political issue that we cut our teeth on. It was what made us understand that we weren't just in this for human rights. There was a political world that went beyond black and white relationships . . . where people had positions and it had to do with history."[43]

Shifting Focus

While fundraising and relations with SCEF and the Bradens dominated SSOC's attention in the fall of 1964, the young whites rightly understood that winning financial support and resolving the situation with SCEF would be irrelevant if SSOC did not begin to grow. Thus, campus traveler Archie Allen spent much of the fall on the road, introducing SSOC to hundreds of students throughout the region and, most important for SSOC's immediate future, encouraging them to attend SSOC's Southwide conference in November in Atlanta.

The conference was a critical moment in SSOC's early history, for it was the group's first opportunity to demonstrate that it had support across the South. Indeed it did, as 144 students from 43 schools in 11 southern states attended the gathering. Additionally, the conference revealed a new commitment by the group to organize on both white and black southern campuses in an effort to become the leading student organization in the South. And while civil rights remained a vital issue to the activists, SSOC urged students to take an interest in other issues that were no less important for bringing progressive change to the South, such as anti-poverty work and the peace movement. SSOC's effort to transform itself into a biracial, multi-issue organization represented a dramatic shift from its founding as a group of white southerners focused almost exclusively on the civil rights movement.

Movies, meetings, and, especially, speeches filled the Southwide conference agenda. Among the speakers who addressed the students were James Forman, SNCC's executive secretary; Ed King, the Tougaloo chaplain and national committeeman of the MFDP; Don West, the radical Georgia poet, labor organizer, and co-founder of the Highlander School; and writer Lawrence Goodwyn, formerly of the Texas Democratic Coalition. Looking back on the conference, Goodwyn recalls being "surprised at the fragile knowledge of half the people in the room." In his estimation, most of the attendees were "very green, earnest college students who didn't know very much about anything except that they were against segregation." Nonetheless, the eclectic group of speakers testified to the students' concern with numerous issues other than segregation. In addition to drawing the young whites into the movement, the drive for black rights awakened SSOC students to injustice throughout southern society. In other words, joining the battle against segregation initiated a process of radicalization for these young whites. Recalling how involvement in the civil rights struggle had altered his view of the world while he was at the University of Georgia, Nelson Blackstock later explained, "I started off worrying about segregation—segregation is a problem; how do you get rid of segregation? And from that . . . you would become very radicalized . . . and begin to question everything."[44]

By the fall of 1964 the movement's successes in fighting legalized segregation convinced many in SSOC that the time had come to confront other injustices more directly. To the young activists, enactment of the Civil Rights Act had mortally wounded Jim Crow. Because this legislation "eliminated most of the necessity of continued action in the form of demonstrations," Archie Allen concluded, "socially conscious students began looking for ways to eliminate the interrelated and more basic problems such as employment opportunity, causes of poverty, new ways of working for civil rights, and examining world issues." That the SSOC activists wanted to emphasize these subjects became evident prior to the conference, as the group prepared the October edition of the *New Rebel*, the first issue of the fall term. The group's leaders did not want civil rights to dominate the newsletter. As Sue Thrasher explained to one colleague, "The paper will probably have quite a bit about Mississippi summer but I want to make sure that it's not overloaded with civil rights. I would like to have some good articles on the fall election and perhaps something on Viet Nam." Thus, the newsletter contained an essay on the military draft by Allen, a reprint of Robb Burlage's 1962 pamphlet "The South as an Underdeveloped Country," and a drawing by Blackstock of the Grim Reaper standing over a field of crosses labeled "Vietnam."[45]

The most important development at the Southwide conference was SSOC's decision to transform itself into a biracial student organization. Throughout the weekend, numerous students advocated trying to recruit black students into SSOC. Some of the students believed that SSOC's progressive goals necessitated that it develop as a racially diverse organization. To conference participants who previously had been involved in an interracial student movement, it was counter-intuitive to reach out only to white students. To the majority of students at the conference who did not have a history of interracial activism on which to draw, the absence of a significant number of black students at the meeting was a severe disappointment. Typical was the student who remarked that after driving all night to get to the conference he had to fight the temptation to head back home upon finding so few blacks at the meeting. Far from being upset by the students' reaction to the racial composition of the meeting, many of the group's leaders agreed with this critique of SSOC. In a letter to a friend in SNCC about the conference, Connie Curry explained that "this desire to be a group for Negro and white students came from the group as a whole with no manipulation from the executive committee, most of whom, however, were glad—overjoyed—to see this happen." As Ed Hamlett told the students, "we are very pleased with your discontent."[46]

The changing focus of SNCC's work provided SSOC with an even greater impetus to turn its attention to black students. SSOC originally had focused on white students, according to the SSOC newsletter, because "The need for working on Negro college campuses had not been felt so keenly since SNCC was considered working with such groups." But in the wake of Freedom Summer and the spurning of the MFDP by national Democrats, SNCC had begun to reorient itself away from black colleges and toward the black community. In fact, the week prior to the Southwide conference, the SNCC staff gathered for a retreat in Waveland, Mississippi, to talk about SNCC's long-term goals and the group's internal problems. Several SNCC members who went from the retreat to the SSOC meeting, including Dona Richards, Ivanhoe Donaldson, and Emmie Schrader, strongly hinted that SNCC was leaving campus organizing behind. The white students believed that this created an historic opportunity for SSOC to reach out to black students even though the group had little experience working with African Americans. The logic behind such thinking was simple; as Hamlett later explained, "SNCC was off the campuses. We were on the campuses." In the words of one SSOC activist, SNCC's move away from the campus meant that "the door is now open for an interracial, intercollegiate student group in the South."[47]

The decision to move onto black campuses made the group's overt symbols of southern consciousness problematic, particularly the name of the newsletter—the *New Rebel*—and the emblem of clasping black and white hands superimposed over the Confederate battleflag. At a long, emotional business meeting on the last morning of the Southwide conference, the students clashed over whether the historical baggage these symbols carried would inhibit the group's ability to attract black students. Many of the activists argued that SSOC's use of images and rhetoric harkening back to the Confederacy conveyed the message that the group was, at best, insensitive to, and, at worst, organized against, the interests and needs of African Americans. The symbols, they insisted, never could be disassociated from the historical context from which they emerged; black—and many white—students always would interpret them as symbols of slavery. The hands-over-flag emblem came in for especially sharp criticism, particularly from the handful of black students in attendance. Explaining to their white counterparts that the battleflag had only one meaning to them, the black activists stressed that no matter how SSOC altered the flag's image, African Americans forever would see it as a symbol of racial oppression. As Ed Hamlett recalls, a black woman from Spelman College spoke for many when she declared, "I'm not having that on any thing that represents any organization that I'm going to be part of."[48]

Significantly, numerous white students at the meeting also expressed their disapproval of the battleflag's incorporation into SSOC's emblem. Dan Harmeling and Nelson Blackstock were appalled by the symbol because they saw themselves as part of a larger interracial movement. Other whites linked principle with pragmatism to condemn the emblem, arguing that using the Confederate battleflag was both morally offensive and counterproductive. In the weeks before the conference, Archie Allen, who had been traveling the campuses of the region, argued that SSOC should drop the emblem because it only would serve to alienate otherwise sympathetic white students. The emblem "is not SSOC," he explained in a letter to Sue Thrasher. At most, he noted, the emblem appealed to 20 percent of his contacts. "The idea of losing support and never getting it because of a romantic, emotional attachment to a symbol is sheer folly," Allen continued. "I don't mind opposition, but I don't want to be destroyed by such a petty factor of SSOC's own creation."[49]

Supporters of the emblem were equally passionate. These activists contended that far from symbolizing racism and slavery, the emblem served to subvert the traditional meanings ascribed to the battleflag. Ed Hamlett was disappointed that the emblem had engendered such opposition because he believed that "it would be a good symbol for us." Anne Braden recalls Hamlett arguing that the flag was part of his heritage, part of the heritage of poor white southerners. And rather than seeing the flag merely as a symbol of oppression, he argued that it signified the bravery and fortitude of his ancestors. Hamlett, Braden remembers, was "almost in tears talking about his grandfather who fought at Shiloh." SSOC chairman Gene Guerrero found much to like about the emblem as well. From his perspective, the shaking black and white hands transformed the battleflag into a symbol expressing "the commonalty of problems across race lines that people in the South face; the fact . . . that race had been used to divide people and to keep people from uniting together to solve problems." Despite his affinity for the emblem, though, Guerrero was not blind to the fact that most of the conference attendees disliked it. Recognizing that the emblem would continue to divide the group if

SSOC retained it, Guerrero memorably concluded, Anne Braden recalls, "let's flush the damn thing." As a result, the activists formally decided to drop *New Rebel* as the name of the newsletter and to jettison the hands-over-flag emblem, necessitating the disposal of hundreds of buttons, stickers, and brochures bearing the symbol and thereby making them instant collector items.[50]

In the days after the conference, SSOC's leaders expressed mixed feelings about the meeting. Ron Parker, who thought the meeting was a success, was encouraged by the group's vow to become interracial and pleased by its commitment to tackle a wide range of interrelated issues. "Perhaps the greatest value of the conference," he suggested, "lies in the fact that it did serve to broaden SSOC's vision of the task that lies before it." Sue Thrasher took a less sanguine view of the proceedings, as the conference had not come close to meeting her expectations. After finding herself writing excessively upbeat reports on the conference to send to prospective donors, she fired off a letter to Anne Braden complaining that the gathering failed to serve as a forum for setting goals and devising local action programs. Instead, organizational matters consumed most of the students' time, and the discussions that did take place were unfocused and became "bogged down in a lot of emotional mish-mash." The conference also failed to address what she considered one of the group's most serious problems, its lack of a clear sense of direction. She was disappointed that the activists had not spent more time talking about specific projects or the organization's long-term goals, and she sarcastically remarked that "we need to have a clear idea in our minds what it is we want to accomplish (other than getting a lot of liberal southerners together to congratulate each other on their liberality)." Discounting SSOC's participation in the White Folks Project, she lamented that the group had yet to initiate any activities. In its first seven months, she wrote distressingly, SSOC "has existed as a coordinating committee and a conference. This in summary is all we have been and done."[51]

Though her letter exaggerated the organization's problems, Thrasher hit on a widely shared sentiment when she complained about the group's lack of activity. Ironically, while they believed SDS devoted too much time to discussions and not enough to initiating action, the SSOC activists themselves had developed a penchant for holding interminable meetings and long, unending debates, many of which focused on procedural matters. Hayes Mizell, the head of the NSA's Southern Project and a frequent participant in SSOC meetings and conferences, considered this the group's most frustrating characteristic. As he wrote a colleague after one such session, "These meetings consume a great deal of time mainly because there is so much anarchy and discussion as to how things should be done." Typically, the students engaged in "a vague and general discussion" about a subject before ultimately "making no decision as to what it wanted to do in this area." Such was the price of trying to operate the committee by consensus. Because SSOC leaders sought to unite everyone on the Executive Committee behind the same position, protracted discussions frequently occurred.[52]

Shortly after the Southwide conference, SSOC's leaders took steps to involve the group in more activities. Working with COFO, the newly elected Executive Committee put in motion plans for a Christmastime project that Tougaloo chaplain Ed King first had proposed earlier in the fall. Just as in the White Folks Project of the summer months, the Mississippi Christmas Project involved the young activists in community work in Mississippi. But reflecting SSOC's new orientation, this project differed in two crucial respects from the work of the summer: rather than bringing white students into the white community, the holiday project entailed an interracial group of students working in the black community of three Mississippi towns, Laurel, Hattiesburg, and Meridian. Held over two four-day periods—one before and one after Christmas—the project arranged for 38 black and white students from 17 southern colleges and universities to participate in voter registration drives and to assist in the painting and remodeling of COFO staff houses and burned-out community centers. Sympathetic white students who were unable to travel to Mississippi found ways to lend their support to the project as well. At the University of Virginia, activist students held a food and clothing drive for the black families evicted from their farms for attempting to register to vote. While detractors had graffitied "KKK" on the trailer the Virginia students brazenly had parked on the main campus thoroughfare to collect donations, the activists were able to send a full trailer of supplies to Mississippi over the holidays.[53]

In Mississippi, the students concentrated most of their efforts on repairing and rebuilding the damaged or destroyed community centers. In Hattiesburg, they devoted most of their time to working on the Dewey Street and Palmer Crossing community centers, while in Meridian activity swirled around the COFO Community Center, which Freedom Summer murder victim Mickey Schwerner and his wife, Rita, had helped to establish. Although the black community in each of the towns welcomed the activists, local whites utilized both legal and extra-legal means to disrupt the project. In Laurel, several of the students spent their Christmas holiday in jail after local authorities arrested them for holding a sit-ins at the Traveler's Inn Restaurant and Pinehurst Coffee Shop, establishments that remained segregated despite the passage of the Civil Rights Act. On the first day of the project in Hattiesburg, the owner of Polk Hardware attacked Ed Hamlett as he accompanied two black youths to purchase building supplies. The storeowner, Hamlett wrote in a letter to John Doar, a lawyer in the Justice Department's Civil Rights Division, had become enraged upon learning that he was one of the civil rights workers. After beating Hamlett to the floor and kicking him in the head, his attacker ominously warned, Hamlett recounted, "that I had better not go to the FBI . . . and that I was up against a well organized group of people who would see me dead by that night, who could see me dead by that night."⁵⁴

The Christmas Project breathed life into SSOC's new emphasis on racial inclusiveness. But the project could not mask the fact that SSOC's effort to become biracial had created serious and potentially debilitating organizational problems. The shift toward biracialism at the November Southwide meeting immediately raised questions about SSOC's relationship with SNCC. Although the groups had been close allies since SSOC's founding, they had yet to formalize their relationship. In the months before the November meeting, the SSOC activists had come to see the two groups as working toward similar and complementary goals—SNCC worked with black students, SSOC organized white students. SSOC's decision to reach out to both white and black collegians, however, along with SNCC's move from the campus to the community, forced SSOC to reconsider the groups' relationship. Were their goals still complementary? Would the organizations remain close? Or would the two groups drift apart, as SSOC concentrated primarily on campus issues and SNCC worked only in the black community? Without a formal agreement to bind them together, each group's shifting agenda called into question the nature of their relationship. Additionally, SSOC's desire to draw black students into the group profoundly altered the organization's raison d'être; organizing white southern students ceased to be SSOC's primary objective. What, then, would take its place as the organization's main goal? Given SSOC's new orientation, would it now try to draw all southern students into the movement? This, however, would be a difficult assignment for SSOC since, despite its commitment to biracialism, SSOC was—and would remain—overwhelming white.

Complicating efforts to reassess its role in the movement was the fact that SSOC focused so much attention on organizational and structural issues. As the proceedings of the Southwide conference made clear, discussion of larger, more substantive issues was given short shrift. Afterward, Clark Kissinger, SDS's national secretary, blamed SSOC's problems on the absence of strong leadership and a developed political ideology: "I guess what SSOC really lacks at this point is leadership. There was no one who came to the SSOC Conference with a burning VISION. No one who got up and laid out an analysis of the problems of the South, a political framework or a set of social values, and a program. . . . [SSOC] will continue to suffer from a sort of intellectual malaise, I think, until someone or something new enters the picture." Consequently, Kissinger explained, "the Conference got bogged down in questions of buttons, structures, etc."⁵⁵

It is not surprising that a leading figure in SDS considered SSOC politically soft and suffering from an absence of strong leaders—that is, for not being more like SDS. Nonetheless, Kissinger's criticisms resonated with others. Hayes Mizell, for example, noted that only vague and general statements could win the support of all the activists, though such statements were not concrete enough to build a program of action around. "At this point," he explained in early October, SSOC's "only aspect of unity is the desire to reach the unreached and to get them to do something substantive."[56]

For an eight-month old organization, there were worse problems to have. Even Clark Kissinger believed SSOC had a bright future. Despite chastising the group for its "intellectual malaise," he noted, "This is not to say that SSOC is in any sense 'doomed.' Quite the contrary—SSOC operates within such an enormous vacuum (the Southern campus) that it couldn't die if it tried." Sue Thrasher did not consider SSOC's floundering to be all bad either, despite her criticisms of the Southwide meeting. While she bemoaned the fact that "Very few people [at the Southwide conference] had any clear-cut idea of what the group should be or do," she also believed such uncertainty could benefit SSOC. "I hope we will always be struggling to find our role," she wrote to Archie Allen earlier in the fall. "For as long as we are doing this, we are probably involved in some kind of creative action."[57]

By the end of 1964, the creative action Thrasher spoke of had begun to take place on predominantly white campuses throughout the South. Locally organized and oriented, the activism of white southern college students gave Thrasher and other SSOC leaders hope that SSOC could play an essential role in coordinating the activism of young white southerners. However, considering that the Mississippi Christmas Project was the only action other than the organizing of conferences that SSOC had taken since its founding, the group had its work cut out for it. But SSOC's leaders believed they were up to the challenge. In eight months they had created a solid foundation for the group, learned hard lessons about red-baiting and the political realities of the movement, and, unlike any other activist group of the day, taken the movement into white communities deep in Mississippi. Their enthusiasm tempered by these experiences, yet still possessing the idealism and arrogance of youth, the young whites believed SSOC would soon become the leading student organization in the South. Expressing the feelings of many in SSOC as the calendar turned to 1965, Sue Thrasher explained,

> If I ever thought of SSOC in a light and shallow vein, I gave that up long ago. For me, the job that we have set out to do is deadly serious. It means working extra long hours for little pay, it means sitting up night after night in "gut sessions" discussing what is really meant by civil liberties, and wrestling ourselves with the problems that confront the South on a much larger scale. I am staying with SSOC because I still believe in that idea, and because I think that right now SSOC has the opportunity of becoming a very dynamic student group that can help to change the South.[58]

Chapter 4

SSOC and White Student Activists at Mid-Decade: The Agenda Grows

> In the South, the transition from civil rights to the war and campus issues was continuous and direct. Students who participated in the civil rights movement were radicalized by their experiences in it. . . . No one started out . . . on radical footing. It was on-the-job training for all of us.
>
> —Marshall B. Jones[1]

> The last five years of southern history has stripped the "southern way of life" of its magnolia blossoms and southern belles. . . . We are living in a region where there is an obvious need for broad secular changes if the "New South" is to be a well balanced and just society.
>
> —1965 SSOC Conference Report[2]

Not long after the first Southwide SSOC conference in November 1964, student activists at Maryville College, a small Presbyterian school nestled in the mountains of east Tennessee, launched a program on campus to advocate progressive causes and to help raise funds for SSOC. Although far from the centers of movement activity in the South, Maryville students had been especially staunch supporters of SSOC during its first year: Maryville was one of the 15 schools represented at the founding SSOC meeting in Nashville, and 8 Maryville students, more than from almost any other institution, had attended the recent Southwide conference in Atlanta. Energized by that gathering, the Maryville students returned to school and, working through their local group, the United Campus Christian Fellowship, quickly organized a "Fast for Freedom," an event that organizer Sandy Chittick explained "involved students in giving up their Sunday evening sack suppers so we could donate the money to a worthy cause—SSOC of course!" The Maryville activists, however, did not consider the black freedom struggle to be the primary freedom for which the students fasted. "Civil rights," Chittick noted in a letter to the SSOC office, "was not emphasized as much as general American and world ideals of democracy, freedom from poverty, fear, etc." The event was a success, as the activists made their peers aware of SSOC and its agenda and convinced 452 of the 700 students to forgo their supper. In early December, they sent the proceeds from the fast, $158.70, to the SSOC office along with a personal check for $5 from Maryville's president, who, Chittick gleefully stated, "was an enthusiastic supporter."[3]

Although Maryville's "Fast for Freedom" was a unique event on a small, remote campus, its organizers were among the growing number of students at the South's predominantly white colleges and universities to become involved in progressive reform movements between 1964 and 1966. SSOC leaders were enthusiastic about this burgeoning local activism since they envisioned the group as a coordinator and facilitator of local activities. The SSOC leadership also worked diligently to draw campus activists to SSOC in an effort to establish a presence for the group at schools across the South. At Maryville, for instance, a visit by SSOC campus traveler Archie Allen had prompted Sandy Chittick and friends to attend the Southwide SSOC conference, which, in turn, had motivated them to stage the "Fast for Freedom." SSOC then publicized this event in its newsletter in hopes that it would encourage students elsewhere to become active on their campuses and in SSOC.[4]

SSOC's focus on local action created a series of opportunities and challenges for the group. On the one hand, white activists on campuses throughout the region readily saw SSOC as a source of organizational support, as the natural group to turn to for advice, resources, and guidance, thus giving it the chance to extend its reach and to win new supporters. On the other hand, since SSOC was a coordinating committee with affiliates rather than a membership group with chapters, the group could influence only indirectly the local organizations aligned with it. Each campus-based group determined for itself which issues to focus on according to the local context and the nature of the student body. The dynamic on each campus was different. Issues that resonated at one school might have fallen flat elsewhere, and actions that earned one group widespread campus support might have alienated potential supporters in other locales.

SSOC's leaders were untroubled and even encouraged by the local orientation of these groups because they believed their organization's job was to support campus activists regardless of which particular progressive causes they embraced. And during SSOC's first two years, local activist groups often focused on issues previously unexplored by SSOC. While SSOC initially devoted its attention to civil rights, the issue that had inspired the activism of most of its founders, campus activists, like the "Fast for Freedom" organizers at Maryville, frequently mobilized around academic freedom, university reform, the war in Vietnam, and other concerns that were more relevant to their local situation. As time passed, the local groups' interest in a broad range of subjects helped to push SSOC to expand its focus beyond civil rights. Though the black freedom struggle always would receive strong support from SSOC, by early 1966 the broad-based activism of white students across the South played an important role in transforming SSOC into a multi-issue organization before the end of its second year of existence.

Student Activism I: Nashville, Auburn, Athens, Gainesville

Between the spring of 1964 and the summer of 1966, SSOC steadily gained supporters and grew in popularity. Crucial to the group's rise was the strength it drew from the newly emergent campus movements. On some campuses, activist organizations were brand new. On others, their existence pre-dated SSOC's founding. In both cases, though, SSOC worked to establish connections with the local activists and bring them into the regional group's orbit. SSOC supported these groups in numerous ways, plying them with printed materials, sending representatives to their campuses, and, most significantly, gradually adopting many of the issues that animated the local activists.

While civil rights was not the overriding concern of many of these campus groups, it was a top priority among some activists. In Nashville, white activists focused on civil rights due to the vibrancy of the local movement. Initiated with the 1960 sit-ins undertaken by black students trained in the techniques of nonviolence by James Lawson, the Nashville Movement still was pressing for desegregation, and garnering headlines for doing so, four years later. Although the movement had forced much of downtown Nashville to open its doors to blacks, a handful of establishments steadfastly had refused to desegregate. As time passed, growing numbers of whites had joined with the black activists to picket and demonstrate against these last holdouts. Most of the whites were associated with Vanderbilt University, Peabody College, and Scarritt College, institutions that bumped up against each other in the central part of the city. The ongoing black protests downtown, as well as the white opposition that greeted each demonstration, had played an important role in galvanizing Nashville whites to act. Equally significant was the indifference and occasional hostility to civil rights they encountered at their schools.

At Vanderbilt, for instance, the state's most prestigious institution, students and administrators alike resisted efforts to open the school to African Americans and sought to isolate the campus from the discussion of civil rights, let alone civil rights activism. In 1960 the administration had precipitated a crisis by expelling James Lawson from the Divinity School for his activism, and in 1963 student government voted to withdraw Vanderbilt from the National Student Association because of the NSA's advocacy of integration. The anti-civil rights sentiment on campus prompted several Vanderbilt whites to seek involvement in the Nashville Movement. With no opportunity to work for civil rights on campus, the downtown campaign provided an outlet for their activism. The same held true for like-minded individuals at Peabody and Scarritt, where, initially, students also reacted coolly to the local civil rights struggle. The fall 1963 protest at the nearby Campus Grill represented the first attempt by whites from the three institutions to challenge segregation in their own community. Besides organizing SSOC's founding meeting, the Nashville activists banded together to establish a local civil rights organization, the Joint University Council on Human Relations.[5]

Sixty students from Vanderbilt, Peabody, and Scarritt founded the JUC, as it was called, on April 21, 1964, two weeks after the creation of SSOC. Though an independent organization, the JUC affiliated itself with SSOC almost immediately. Not surprisingly, many of the same individuals were involved with both groups, including Sue Thrasher and Archie Allen of Scarritt, and David Kotelchuck and Ron Parker of Vanderbilt. Barely a week after its founding, the JUC joined with the Nashville Leadership Council and the local SNCC chapter to inaugurate a new push to desegregate downtown restaurants. The campaign targeted some familiar names, including Krystal's, the popular hamburger chain, the Tic Toc, where Kotelchuck had been attacked while picketing in 1962, and Morrison's Cafeteria, the well-known eatery on West End Avenue, close to the schools. In the ensuing weeks, the campaign resulted in the arrest of more than 300 protesters and precipitated several confrontations with counter-protesters, including one in which Scarritt student William Barbee was hospitalized after a baton-wielding police officer knocked him unconscious.[6]

For the white activists of the JUC, the high point of the campaign was a sip-in at Morrison's on May 3, an event timed to coincide with Martin Luther King's visit to the city. While 70 black protesters picketed outside, 200 whites occupied all of the cafeteria's tables, lingering over coffee for more than two hours as a means of shutting down the restaurant because of its segregation policy. When they departed, each protester handed the cashier a small card declaring "my dissatisfaction . . . [with] your present policy of discrimination. . . . As a white citizen of Nashville, I see no reason for this blight of segregation to continue and urge you to refrain from disgracing your city by your policy." The sip-in created mass confusion among the cafeteria's managers because, said one participant, "They didn't know who to keep out. They couldn't tell who was a sip-inner and who was a real customer or a prominent Nashville citizen." Pleased that the sip-in had disrupted the cafeteria's operation and captured the attention of the local media, the JUC sponsored a second sip-in at Morrison's a short time later, drawing more than 300 whites to the cafeteria to protest its racial policy.[7]

The JUC continued to focus on civil rights after the sip-ins. Later in the year, it helped to plan an interracial march downtown to protest segregation, developed a tutoring program for black schoolchildren, and took to the streets of black Nashville to distribute flyers on the civil rights movement. The JUC also worked closely with SSOC. Given the strong ties between the groups, SSOC maintained closer relations with the JUC than it did with its other affiliates, and, over time, the JUC began to assist in the SSOC office by sending out mailings, printing literature, and even raising money for its sister organization.[8]

Student activists on a few other campuses also emphasized civil rights. At Auburn University in Alabama, students advocated black equality through the Young Democrats Club. By the end of 1964, the group, whose 50 members included two African Americans, was the liberal base on campus. Most of the club's members, though, were unwilling to take part in civil rights actions, instead cheering on the movement from the sidelines, such as when it passed a resolution expressing support for the Mississippi Freedom Democratic Party.[9] Students who actively participated in civil rights efforts did so through the Auburn Freedom League. Founded in early 1965 as a community civil rights organization by the minister of the White Street Baptist Church and Tom Millican, an Auburn student and Alabama native who was SSOC's main contact in the state, the Auburn Freedom League organized desegregation and voting rights campaigns throughout Lee County, where Auburn is located. Millican was well known on campus as an outspoken advocate of civil rights, and his organization of a protest against an appearance at Auburn by Governor George Wallace only added to his notoriety. While Millican's activism alienated him from most Auburn students, it did win him the support of a small band of students, many of whom quickly joined the Auburn Freedom League. These students were among the protesters who marched on the Lee County Courthouse in Opelika in September 1965 to protest local authorities' efforts to prevent blacks from registering to vote.[10]

At the University of Georgia in Athens, civil rights competed with other issues for attention from student activists in 1964. For example, the widespread disdain that had developed among students for the special rules that governed women students offered activists a chance to build support for challenging these regulations and restrictions. Additionally, the sudden appearance on campus in October 1963 of peace activists taking part in a Quebec-to-Guantanamo Peace March gave students the opportunity to take a critical look at American foreign policy and war-related issues. But neither women's rules nor peace and war received significant attention from campus activists. Instead, the activists concentrated on civil rights, the issue that resonated most loudly on campus owing to the still-fresh memories of the bitter 1961 struggle to desegregate the university. Indeed, desegregation remained a divisive issue at the university long after the crisis had subsided, as evidenced by the enthusiastic reception students gave Mississippi's rabidly segregationist governor, Ross Barnett, during an appearance on campus in 1963.[11]

Georgia activists, thus, made the struggle for black equality their primary concern. In early 1964, they joined with their peers from several Atlanta area schools, including Emory, Georgia Tech, and Oglethorpe, to create the Georgia Students for Human Rights, a white student organization dedicated to speeding the desegregation process in Atlanta and around the state. One way the group hoped to accomplish its goals was by drawing more whites into the movement. University of Georgia students' involvement in the new group caused an uproar on campus. In one incident in February, Nelson Blackstock, who two months hence would represent the University of Georgia at the founding SSOC conference, created a crisis in his dormitory by posting flyers calling for students to join the protests in Atlanta. This enraged his dormmates, who immediately shredded the flyers and spent the better part of the evening hurling epithets at him and threatening him with violence, thus forcing him to take refuge in his room. Within a few days, he was summoned to meet with the dean of housing, who cautioned him not to fan the flames of racial discord on campus. Shortly thereafter, the powerful and respected dean of men, William Tate, asked to confer with Blackstock. As he wrote the young student, "Often I like to talk to boys about their actions in terms of student opinion and attitude, and . . . I would like to talk to you about the material that you have been placing on the bulletin boards. . . . I am primarily interested in the practical and theoretical aspects, both complicated, that this action implies." But these were not his only reasons for calling the meeting. As their conversation revealed, Tate was trying to learn more about Blackstock in order to assess how serious a threat he—and the new group—was to the school's stability. That the posting of flyers had inspired student harassment and led to a meeting with two deans indicated to Blackstock and his colleagues that civil rights was a potent issue on campus.[12]

The Georgia activists, like their counterparts at Auburn and in Nashville, made civil rights their top priority because of its power as a local issue. Elsewhere in the South, the local calculus was different, as civil rights frequently took a back seat to other, more locally pressing issues. This was particularly true at the University of Florida in Gainesville, the flagship school of the state's higher education system. Florida activists initially focused on civil rights, but as time passed they devoted greater attention to academic freedom, university reform, and the Vietnam War, issues that were of immediate concern to the Gainesville activists. With its multi-issue focus, local orientation, and small but committed core of supporters, SSOC hoped that the Florida movement could serve as a model for student activists throughout the South. SSOC leaders believed that if other young whites could duplicate the Florida activists' passion for progressive causes and eagerness to engage in direct action, southern students would play a leading role in transforming the South.

Propelled to action by the anti-segregation activities undertaken by the young blacks on the Gainesville NAACP's Youth Council, approximately

25 whites on the Florida campus came together in the spring of 1963 to create the Student Group for Equal Rights (SGER). Although called the *Student* Group for Equal Rights, faculty members Austin Creel, Ed Richer, and Marshall Jones played important advisory roles in the group. Richer and Jones were especially influential with the student activists. Richer, a member of the humanities department, was the most outspoken of the faculty activists—as student activist Marilyn Sokalof puts it, "Richer was just out there saying everything there was to say"—while the more reserved Jones, a psychiatry professor, and his wife, Beverly, became mainstays of the Gainesville movement. Among the students who founded the SGER were several who would go on to play leading roles in the campus movement at Florida, including Judy Benninger and brothers Jim and Dan Harmeling. The rigid segregation they encountered in Gainesville had outraged these students and prompted them to seek ties with black activists. Dan Harmeling, for instance, began to frequent NAACP Youth Council meetings, and shortly thereafter both brothers as well as Benninger were walking picket lines to push for desegregation.[13]

When classes resumed in the fall of 1963, SGER focused its attention on the restaurants near campus that regularly refused to serve the few black students who attended the university. After failing to win the desegregation of the eateries through negotiations, SGER organized the picketing of two of the restaurants, the Goldcoast and the College Inn. The group also participated in civil rights actions beyond Gainesville. Later in the fall, a dozen SGER activists joined an anti-segregation demonstration in nearby Ocala, where police quickly identified them as outsiders and arrested them for disturbing the peace. Shortly thereafter, the group expanded its civil rights work by participating in voter registration drives in the black community and developing the Gainesville Tutoring Program to bring white collegians into contact with local black students. By 1964, the group, with 150 members, had become a vocal proponent of black equality on campus. In Marshall Jones's estimation, "The advent of SGER broke the doldrums at Florida."[14]

Nevertheless, SGER was an organization of limited vision, as it was reluctant to broaden its agenda beyond civil rights or to challenge aggressively the powers-that-be in the community. Anticipating that social change would come slowly to Gainesville and the university, most SGER members, including faculty advisers Jones and Creel, believed that patience and diligence eventually would lead to the destruction of Jim Crow. A minority of the group's members, however, did not share this optimism about the future nor this faith in the efficacy of non-confrontational tactics. Jim Harmeling, Dan Harmeling, Judy Benninger, Ed Richer, and several others concluded that only by directly challenging the opponents of progressive social change, from segregationist merchants to pro-censorship administrators, could they achieve their goals. As time passed, they grew increasingly frustrated with the other members' acceptance of the slow pace of change and their unwillingness to adopt more confrontational tactics. In March 1964 they withdrew from the group, and shortly thereafter SGER fell into decline. The splintering of SGER liberated these activists from the bonds that membership in the group had imposed and allowed them to engage in direct action on a variety of issues without fear of alienating reluctant SGER members. Organizational loyalty no longer was paramount. In the succeeding months, SSOC was the only group to which the Gainesville activists maintained close ties. Dan Harmeling represented the University of Florida at the founding SSOC meeting, and the connection with SSOC kept the Florida activists informed about the developing student movement in the South.[15]

No longer constrained by SGER's civil rights orientation, the activists embraced a range of other causes, particularly university reform. To student activists in the 1960s, university reform was a broad and flexible term that encompassed everything from calls for a relaxation of the rules governing student life—the hated in loco parentis policies—to demands that the university change its image as a subsidiary of the federal government and big business. For most students around the nation, the well-publicized 1964 Free Speech Movement on the Berkeley campus of the University of California marked their introduction to the issue. The Free Speech Movement, which had its genesis in the Berkeley administration's refusal to allow outside political organizations to distribute materials on university grounds, made an especially strong impression on student activists at Florida because it threw into relief the connection between university reform and off-campus political activism, the issue that the Florida activists focused on when they turned their attention to university reform in late 1964. The school's in loco parentis policies also inspired the students to call for reform. Marilyn Sokalof recalls that the numerous detailed rules that governed the daily lives of women dormitory residents was a particular point of contention. While some female students perhaps saw these issues as gender-specific concerns, many others, like Sokalof, whose "feminist consciousness really wasn't tapped into yet," viewed them through the prism of university reform. Equally significant in pushing the Florida activists to advocate university reform was their desire to attract more students to progressive causes. They were convinced that raising university-related concerns would motivate many previously inactive students to speak out and thus would constitute a first step toward drawing them into the larger movements of the day. The activists' goal, Jim Harmeling noted, was to reorient students away "from teen-age trivia" by exposing them "to the real social problems of campus and Florida life."[16]

To work for university reform, the Florida activists founded a student political party, the Freedom Party, and ran a full slate of candidates in the February 1965 student government elections. The party's platform called for the abolition of compulsory R.O.T.C., the scrapping of curfew hours for women students, the imposition of rent control on student housing, the end to racial and religious discrimination in university organizations, and the removal of administrative veto power over outside speakers brought to campus by student groups. More generally, the Freedom Party sought to loosen university control over students, and, in one of its nominee's words, "strive toward a radical re-orientation of the basic campus relationships between students, faculty, and administration." While the other presidential candidates spoke of safety concerns and student social life when queried at a debate about the most important issue facing student government, Freedom Party nominee Jim Harmeling answered that guaranteeing student government's independence from the all-powerful administration was of paramount importance. As he emphatically declared, student government, long dominated by fraternity- and sorority-based parties, has been "an echo of the administration and the deans. The only way to alleviate this is through revolution."[17]

When election day arrived, few Florida students heeded Harmeling's call for revolution; he finished third in the presidential race, and only one of the Freedom Party's 23 other candidates won office—Chris Benninger, Judy's brother, was voted onto student council. The election results were a sign, Professor Walter T. Rosenbaum believed, that "this campus isn't inclined toward radical action. . . . The Freedom Party . . . ran with the type of people that are at Berkeley. It fell flat on its integrated face." Yet despite the electoral disappointments, the Freedom Party made important inroads at the university. It not only had transformed university reform and community action into legitimate topics of debate on campus, but its campaign represented the first challenge to the Greek system's stranglehold on political power, a challenge that culminated in 1967 with the capture of student government by a non-Greek party that campaigned on a university reform platform. Although many of the original activists had moved on from Gainesville and the Freedom Party had ceased to exist by 1967, both had left their mark on the University of Florida.[18]

They had left their mark on SSOC as well. In the four months after the November 1964 Southwide conference, SSOC not only reached more students—it tripled its number of campus travelers from one to three—but also began to promote a range of new issues, including poverty, labor organizing, and especially university reform. The Florida movement had deeply impressed SSOC leaders, and they consistently spotlighted activities on the Florida campus as a means of encouraging students elsewhere to become involved in progressive causes. In 1965, the Florida students' success in establishing the Freedom Party and raising concerns about university rules and regulations inspired SSOC to develop a program built around university reform.[19]

SSOC first trumpeted university reform at its spring Southwide conference, where it was the subject of two workshops and the focus of one of the featured speeches. Florida activists were conspicuous at the meeting.

Marshall Jones was a panelist for a session on local campus actions and Ed Richer spoke on university reform, student politics, and the Freedom Party. That Richer joined such luminaries as Andrew Young of the Southern Christian Leadership Conference, Stokely Carmichael of SNCC, and Leslie Dunbar of the Southern Regional Council in making a plenary speech testified to the importance SSOC now attached to university reform.[20]

After drawing attention to university reform at the conference, SSOC sponsored a region-wide tour on university reform by Steve Weissman, a Florida native and veteran of Berkeley's Free Speech Movement, and Heddy West, Don West's daughter, who had become a popular singer of Appalachian music. At each stop, West performed songs to raise funds for SSOC, and Weissman spoke about university reform and the Free Speech Movement. The tour traveled to campuses throughout the region, including predominantly white schools such as Virginia, Duke, Emory, Vanderbilt, and the University of Kentucky as well as a handful of historically black institutions, among them Virginia State, North Carolina A&T, and the Atlanta University Center. The tour visited both schools where students already had begun to speak out about university reform issues, such as Marshall University in West Virginia, where activists recently had founded a new political party calling for an end to women's hours and the establishment of a "free speech podium" on campus, and schools where it hoped to fuel student discontent with university regulations. At the University of Arkansas in Fayetteville, the tour helped galvanize students interest in the issue, as in subsequent months activists on campus began to publish "Scuse Me," an alternative newsletter, and hosted a SSOC-sponsored conference on university reform. By the end of 1965, many of the students associated with SSOC had come to support university reform, and SSOC leaders had committed the organization to the issue.[21]

Student Activism II: Charlottesville

Despite its prominence within SSOC, university reform supplemented rather than replaced civil rights in the group's agenda. Although university reform received considerable attention at the spring conference, civil rights still dominated the meeting's sessions, workshops, and speeches. The conference also furnished evidence that SSOC had not wavered in its commitment to biracialism. Giving life to its decision at the November 1964 meeting to build a more integrated organization, the group elected two black students to leadership positions, choosing Howard Spencer of Rust College in Holly Springs, Mississippi, and Herman Carter of Southern University in Baton Rouge, Louisiana, as the organization's vice-chairman and secretary, respectively. With white students filling the other two leadership posts—Howard Romaine of the University of Virginia succeeded Gene Guerrero as chairman and Roy Money of Vanderbilt was the new treasurer—SSOC was integrated at its highest level. In the months after the conference, the group's continued commitment to civil rights and its desire to support biracial efforts converged in its backing of a community project in rural Virginia undertaken by the interracial Virginia Students' Civil Rights Committee (VSCRC). SSOC lavished attention on the VSCRC because the group, by concentrating on affairs beyond the university, sought to bridge the gap between the campus and the community.[22]

VSCRC was born in late 1964, the creation of a small group of black and white students from across Virginia who sought to speed the process of civil rights reform in the state. Prominent among the founders were whites from the student movement that had emerged at the state's premier institution of higher learning, the University of Virginia in Charlottesville. Virginia seemed an unlikely setting for student activism given that in the early 1960s the university remained an elitist bastion of white male privilege. Few blacks or women students could be found on the university's grounds, and the vast majority of students, faculty, alumni, and administrators wanted to isolate the school from dangerous ideas and people that might force social change upon it. Yet at the university of serpentine walls and Virginia Gentlemen, a small band of students came together to promote desegregation. As time passed, these students deepened their commitment to progressive causes, expanding their focus beyond civil rights and making connections with outside activist organizations, including SSOC. Although the Virginia activists had started, as one of their numbers remembers, as "a pretty tame bunch of liberals who were trying to desegregate things," they eventually became leading members of the southern antiwar movement and important figures within SSOC; for three consecutive years SSOC's chairman hailed from Jefferson's academical village.[23]

In the spring of 1961, liberal leaning students founded a university chapter of the Virginia Council on Human Relations (VCHR), a liberal statewide organization which, like its sister organizations across the South, actively supported desegregation efforts. For the next three years, the Jefferson Chapter of the VCHR was the primary vehicle through which pro-civil rights students worked for desegregation. The group's top priority was to educate the university community on the importance of civil rights, and it shied away from direct action on the grounds that such "an extremist approach . . . may lead to confusion or animosity among individuals rather than solution of . . . problems," its newsletter stated.[24] Several of the few black students at the university participated in the chapter, including Wesley Harris, the first African American given the privilege of living on the Lawn, the row of dormitory rooms, faculty residences, and classrooms that comprised the original campus. Harris became the president of the group in the fall of 1963. Among the white activists, many were drawn to the group by its faculty advisors, history professors Thomas Hammond and Paul Gaston. Gaston, a warm and affable

Alabaman, especially fired the students' interests in civil rights. Many were impressed that as a native southerner he had chosen to remain in the South to teach about the South instead of fleeing the region for a plush appointment at a northern university. His outspoken support for civil rights and his active participation in the VCHR made him a role model to these students. For white students like Bill Leary, the chapter's president in 1964 who considered himself a moderate, Gaston was the "perfectly credentialed person for playing the role he was playing—a southerner with the right trace of a southern accent . . . the embodiment of the Virginia Gentleman, polished mannerisms. . . . And the fact that somebody like Paul Gaston also [supported civil rights] showed that people who were not alienated from normal convention nevertheless could be committed to this kind of activity."[25]

By 1964, some of the students associated with the Jefferson Chapter, including undergraduates Roger Hickey and David Nolan, and graduate students Anne Cooke and Howard Romaine, had grown impatient with the group's non-confrontational style and its moderate outlook; they were appalled, for example, that some of the group's members did not support the 1964 Civil Rights Act. They wanted the VCHR to more aggressively promote civil rights, and they believed that the group should advocate racial reform not just at the university and in Charlottesville but around the South. Above all, these students craved action. They believed that one of the chapter's most common activities, sponsoring talks by prominent civil rights supporters, such as James Farmer, Norman Thomas, and Martin Luther King, Jr., was important but would not, by itself, lead to desegregation. Only bold and confrontational tactics, they argued, would bring lasting change to the university, Charlottesville, and the South.[26]

Romaine, in particular, played an important catalyzing role. Romaine came to Virginia from Mississippi, where he had participated in the White Folks Project and briefly had shared a house in Jackson with Mario Savio, who shortly thereafter led the Free Speech Movement at the University of California. Having taken part in the most dynamic civil rights project of the summer, Romaine refused to settle comfortably into VCHR-style activism. Talking incessantly about Freedom Summer, Mario Savio, and civil rights, he pushed for the group to adopt a more activist agenda. To the students who shared his impatience, Romaine's appearance at VCHR meetings was, David Nolan remembers, "a real shot in the arm." In the fall of 1964, Romaine and the others found support for taking a more aggressive stance from SSOC. In October, Archie Allen visited Virginia and encouraged the students to become more active. Then in November, several of them attended the Southwide SSOC conference in Atlanta, where they met activists from campuses across the region. Inspired by the range of activities taking place elsewhere, they were convinced that the Jefferson Chapter needed to modify its program and goals. When they failed to persuade the chapter to develop a community action program to challenge segregation both at the university and in Charlottesville, they spontaneously organized a new student organization dedicated to developing action-oriented programs in support of progressive causes. At Howard Romaine's suggestion, the activists named the group Students for Social Action (SSA) and immediately elected him its first president. Soon after, the group officially affiliated with SSOC, becoming the first local organization to establish formal relations with SSOC.[27]

SSA's first major action—to collect food and clothing in support of SSOC's Mississippi Christmas Project—was undertaken in cooperation with the Jefferson Chapter. But while the VCHR students limited their involvement to gathering the materials to ship south, SSA members Sidney Kamerman, Leo Bowden, Anne Cooke, and Howard Romaine, along with three students from nearby Mary Washington College, traveled to Mississippi to help repair a vandalized community center in Meridian. The group also joined with the VCHR and the local NAACP chapter in March 1965 to organize a "Sympathy for Selma" demonstration in the wake of the "Bloody Sunday" beating of civil rights marchers as they attempted to cross the Edmund Pettus Bridge. Singing "We Shall Overcome," 300 students and faculty marched to the Rotunda, the university's historic central building, where they rallied with the NAACP members who had marched down Main Street to the university. Several SSA members, including Alan Ogden and Bob Fisher, then went to Selma to take part in the successful march across the bridge.[28]

The cooperation between the SSA and the VCHR masked the growing irrelevancy of the Jefferson Chapter to student activists at Virginia in 1965. While the rest of the university community might have considered the groups to be nearly identical since they both advocated desegregation, the action-oriented SSA had eclipsed the VCHR as the most dynamic student organization on campus. SSA also distinguished itself from the Jefferson Chapter by its willingness to speak out on issues not directly related to civil rights. Like activists at the University of Florida, SSA members had no compunction about bringing attention to free speech, academic freedom, and other university reform-type issues. The group, for instance, criticized the university's hallowed honor code, charging that it was absurd for a segregated institution, on the one hand, to profess allegiance to a code of honor that called for the expulsion of students for such offenses as cheating or plagiarizing and, on the other hand, to use racial criteria to regularly deny admission to a whole class of individuals. And in fall 1965, in a flyer it distributed to all incoming students, SSA attacked compulsory R.O.T.C., the honor system, and a conformist "miseducation" system that transformed an inquisitive young person into "a neat little machine boxed in coat and tie."[29]

Additionally, SSA differed from the VCHR by eagerly taking part in civil rights campaigns off campus. In 1965, SSA members played an instrumental role in the Virginia Summer Project, a community action program in six black-belt counties in Southside Virginia, the mineral-rich, tobacco-growing land that lay south of the James River. Although both SNCC and SSOC supported the project, it was organized and led by the Virginia Students' Civil Rights Committee, an independent organization of black and white students from across the state founded in the aftermath of a SNCC-sponsored conference at Hampton Institute in December 1964. Inspired by the volunteers who had gone to Mississippi the previous summer, these students, among whom were SSA members Howard Romaine, Anne Cooke, David Nolan, and Carey Stronach, sought to develop a mini-Freedom Summer in communities closer to home. As Hampton Institute student and VSCRC chairman Ben Montgomery remarked, "We decided we didn't need to go to Mississippi to find work that needed doing. We had problems right here."[30]

On June 1, fifteen whites and five blacks from seven Virginia colleges and universities set off for Southside with the intention of organizing themselves out of a job. Because segregation was a defining feature of black life in the counties, much of their work centered on challenging the enforced separation of the races. In the Nottoway County town of Blackstone, VSCRC encouraged residents of the Westend, the segregated section of the town, to demand better city services, and they responded by organizing a petition drive and then threatening to hold demonstrations in a successful effort to force the town's white leaders to install sewer lines in their neighborhood. On more than one occasion VSCRC students organized integrated groups to seek service from segregated businesses, and in Powhatan County several of the activists went so far as to picket one such establishment, the Plainview Service Station.[31] Voter registration was another important facet of the volunteer's work. VSCRC encouraged African Americans to register in a variety of ways: they held small workshops on registering; they spoke to congregants before, during, and after church services; they conducted registration drives; and, most notably, they organized public demonstrations. In Lunenburg County, for example, VSCRC organized a march to the county courthouse, after which 175 people registered to vote for the first time.[32]

In several counties, the volunteers helped local residents develop community poverty programs or pushed for the inclusion of more blacks in existing programs controlled by local whites. The volunteers worked as well to expand job opportunities for Southside's black citizens. To protest the discriminatory hiring policies of white businessmen, the activists joined black community leaders in Victoria and Lawrenceville, the commercial centers of Lunenburg and Brunswick, respectively, in organizing a boycott of white-owned stores to protest the businesses' refusal to hire blacks. The "selective buying campaigns" were not trivial events; both the Lawrenceville boycott, which occurred during the Christmas buying season, and the Victoria boycott, which began in the fall and lasted well into 1966, severely hurt local merchants since blacks comprised a significant portion of their clientele. Despite appearances by the Klan, assaults of boycott supporters, and initiation of a $1,000,000 conspiracy suit against the volunteers and black community leaders in Lunenburg, local whites were unable to break the boycotts.[33]

The boycotts were the high point of VSCRC's work in Southside. They won widespread support from the black community and inflicted financial pain on many white storeowners. That the boycotts took place in the second half of 1965 signaled that VSCRC remained committed to working in Southside long after the summer had ended. While most volunteers returned to school at the end of the summer, they visited Southside on weekends, publicized the campaign on their campuses, and helped raise funds for the project. A few VSCRC volunteers, including David Nolan and Nan Grogan, decided to leave school altogether after the summer in order to continue organizing on a full-time basis in the counties. The following summer they were joined by a new, larger group of volunteers, as VSCRC organized a second, expanded summer project. VSCRC, however, began to break apart in the summer of 1966, as the organization was wracked by a host of internal problems that worsened over time, from financial difficulties to personality conflicts to organizational problems. Equally significant was the fact that the Vietnam War had begun to displace civil rights as the primary issue of concern among some of the volunteers, particularly those from the University of Virginia. By 1966, these Virginia students, like growing numbers of activists on other southern campuses, had come to see the war as the defining national issue of the day.[34]

Discovering the Vietnam War

Opposition to the war in Vietnam grew slowly among southern activists. Before 1965, the vast majority of activist white students in the South were reluctant to focus on the war. While student activists could cast their opposition to compulsory R.O.T.C. in terms of university reform by arguing that making it mandatory was unfair to students, they could not focus on the war without directly considering American foreign policy. Many simply were unwilling to do so. Desegregation and black rights had motivated them to become socially active, and they had no desire to turn their attention to a new subject. In addition, many saw no reason to dissent from the war, believing either that it was their patriotic duty to support the war effort or, more positively, that the United States was right to have entered the conflict. Nor were those students who were unsympathetic to the government necessarily inclined to speak out about the war. To them, the war was a perplexing issue. Whereas southern activists instinctively knew that supporting civil rights was morally correct, they were uncertain who was in the right in Vietnam. The war was confusing, far away, and hard to make sense of, everything the civil rights struggle was not. Consequently, Marshall Jones, the activist professor at the University of Florida, explained that it took time and effort to arrive at an understanding of the war. "Viet Nam was on the other side of the earth and all the information we had concerning it came to us through intermediaries. Adopting a position on the war required an extension of faith, a sifting of informational sources, an imaginative leap of continental proportions."[35]

But as 1965 progressed and the United States became more entangled in the war, an increasing number of southern activists made the leap. In the second half of the year, opposition to the war coalesced at several schools. In October 1965, students at Emory University in Atlanta, including former SSOC chairman Gene Guerrero and Jody Palmour, who soon would play an important role in the group, organized "Conversation Vietnam," the first major teach-in on the war in the South. The students were encouraged in their efforts by Thomas J. J. Altizer, the well-known Emory theologian who had participated in the nation's first teach-in on the Vietnam War in March 1965 at the University of Michigan. The event was a huge success, attracting an audience of more than 1,000 people to the school on October 29, 1965. Eugene Patterson, the editor of the *Atlanta Constitution*, and Norman Thomas gave the keynote speeches for and against American policy in Vietnam, respectively. After they spoke, several panelists further debated the issue: Georgia Representative James Mackay, Morehouse College professor Robert Brisbane, Young Republican's officer Arthur Collingsworth, and New York University professor Ernest Vanderhaag, all of whom offered support for the administration's current policy, were rebutted by radical Yale professor Staughton Lynd, Emory professor Louis Chatagnier, *Viet Report* associate editor John McDermott, and antiwar activist Nanci Gitlin, who opposed American involvement in the war.[36]

Similar events took place later that fall at the University of Florida and Vanderbilt University. At Florida, an audience of over 1,000 students turned out to watch professors Marshall Jones and John Spanier debate American policy on Vietnam, and at Vanderbilt, David Kotelchuck, the young professor who had been active in local civil rights actions, argued the antiwar position in a debate on campus about Vietnam with political science professor John T. Dorsey. Likewise, activists at both schools founded antiwar organizations, the Gainesville Committee to End the War in Viet Nam and the Nashville Committee for Alternatives to the War in Vietnam. Both groups organized marches, led demonstrations, and staged additional teach-ins.[37]

The war also emerged as an important issue at the University of Virginia. Vietnam contributed to the fragmentation of the movement on campus, as the activists aligned with Students for Social Action sought to work against the war while most of those in the Jefferson Chapter of the Virginia Council on Human Relations remained focused on civil rights. So too did the VCHR's faculty advisers, Thomas Hammond and Paul

Gaston. As a result, relations between Gaston and Hammond and the SSA students became strained. Tom Gardner recalls that "we still related to Paul [and Tom] around . . . civil rights stuff, but there was a little bit of a distancing around the war." Gaston concurred, later recalling an episode when Howard Romaine "berated me . . . for not seeing the light" on Vietnam. Although Gaston eventually turned against the war, the fact that he, in his own words, "took a little time to become [an] antiwar activist" prompted the students to search for new faculty mentors. They eventually found support for their views on the war from sociologist Richard Coughlin, who had served as the U.S. vice-counsel in Vietnam in the late 1940s, and East Asian historian Maurice Meisner, "people who," Gardner notes, "had really no grounding in . . . civil rights but who were Southeast Asian experts."[38]

With the support of these professors, SSA made the war one of its top priorities in early 1966. On February 12, SSA held an hour-long protest on the steps of Alderman Library to show its opposition to the increased U.S. military activity in Vietnam. The 23 SSA members who gathered for the demonstration, however, were unprepared for the hostility directed at them by their fellow students, as more than 300 spectators jeered and pelted them with snowballs. Although SSA did not organize another antiwar protest that spring, opposition to the war continued to grow within the group. Of particular significance was that by the summer most of the Virginia students involved in VSCRC's Southside project had begun to shift their focus from civil rights and community work to the war and campus organizing. In March 1966, several VSCRC volunteers were among the 43 people who demonstrated at the Federal Youth Reformatory in Petersburg, Virginia, to protest the incarceration of three draft resisters. Frustrated by the petty disputes that were undermining VSCRC, the volunteers wanted to reorient themselves toward working with their peers. "Instead of our being whites organizing blacks," volunteer Bob Dewart offered, "I would suggest that we be Virginia students organizing Virginia students." And believing that the war had become the most salient national issue, they wanted to return to campus to help organize student opposition. In their view, the white university, rather than the black community, was the obvious locale for building opposition to the war. The Southside, veteran VSCRC organizer David Nolan noted, "was not the most effective place to work to change American policy in Vietnam."[39]

The antiwar activities undertaken by students at Emory, Florida, and Virginia indicated that by late 1965 and early 1966 the Vietnam War had begun to capture the attention of activist students in the South, including SSOC's leaders. Though it had assumed a leading role in advocating civil rights and in encouraging white southern students to take part in the movement for black equality, the group was conspicuously silent about the war. By mid-1965, a consensus had yet to form within SSOC about

Vietnam, a fact underlined by the group's refusal to take a stand on the conflict during a summer workshop. But SSOC's leaders recognized that as the war began to escalate and dominate the national news, so, too, would it command greater attention from college students. The success of Emory's "Conversation Vietnam" suggested to the SSOC leadership that the war was not merely a faddish issue or one in which student interest would soon dissipate. In the group's newsletter, SSOC executive secretary Sue Thrasher acknowledged that the group would have to discuss the war more thoroughly in the near future. Pleading with SSOC's supporters to share their views on the conflict, she wrote, "discussion about Vietnam is . . . desperately needed in the newsletter—discussion means that we need to hear your comments." In the coming months, SSOC would commit itself to opposing American policy toward Vietnam as well as struggle to reorient its goals and activities toward the war, an issue which, unlike civil rights and university reform, required SSOC to focus on affairs both beyond the campus and outside the South.[40]

Chapter 5
Shifting Ground: From the Campus to the Community

I believe they, more than anyone else in the South, started and made possible some genuine anti-Vietnam protest in the region. Before SSOC there was nothing. After SSOC some things began to happen. Now I know that there were a number of other historical developments coming along at the time. But I would still insist that this little group sparked the thing.

—Will D. Campbell[1]

What do you do when the radical black activists, the ones who share your . . . ideology and politics, are saying "we need a black organization?" I mean, [do] you say, "yeah, that's fine, we totally support your rights to do that but loan us a few of your best activists to be part of our white organization?" It's a weird position to be in.

—Tom Gardner[2]

By the first anniversary of its April 1964 founding, SSOC had established itself as the leading organization of white students in the South. Over the course of the year, the group had developed programs, written a constitution, elected leaders, and solidified its organizational structure. With a mailing list of more than 1,500 names, SSOC's reach extended into every southern state. The previous six months were particularly good for the group. The Mississippi Christmas Project won SSOC widespread support and proved that the group could organize a major project. The Southwide conferences in November 1964 and March 1965 attracted new students to SSOC and enhanced its reputation as a progressive organization. And the election of two black students to leadership positions at the March meeting bolstered SSOC's credentials as an interracial organization. While these developments pleased the group, SSOC executive secretary Sue Thrasher found in them cause for concern. Thrasher worried that SSOC's narrow focus on civil rights jeopardized these recent gains. She believed it was vital for SSOC to expand its program in order to serve students galvanized by such issues as university reform and the Vietnam War; if the group remained wedded exclusively to civil rights, it risked making itself irrelevant on southern campuses. "I guess," she wrote Anne and Carl Braden near the start of SSOC's second year, "that our 'program' . . . [is] quite limited."[3]

Less than 18 months later, the notion that the group's program was limited would seem quaint. The spring of 1965 to the fall of 1966 was a transitional period for SSOC, as it evolved from a civil rights-oriented, single-interest organization into a multi-issue group with a broader perspective and an interest in a wide array of issues. The group's growing financial resources facilitated this transition; with an annual budget topping $35,000 in 1966, SSOC could afford to undertake a host of new programs.[4] More significantly, the development of vibrant, campus-based student organizations that focused on issues other than civil rights impelled SSOC to embrace several new causes. For instance, with opposition to the Vietnam War building on campuses throughout the South, by early 1966 SSOC actively began to support the antiwar movement, eventually becoming the movement's leading white voice in the South. Additionally, beginning in late 1965 the group moved away from its earlier commitment to biracialism, as the white activists' desire to work exclusively with other whites as well as SNCC's increasingly separatist rhetoric, rendered efforts to make SSOC more inclusive both less appealing and less successful. The group's renewed interest in white community organizing, an interest that had lain dormant since the White Folks Project of 1964, prompted the group to reach out to non-student white southerners starting in 1966. Finally, that summer the activists recreated SSOC as a membership organization in order to move it closer to the emergent campus groups and to give the local activists a greater sense of investment in the organization.

But if these developments ensured that SSOC would remain both in close contact with its supporters and focused on issues relevant to local activists, they also created a new set of problems for the group. Some SSOC activists, for example, disliked the group's new focus, especially its emphasis on the Vietnam War, and they gradually drifted away from the group as these other issues displaced civil rights from the forefront of the group's agenda. Similarly, SSOC's multi-issue perspective made it more difficult for the group to shape a coherent program of action. While the vision of a freer, more caring, and peaceful society underlay the group's activism, SSOC struggled to explain how the new issues and causes it advocated reflected its vision of a better world. It was one thing to take a stand against the Vietnam War or to oppose in loco parentis policies and quite another to tie these issues to SSOC's broader goals. Throughout its life, SSOC labored to make these connections.

Opposing the War

American involvement in the conflict in Vietnam grew rapidly in 1965. Early in the year, President Johnson announced that American bombers had started dropping their payloads over North Vietnam, and in the summer he authorized the first use of American ground troops in the conflict; by the end of the year 184,000 American soldiers were stationed in Vietnam, more than 11 times as many were there at the time of John Kennedy's assassination.[5] In the United States, these military developments fueled the antiwar movement. In April Students for a Democratic Society sponsored its, and the nation's, first major antiwar demonstration in Washington, D.C. The Americanization of the war also captured SSOC activists' attention, and over the course of 1965 they too devoted increasing time and energy to protesting the war. As the decade progressed, increasing numbers of white southerners, especially students, spoke out against the war, and SSOC was often the organization through which they registered their dissent. Just as SSOC's civil rights activism shattered the myth that southern whites stood as one against desegregation, the group's work against the Vietnam War rendered obsolete the view of a white South united in support of the escalation of the Vietnam War, in particular, and Cold War American foreign policy, in general.

Despite the group's growing opposition to the war, antiwar sentiment was not universal in SSOC. For some student activists, particularly those at the region's smaller or more isolated schools, such as Maryville College, Auburn University, and Davidson College, SSOC's involvement in the antiwar movement was especially disillusioning. Although the group was the lifeline to the broader movement for these students, more than a few of them severed their ties to SSOC as the group moved to make the antiwar cause its own.

White activists' lack of enthusiasm for the antiwar movement sprang from several sources. Some avidly backed both the military campaign in Vietnam and American foreign policy. Having come of age during the most tension-filled years of the Cold War, when nuclear annihilation seemed an horrific possibility and not merely a remote abstraction, they, like most everyone else around them, agreed that the threat of Soviet expansionism required the nation to be ever-vigilant in protecting its interests worldwide. From their perspective, international communism was as pernicious and as dangerous to the nation's health as the racism against which they fought at home. Tom Millican, the leader of the handful of civil rights activists at Auburn, was one activist who held these views. Millican was an unabashed supporter of the war effort, and he made his feelings known at a SSOC meeting when, Howard Romaine later recalled, first he pleaded with the others that " 'you gotta fight for your country,' " and then accused them of being communists for opposing the war. As opposition to the war grew within SSOC, Millican gradually reduced his involvement in the group until he drifted away from the organization altogether. Other SSOC activists believed the group's primary focus should remain its work for racial reform at home rather than agitation for an end to a military campaign halfway around the world. One white student at Morristown College in east Tennessee who had taken part in local civil rights actions was so incensed by SSOC's growing interest in the war that he demanded that his name be removed from the mailing list. The final straw, he explained, was the increasing discussion of the war in the newsletter, thus making it "another organ that juxtaposed the racial integration movement (highly admirable) with the pacifist movement (highly dubious)."[6]

But as SSOC's involvement in the civil rights movement deepened, most activists came to recognize the interconnectedness of the war and the black struggle. They were especially sensitive to the war's pernicious effect on the movement. Ed Hamlett recalls thinking at the time that "You couldn't fight poverty, you couldn't fight racism, you couldn't do the kinds of things that we needed to do domestically 'cause everything was being poured into—emotionally, fiscally—everything was being poured into this damn war over in Vietnam." Many others shared his concerns. Bill and Betsy Jean Towe, leading figures in the Virginia Students' Civil Rights Committee's project in Southside Virginia, were equally blunt in suggesting that the war was inextricably linked to a wide range of domestic causes. As they noted with consternation in 1967, "The war is diverting our attention and funds from the problems caused by the pollution of our water and air, from solving the growing crises of our cities, from solving the problem of poverty, [and] from solving our racial problems."[7]

Sue Thrasher also concluded that the war and civil rights were interconnected. But although she sympathized with those who argued that SSOC should protest the war because it impeded the black struggle, she instead stressed how the principles that animated the civil rights movement motivated her to oppose the war. SSOC was obligated, she believed, to speak out against the war because it, like the civil rights movement, was fundamentally about people's right to control their own destinies. Just as Jim Crow laws had prevented black southerners from exercising full control over their lives, so too did the American military deny the Vietnamese people the right of self-determination. Archie Allen wholeheartedly agreed with Thrasher, and he castigated civil rights activists who failed to see the connection between the issues. "It distresses me," he complained in the SSOC newsletter in early 1966, "when persons who, for a few days, risked their lives in voter registration projects in Mississippi cannot envision the pursuit of democracy and self-determination for the people in the Mekong Delta in Vietnam. It is inconsistent that those who abhor violence in Selma and Birmingham should not abhor and just as vigorously protest violence carried out by U.S. forces in Vietnam."[8]

For some white activists, the realization that the most vocal opponents of desegregation and black rights frequently were staunch advocates of an aggressive Vietnam policy revealed the connection between civil rights and the war. Bruce Smith, a Lynchburg College student who had helped to organize civil rights protests at the school and who was present at the founding SSOC meeting, quickly realized that most of the South's leading politicians, such as Richard Russell of Georgia and James Eastland of Mississippi, not only opposed extending civil rights to black Americans but "were the biggest cheerleaders for the war." "These horrendous racists were saying it was our duty to go over to Vietnam and defend democracy," Smith bitterly recollects. "We knew from dealing with racists that was a total lie, that they were not interested in democracy.... If there had been Californians and Coloradans and Ohioans hollering about Vietnam it wouldn't have hit us so hard or so early. But what made me look at it was these southern, white, racist guys and the language they used and knowing they were my enemy at home. They were my enemies. And anything they were for couldn't be [at] all good."[9]

Civil rights more indirectly inspired the antiwar activism of others in SSOC. For them, the passage of the 1964 Civil Rights Act and the 1965 Voting Rights Act were signal achievements. While they did not naively think that these acts, by themselves, would secure civil rights for African Americans, they believed that their enactment represented great victories for the Movement. Consequently, SSOC activists could turn their attention to other, equally pressing issues. "Socially conscious students," Archie Allen noted near the end of 1964, now "began... examining world issues."[10]

Among these "socially conscious" students were those whose opposition to the war was rooted in a critique of Cold War American foreign policy, not a recognition of the links between the war and civil rights. For University of Virginia student Bob Dewart, for instance, the war was wrong on its own "merits," a position he had adhered to since John F. Kennedy had vowed to end the "chaos in Laos" in the 1960 presidential campaign. From Dewart's point of view, the war should be opposed because it was wrong, misguided, immoral, and dehumanizing. In early 1966, as a member of Students for Social Action at Virginia, he was one of the primary organizers of the first antiwar demonstration at the university. And after he joined the SSOC staff and became editor of the *New South Student* later that year, he frequently published dissenting articles on the war.[11]

The growing black separatist sentiment within SNCC was another important factor that impelled some SSOC students to turn against the war. Lyn Wells, who soon would become SSOC's most popular and effective organizer, was particularly swayed by the anti-imperialist aspect of SNCC's Black Power message while, as a 15-year-old, she worked in the SNCC office in her native Washington, D.C. Wells vividly recalls that at the last staff meeting at which whites were present, Ivanhoe Donaldson made a presentation on imperialism that implied, among other things, opposition to the Vietnam War and support for the Palestinian quest for a homeland. Black Power led other white activists to oppose the war as well, but in a more indirect manner. In their view, SNCC's black separatist turn forced whites to search for new issues on which to focus. For many, the war became that issue because its escalation coincided with the changes taking place within SNCC. " 'Instead of organizing the black community, let SSOC do the war' " Howard Romaine remembers people in the group arguing.[12]

An upturn in antiwar activism on predominantly white southern campuses also pushed SSOC to treat the war as a serious and important issue. That such leading SSOC activists as Gene Guerrero were among these students ensured that the war would become a subject of discussion within the organization. During the last months of 1965, as the war intensified and the national antiwar movement picked up steam, SSOC devoted space in its newsletter to reports of antiwar activities in the South and discussions about the growing opposition to the war. Between October and December, the newsletter informed readers about teach-ins at Emory and the University of Kentucky, reported on the creation of the Atlanta-based Southern Committee to End the War in Vietnam, and presented thought pieces on the tactics utilized by antiwar activists and the general student response to the war. Sue Thrasher pointed out in one article that antiwar sentiment among white southern students had forced SSOC to recognize the war's relevance to the group's mission. As she explained, SSOC's commitment to building a truly democratic society, ending "man's inhumanity to man," and exposing southern students to "the most vital issues of the nation and the world," required the group to focus on the conflict in Vietnam.[13]

While white southern activists' interest in Vietnam encouraged SSOC to oppose the war, the group moved slowly to translate its antiwar sentiments into action. Indeed, in 1965 the group hardly was the voice of southern white opponents of the war that it later became. Instead, its opposition was tentative and often indirect. At its spring 1965 conference, the group declined to send an official delegation to SDS's upcoming "March on Washington to End the War in Viet Nam," the first national demonstration against the war. And after a protracted debate at a workshop in the late summer, SSOC leaders decided against taking a formal position on the war despite their personal opposition to the conflict.[14]

SSOC's initial reluctance to actively oppose the war reflected the leadership's lingering doubts over how extensively the organization should immerse itself in antiwar work. In a lengthy letter assessing the challenges facing the group published in its newsletter in November 1965, SSOC chairman Howard Romaine argued that the group's prior concentration on civil rights, though unsatisfying to those who advocated the development of a more far-reaching program, had the happy consequence of ensuring that SSOC was known for its support of black equality. Broadening its message risked muddling the public perception of the group and creating confusion about which issues concerned the organization. Moreover, he worried that devoting greater attention to the conflict would alienate the significant if uncertain number of SSOC supporters who staunchly backed American involvement in Vietnam or, at the least, were ambivalent about the war. He also was unable to resolve the central contradiction of antiwar work for SSOC: why should a southern-oriented organization devote time and energy to an international issue? Despite his personal opposition to the war, Romaine suggested that SSOC should remain focused on the most important domestic issue of the day, civil rights. "Perhaps one ought to make the distinction between issues about which one feels morally outraged but can do little about," he explained, "and those issues which because of one's time, place, and personal identity, one can do a good deal about. By the latter I mean, of course, changing the South."[15]

Gradually, Romaine and the rest of the leadership overcame their hesitancy to organize against the war. Working with the newly organized Southern Coordinating Committee to End the War in Vietnam, SSOC helped plan the Southern Days of Protest, billed as the first Southwide protest against the war. Its high point was a series of events held around the region on February 12, including the gathering of 500 people at Vanderbilt to hear SDS leader Tom Hayden talk about his recent visit to North Vietnam; the coming together of more than 125 people in Richmond to listen to antiwar speakers at a Vietnam Forum; and the picketing through the rain by approximately 40 protesters of the pro-war Affirmation Vietnam Rally in Atlanta, whose speakers included Secretary of State Dean Rusk and Georgia senator Richard Russell.[16]

SSOC assumed a more visible role in the antiwar movement after the Southern Days of Protest. During SSOC's spring meeting in Atlanta, the students participated in the first antiwar action that SSOC organized by itself. On Easter Sunday morning, the second anniversary of the group's founding, SSOC activists held a peace vigil and distributed antiwar literature at a sunrise service in an Atlanta sports stadium.[17] SSOC also targeted the military draft. Working closely with SDS, it helped to coordinate southern demonstrations against the Selective Service College Qualification Test that high school seniors and college students were required to take to determine if they were eligible for student deferments. The exam took on greater significance in 1966 as a result of Selective Service System Director Lewis Hershey's announcement that the nation's military needs necessitated an end to the policy of granting automatic deferments to students. The idea of using the exam to determine draft eligibility repulsed antiwar activists. While some, no doubt, feared losing their own deferments, many agreed with an SDS pamphlet that asked, "SO WHAT DOES GEOMETRY HAVE TO DO WITH THE VIETNAM WAR? If the government intends to train us to be killers if we fail its test, why doesn't it ask us instead what we think this war is all about?" SDS decided to do just that by distributing its own "exam" during demonstrations at test centers. While the government exam queried students about their math and science skills, SDS created a 25 question exam that tested students' knowledge of the war. SSOC activists opposed the government's exam for many of the same reasons as their northern colleagues. But the group's southern orientation also informed its views, as it argued that southern students were poorer and less well educated than their northern peers and thus more likely to fail the test and become draft eligible. This perception of the draft was closely related to the increasingly common view among antiwar activists around the country that the selective service system was a tool to force lower-class youths into the military. SSOC simply gave a regional twist to this class perspective on the draft since the South had a larger proportion of poor and working-class young men who were vulnerable to the draft. On the three dates the exam was given in May and June, SSOC took a lead role in dispensing the alternative exam. At 80 southern test centers, including those on the campuses of Memphis State, the University of Tennessee, Randolph Macon College, and the University of South Carolina, SSOC activists distributed the SDS exam to test takers and organized demonstrations against the war.[18]

The Southern Folk Festival

Although SSOC increasingly focused on the Vietnam War in 1965 and 1966, it was not the only issue of concern to the white students. True to the organization's roots, SSOC remained committed to the cause of black equality. In the fall of 1965 and the winter of 1966, SSOC initiated organizing projects in several impoverished southern black communities. In the North Nashville Project, local SSOC activists joined with students from the area's historically black colleges in an effort to establish what Howard Romaine termed "an experimental urban organizing project" among poor Nashvillians. SSOC also helped to fund a project led by Herman Carter, a student at Southern University in Baton Rouge and SSOC's vice-chairman, in which the students worked to establish a community center and open an African American library in Scottlandville, a black ghetto on the outskirts of New Orleans. And Ray Payne, a black University of Alabama student, coordinated a similar effort in Atlanta's Vine City project, a SNCC-led campaign to organize black residents of this slum area.[19]

These projects represented an effort by SSOC to involve interracial groups of student activists in black community organizing. As such, however, they were of limited effectiveness: relatively few SSOC activists took part in them, and the intense focus on one particular black neighborhood that characterized each campaign meant that the projects only reached a small number of people. In early 1966, in an effort to achieve a broader impact, SSOC began planning the Southern Folk Festival, a musical tour of both black and white southern communities and campuses by an interracial group of folksingers. While most of the performers were not well known, by the end of the decade legends Pete Seeger, Johnny Cash, and Bill Monroe also had played on the tour. Thanks to its musical format and its frequent travels around the region, the festival became SSOC's most popular interracial endeavor. More than that, the festival blossomed into one of the most visible and successful projects initiated by SSOC during its five-year existence. As Gene Guerrero fondly remembers, the festival was "one of the most productive things SSOC did."[20]

It is not surprising that music was at the heart of such an important SSOC project since the young whites, like activist students elsewhere, found inspiration for social change in the new folk music of the day. The songs of artists such as Joan Baez, Bob Dylan, and Phil Ochs made a lasting impression on the SSOC students. When Dylan sang about the manipulation of poor whites in "Only a Pawn in Their Game" or when Ochs mercilessly mocked white liberals in "Love Me, I'm a Liberal," they expressed sentiments that many of the activists shared. That country music also resonated with SSOC activists, however, set them apart from their northern peers. The folk music of the day introduced many of the southern students to country as well as to bluegrass and gospel. Gene Guerrero first discovered this music through the work of Joan Baez. One of her first albums "had some bluegrass on it, which I liked a lot," he remembers, "and [I] started to buy some bluegrass records and . . . then went from bluegrass into country." Others found country music through the Grand Ole Opry. Since SSOC was headquartered in Nashville, the students made frequent trips to the shrine of country music for the weekly radio show, where they heard icons Bill Monroe, Roy Acuff, and the Carter Family perform.[21]

Many of the SSOC activists were drawn to country and bluegrass music by its "southernness," by its ability to pull together, in historian Thomas Connelly's words, "the patchwork quilt pieces of the Southern soul." While northern activists might scorn country and bluegrass as the music of uncouth and unsophisticated provincials, the students in SSOC, as part of an organization that celebrated southern distinctiveness, embraced this music as their own, as a reflection of their cultural heritage. Yet not all southerners shared equally in this cultural heritage. From SSOC's perspective, country and bluegrass were such powerful musical forms because they were rooted specifically in the history, culture, and condition of southern whites. The activists' celebration of this music, like their search for a radical southern past, reflected their effort to locate a cultural tradition and a history with which they could identify. Just as significantly, the young activists believed that country and bluegrass linked them to the poor whites of the South. While spending time in 1964 in Prince Edward County, Virginia, which earlier had gained notoriety for closing its public schools rather than integrating them, Ed Hamlett wrote a friend, "Tonight . . . I'm listening to Johnny Cash singing 'All God's Children Ain't Free' and I think about the 200–300 white kids here in PEC who can't afford to go to the white private segregated academy and whose parents won't send them to the public school." The group also adopted as its unofficial song a country tune about former Alabama governor "Big Jim" Folsom impregnating a poor country girl and then leaving her to a life of prostitution while he went off to the statehouse. The white students attached such importance to the song, Sue Thrasher later explained, because it "made us come to terms with our own backgrounds, which were largely poor and rural, and admit that was where we came in, that was where we had to begin."[22]

The SSOC activists did not always adhere to such a racialized view of southern culture or consider country music to be the province only of southern whites. Many of the activists recognized that African Americans had made important contributions to southern culture and music, which inspired them to believe that genuine racial reform was possible in the South. In 1966, several SSOC staffers, after first attending the 41st birthday celebration of the Grand Ole Opry, went to a Midnight Jamboree at the Earnest Tubb Record Shop, where the mostly white crowd surprised them by giving an enthusiastic welcome to black country singer Charley Pride.[23]

A similar sense of optimism infused the Southern Folk Festival. SSOC activists hoped that the festival, with its interracial concerts of country, bluegrass, gospel, blues, and freedom songs would demonstrate that black and white southerners shared a common culture, and thus help to break down the barriers to interracial cooperation. Additionally, the activists hoped that the festival would raise much-needed funds for the group. While SSOC had supported itself on a handful of small donations during its first year of existence—the largest donation was a $2,000 grant from the Norman Fund in late 1964—the group subsequently solicited and received several large contributions, including $1,000 from the Rabinowitz Fund and $2,500, $5,000, and $10,000 dollars from the Taconic, Field, and New World Foundations, respectively. By early 1966, SSOC had slightly more than $20,000 in the bank. The group's commitment to progressive reform and the recognition that the activists faced an uphill battle in the South had attracted these organizations' support. As an anonymous Field Foundation official approvingly noted upon reviewing the group's grant request, "SSOC, with its idealism and devotion, deserves a chance to exist." The group even managed to win the support of the venerable socialist leader Norman Thomas, who donated $5,000 to SSOC in the spring of 1965. Yet despite the growing list of contributors, SSOC remained in perilous financial health since its expanding agenda and newly developed initiatives quickly consumed the increased funding.[24] It was in this context that, in the midst of a meeting between SSOC and SNCC leaders at the Highlander Center, SNCC's Bob Moses suggested that SSOC appeal to sympathetic folk singers to come to the South to perform a series of benefit concerts for the group. Citing the example of the SNCC Freedom Singers, Moses and others from SNCC testified to music's power to draw people together and to help raise funds. The SSOC activists were enthusiastic about the possibilities for such an undertaking, and Moses's idea quickly blossomed into the Southern Folk Festival.[25]

To coordinate the project, SSOC turned to Anne Romaine, SSOC chairman Howard Romaine's spouse who, like him, was a graduate student at the University of Virginia and part of the core group of activists on campus. A talented guitar and piano player, Romaine happily accepted the challenge of organizing the project because she believed it gave her the opportunity to create an identity for herself within SSOC apart from that of the chairman's wife. Her first move was to seek the advice of Guy Carawan and Gil Turner, folksingers and movement supporters who, by chance, were at Highlander the same weekend as the SSOC–SNCC meeting. They both thought the idea for the project was wonderful. In particular, Turner, who organized the Mississippi Caravan of northern folksingers during Freedom Summer, played a key role in getting the project off the ground. In addition to inviting Romaine to New York and introducing her to Joan Baez and several other luminaries of the folk world, he suggested that she contact Bernice Reagon, a veteran of the Albany, Georgia, civil rights movement and a member of the SNCC Freedom Singers. Romaine did so, and the two women took an immediate liking to one another, discovering that they shared similar interests in history and politics and the belief that music could be a powerful tool for social change.[26]

The two women quickly agreed to organize the project together, but they decided to alter Moses's original idea significantly. Rather than bringing a collection of well-known white, northern folk singers to the South to sing about the movement, they would assemble an interracial group of relatively unknown southern folk singers to tour the region. The singers would play both movement songs and the indigenous music of the South, thereby giving meaning to the name Southern Folk Festival. As Romaine explained shortly before the first concert, the tour would confront white students with the social issues of the day by presenting an integrated group of singers performing the songs of the civil rights movement as well as reintroduce all southerners "to the folk music tradition of the black belt and mountain region of the South," thus demonstrating that the music and culture of black and white southerners overlapped. Moreover, the festival's creators explained that the tour could be an important weapon in combating the view that "corrupt politics, race hatred, [and] Uncle Tomism" define the region's past. "Few people," Romaine and Reagon contended, "seem to know much about the [slave] revolts, the Populist Movement, the labor organizing in cotton mills and coal mines of the thirties, the freedom movement or of the music which these social movements created." Romaine and Reagon believed that exposing southerners of both races to activist-tinged music would inspire them to empathize with one another and be more willing to make common cause over issues of mutual concern.[27]

The festival's inaugural concert was on March 31, 1966, in Richmond, Virginia, the first stop on an 18-city tour of the South. The diverse group of performers included Pearly Brown, a blind black street singer from Americus, Georgia; Hedy West, a performer of Appalachian music and the daughter of radical poet and Highlander Folk School co-founder Don West; Mable Hillary, a blues singer from the Georgia Sea Islands; Eleanor Walden, an Appalachian singer from Atlanta; Len Chandler and Gil Turner, northern folksingers who played on part of the tour; and Romaine and Reagon. To heighten the sense of cultural interaction and contact, the performers sat in a semi-circle on stage and alternated playing songs, thus allowing the audience to see how the various musical forms flowed from one another. After Richmond, the project spent the month of April playing community halls and campus auditoriums in Charlottesville, Durham, Columbia, Atlanta, Tallahassee, Birmingham, Tougaloo, Tuskegee, Austin, New Orleans, Fayetteville, Nashville, and Lexington. Traveling and performing as an integrated group across the South led to a few unpleasant encounters. After their concert in Tuskegee, a carload of whites trailed the performers as they made their way late at night toward Montgomery, at one point brandishing guns at the group's vehicle. In several instances, whites picketed the group's performance. At the University of Kentucky in Lexington, members of the campus chapter of the Young Americans for Freedom protested outside the campus auditorium where the concert took place. More commonly, though, audiences gave the performers an enthusiastic welcome. And many audience members, both students and community people, attended the pre-concert workshops the performers held to discuss the history and roots of southern music and the relationship between music and social activism. As singer Mable Hillary put it, "We're trying to teach people about where they came from. If you don't know where you've been, how can you know where you're going?"[28]

The tour was a great success for SSOC. The performers attracted attention wherever they went, thereby introducing SSOC, the sponsoring organization, to hundreds, if not thousands, of southern students. Moreover, as Bob Moses originally had predicted, the festival helped to fill SSOC's coffers. The tour also was SSOC's most noteworthy project in support of its oft-stated desire to become a racially inclusive group. In that sense, however, the festival was an exceptional event; its very success only served to highlight the fact that SSOC typically was unable to bring to life its commitment to interracialism. Few African Americans participated in the group, and, as 1966 progressed, the group's interest in black community work waned. Thus, when Romaine and Reagon separated the festival from SSOC by making it an independent entity after the second tour in 1967,

SSOC was left without any visible evidence of being an interracial group.[29]

Back to the White Community

In reality, SSOC was not an interracial organization. But the group's fading commitment to biracialism resulted from more than just an inability to attract black students or develop interracial projects. It also represented a response to the growing racialization of SNCC. SNCC's heightened racial consciousness, which culminated in the rise of Black Power and the ouster of the few remaining whites from the group before the end of 1966, prompted the white activists to reconsider and ultimately to abandon the goal of building an interracial group of progressive southerners. SNCC's evolution from a group that promoted racial inclusiveness into one that advocated racial separatism and black nationalism was a result of the bitter defeats and crushing setbacks the organization had experienced since its founding in 1960. The refusal of the Kennedy administration either to condemn forcefully the civil rights violations daily taking place in the South or to act of its own accord to protect civil rights workers; the continual violence visited on SNCC activists across the region; and, especially, the 1964 murder of the Freedom Summer volunteers and the subsequent failure of the MFDP challenge at the Democratic National Convention, all fed SNCC activists' disillusionment with the federal government and convinced them that white liberals, their putative allies, could not be relied upon to work for their best interests. As a result, beginning in 1964 a growing number of blacks in SNCC began to insist that African Americans must lead and control the movement for black liberation, that to continue to try to goad the federal government to action or to work with moderate whites would invite further betrayals. And they stressed that SNCC needed to become an all-black organization so that black southerners would have the opportunity to achieve their own freedom for themselves.[30]

Initially, SNCC's move toward racial exclusivity troubled SSOC activists. Many worried that SNCC's evolution into a black-only organization undercut the goal of building a racially unified movement for a progressive South. Alan Levin, a leading activist in Gainesville, remembers that SNCC's move toward separatism "struck me as offensive and against my universalist ideals. . . . 'We're all one people, right? Isn't that the idea?' " Ed Hamlett, who had worked in SNCC before SSOC's creation, found it hard to believe that the group would spurn white assistance. "Want to talk to you about the reported white purge from projects going on in SNCC," he wrote in March 1966 to Jane Stembridge, another early white activist in SNCC. "They appear to be saying it is detrimental to the movement for whites to work in the black community. . . . I don't know—what have you heard?" Expressing his own disappointment as much as his colleague's, Hamlett told her of a SSOC activist who "went down to a conference in Atlanta at SNCC offices recently and returned completely pushed out of shape—destroyed his dreams for an integrated movement." Similarly, SSOC Chairman Howard Romaine disappointedly wrote in the group's newsletter that SNCC's "move toward black exclusiveness . . . pose[s] serious problems to any visionary who is thinking in terms of an integrated student movement in the near future."[31]

Gradually, though, the majority of SSOC activists grew more comfortable with the rhetoric of racial separatism. The turning point occurred in April 1966 at SSOC's spring Southwide conference in Atlanta, where several SNCC leaders spoke to the white students about black consciousness, the need for an all-black SNCC, and the new role for whites in the movement. The discussion that ensued was as rare as it was important, for not since the days immediately after SSOC's founding had the two groups met to consider such delicate issues. As Tom Gardner recalls, the gathering allowed for "a real open—not always easy, sometimes painful—but open, honest, conversation . . . about how the movement should be organized in terms of race . . . between an organization that represented a lot of white activists and kind of a vanguard organization like SNCC."[32] Led by new chairman Stokely Carmichael, a passionate advocate of black separatism who soon popularized Black Power as both a slogan and an ideology, the SNCC members argued that only black activists could work effectively as organizers in African American communities. Moreover, they insisted that an all-black SNCC would help to instill racial pride in the black activists themselves and to combat the sense of racial inferiority that continued to plague African Americans across the South. While they appreciated the white activists' commitment to civil rights and their past contributions to the movement, Carmichael and the others told the whites of SSOC that the time had come for them to return to their own neighborhoods and to work with their own people. They stressed that this was essential work that only the SSOC activists could carry out. As one SNCC organizer from Mississippi put it, Tom Gardner vividly remembers, "'What we're saying is that someone has got to organize those white guys hanging around the gas station. We can't do it, but you can.'"[33]

In the subsequent weeks and months, SSOC concurred with the view that the changing nature of the movement required racially segregated groups. Its support for SNCC's new direction stemmed, in part, from the activists' unwillingness to break with its close ally on such an important subject. SSOC's origins were in SNCC, and the white activists consistently looked to SNCC for leadership and guidance. As a result, they felt compelled to support SNCC's changing philosophy. Not to do so, Tom Gardner explained later, would have made it appear that SSOC was not "being supportive of the . . . integrity of the black movement." Gardner and the others also genuinely sympathized with SNCC's separatist impulses. In their estimation, the emphasis on black consciousness was an understandable development in the struggle for black equality. Rather than something to be feared, many in the group agreed that racial separatism was, Gardner recounts, "the logical extension of the movement by blacks for self-empowerment."[34]

SSOC's general agreement with such a view had deep consequences for the group. By validating racial separatism as a legitimate organizing principle, the white students rendered SSOC's commitment to interracialism anachronistic and signaled their willingness to reorient the group toward white community organizing. Instead of viewing the group's turn away from interracialism as a disastrous development, most considered this a prudent course of action for SSOC to take given the group's own difficulties in becoming interracial. After all, SSOC never could offer black students a clear rationale for becoming involved in the organization. As Steve Wise, the group's incoming chairman, later asked, "why would any black student join SSOC when they could join SNCC?"[35]

The white activists also recognized that someone needed to reach out to the people who they believed were most hostile to progressive goals: non-student white southerners. Since most SSOC activists were native southerners, they considered it their duty to take SSOC's message into white southern communities, to the individuals "who do the dirty work of bombing and beating," in Ed Hamlett's phrase.[36] Although SSOC still declared that its ultimate goal was to bring blacks and whites together around common concerns, the group realized that would be "impossible as long as the Klan is almost the only group now attempting to organize this large, sometimes hated, and often forgotten minority group." Consequently, the white activists could agree with their friends in SNCC that a greater division of labor would characterize the movement in the future: black SNCC activists would work on historically black campuses and organize in the black community, white SSOC activists would strive to build support for the movement on predominantly white campuses and in the wider white community.[37]

Importantly, none of the white activists who supported the group's reorientation believed that SNCC and SSOC would cease to be close allies or that relations between the black and white activists would dissipate. SSOC's waning biracialism and its reorientation toward southern whites neither were accompanied by a concomitant decline in the group's commitment to civil rights nor signaled that the group had become estranged from the black activists of SNCC. Relations between the groups endured, even as both organizations purified themselves racially. And SSOC's commitment to black equality remained strong, even as the Vietnam War and a series of other issues displaced it from the center of SSOC's agenda. Long after both groups' demise, David Nolan proudly notes, "Wherever SSOC chapters were and wherever there was any kind of black activity, the SSOC people were going to be their most direct allies.... So although the chapter itself would not be black and white together, there would always be contacts, there would always be some link."[38]

SSOC's decision to focus exclusively on southern whites, especially its decision to begin organizing among non-students, won the group accolades from a number of its friends and allies. This support was crucial, for it helped to allay the young activists' concerns that their lack of experience in white community work would hamper their new efforts; until mid-1966, the only previous white community program SSOC had developed was the White Folks Project in Mississippi during 1964's Freedom Summer. The group had devoted so little time or energy to such work because the students knew that advocating progressive goals in the white community was lonely, difficult, and potentially dangerous work; it was commonly known, for instance, that segregationists regularly targeted white civil rights activists for especially brutal treatment since they were seen as "traitors" to their race. For many of the activists, their involvement in the civil rights movement and their open rejection of white supremacy left them deeply alienated from white society. Some became estranged from their families, and many others avoided spending any time in white communities. Thus, the support the group received for its decision to try to make inroads into southern white communities bolstered the activists' confidence and convinced them they were moving in the right direction.[39]

SNCC, of course, approved of SSOC's new goals. Stokely Carmichael, Willie Ricks, Stanley Wise, and other SNCC leaders applauded the group for dedicating itself to building support for the movement in white communities. More importantly, the white students garnered praise and encouragement for their work from the "SSOC elders," older white southern activists who, owing to their personal history of activism and their outspoken criticism of the South's prevailing racial and economic order, SSOC often looked to for advice and wisdom. Experience in social justice movements rather than age was the defining characteristic of these activists; among them were both individuals who had fought in the lonely battles of an earlier generation, such as Highlander Folk School founder Myles Horton and Southern Regional Council and Field Foundation official Leslie Dunbar, and younger whites who, though only a few years older than the SSOC students themselves, already had played an important role in the contemporary struggle to change the South, such as Hayes Mizell, head of the National Student Association's Southern Project, and Robb Burlage, one of SDS's founders and the author of SSOC's founding manifesto, "We'll Take Our Stand." All of them informally had advised the group in its early days, consulted with the student activists as they developed programs and devised strategies, and, in 1966, backed SSOC's attempt to reorient itself toward southern whites. They shared SSOC's view that SNCC's drift toward racial exclusivity made the early months of 1966 a propitious time for SSOC to refocus its energies on white southerners. Moreover, they believed that SSOC's attempt to build support for civil rights among southern whites would help to dispel the notion that, as Robb Burlage wryly remarked, "white people don't organize <u>as white people</u> except for a bad purpose." White organizing would prove that, contrary to popular perceptions, the region's whites were not uniformly opposed to the movement for black rights.[40]

One of the "elders" who considered SSOC's reorientation toward southern whites appropriate was Will Campbell, the iconoclastic Southern Baptist minister who during much of the 1960s vigorously promoted black equality from positions within the National Council of Churches and the Committee of Southern Churchmen. The whites of SSOC particularly respected Campbell for his fiercely independent streak and his willingness to promote racial equality in hostile settings. In the mid-1950s, while serving as Director of Religious Life at the University of Mississippi, Campbell's support for black rights made him an outcast on campus and hastened his departure from the school. More recently, and to the consternation of some civil rights supporters, he had ministered to Ku Klux Klan members in an effort to understand and hopefully soften their racist views. Campbell first became acquainted with the students who founded SSOC through their involvement in the civil rights movement in Nashville, not far from his home in Mt. Juliet, Tennessee. When these whites decided to create SSOC, Campbell gave them his full support, and he soon became a self-described "uncle, chaplain, [and] sometimes restrainer" of the group.[41]

Campbell backed SSOC's efforts to organize among southern whites because he firmly believed that the movement for black equality had to be organized and led by blacks. But, Campbell argued, whites remained vital to the cause because they were uniquely qualified to promote civil rights in the white community. As he later explained, "There were things I could do which at that time a black student could not do by virtue of the color of my skin."[42] And since the white, rather than the black, community was the source of the South's racial problems, true progress on civil rights would not occur until southern whites began to support racial reform. From Campbell's perspective, SSOC's new commitment to white organizing represented an important if all-too-rare attempt by progressives in the South to reach out to un-reconstructed southern whites in an effort to persuade them to accept black equality as both an inevitable and positive development.

Anne Braden also supported SSOC's decision to focus more attention on southern whites. Braden was the older activist to whom SSOC students were closest. Like Campbell, Braden was drawn to SSOC because she believed the group could carry out the important work of organizing among white southerners. Over the course of her long career in social justice movements, Braden consistently advocated that white southern activists organize among the region's white population. Progressive reform, she argued, necessitated that white southerners, particularly poor and working-class whites, join the region's African Americans in an interracial coalition to address the problems that affected the poor of both races. She dedicated herself to trying to convince southern whites that racism was both immoral and antithetical to their own interests and that they could help themselves by allying with blacks around issues of mutual concern, from poverty to political representation. But Braden well knew that southern whites largely were unorganized and that few identified their interests with those of their black neighbors. To bring an interracial alliance into being, therefore, Braden insisted that her fellow white activists reach out to those whites untouched by the movements sweeping the South. As she wrote in May 1966, "white people must get to the task of organizing the poor white people of the South who are also oppressed but have been bypassed by the civil rights movement."[43]

Thus Braden eagerly, if paradoxically, welcomed SNCC's increased racial consciousness because, though she abhorred racial separatism and preached interracial collaboration, she believed SNCC's reorientation would force white southern activists to work in the white community. Consequently, she was pleased by SSOC's new-found interest in non-student whites, believing it represented an important first step toward building an interracial coalition. But though SSOC now had a chance to make real gains in the white community, she also saw in this opportunity the danger that the group might turn to organizing whites completely apart from the black movement. Her greatest fear was that SSOC would organize whites around their own issues but then fail to bring them together with black southerners. "We who want a democratic South," she stressed in 1967, "must organize white people not as white people but around their problems. And we must encourage them to see immediately that they will never solve those problems unless they can unite with black people."[44] She believed that organizing whites apart from blacks would encourage whites to create their own groups which, at best, would view blacks as having different interests and needs from whites and, at worst, believe that African Americans stood in the way of solving the problems that bedeviled the white community. In a lengthy memo she penned to the group in June 1966, shortly after SSOC had indicated its new direction, Braden warned the activists that "if you begin to organize groups of white people without tackling this issue in a very concrete way, I think the problem is much greater than that you will just be wasting your time. I think you may be creating a Frankenstein. I think you may well find that you have organized groups and organizations that become a real danger—to the South and to all you stand for. . . . To put it bluntly, these groups you form could well—and not too far in the future—become active anti-Negro groups." Black and white groups, thus, would come to see each other as adversaries competing for resources, attention, and power, rather than as partners in a common struggle. The challenge for SSOC, she believed, was to awaken whites to the problems they shared with blacks and to lead them into an interracial movement to work for progressive reforms that would solve these problems.[45]

She therefore worried that SSOC's response to the changes within SNCC signaled that the white students had lost sight of this ultimate goal. In her June 1966 memo, Braden made clear that she felt it was wrong to think that one could organize "poor white Southerners completely apart from the Negro movement—and figure that somewhere down the road, maybe several years hence, they'll get together in some sort of coalition." In her view, SSOC needed to make anti-racism an explicit part of its agenda in the white community and to persuade whites of the virtues of extending civil rights to blacks. Braden readily acknowledged that the white students would find this work especially difficult. But she stressed that organizing white support for black civil rights was essential for bringing progressive reform to the South. As she explained a few months later in an article in the *New South Student*,

> I've been a white southerner all my life. I've seen a lot of racism, and I've seen a considerable number of people change their views. Nothing in my experience or my observation has ever led me to doubt a proposition that I consider key: that there is no easy way to get over this hump, that it must be tackled head on, that people must be confronted with the consequences of their racist views and put into a situation where they have to face the truth. Truth is painful, and people usually have to turn themselves inside out to accept it. Once they've done that, things happen.[46]

Because few SSOC students had experience in white community organizing, Braden's communications offered the group a clear purpose and goal for working with non-student whites. Yet SSOC did not entirely follow her advice. While Braden had counseled the young activists to make poor whites the focus of their efforts, the students instead sought to reach out to a broad cross-section of the white South. Echoing the view of the activists who took part in the White Folks Project, many students reasoned that success in the white community required SSOC to work with the white middle class in addition to poorer whites. In their view, the group could not rely exclusively for support on lower-class whites since, in Sue Thrasher's words, "the working-class just isn't a great revolutionary vanguard." Thus in Atlanta shortly after the group's reorientation, SSOC joined with local white activists to develop the Atlanta Project, a pilot program for middle-class organizing that sought to convince middle-class white Atlantans not only that they should support desegregation and black equality, but that they needed to take an interest in the numerous other problems that plagued the city. As the organizers explained, the project was based on "the assumption that problems such as racism, poverty, poor education, and public facilities are not the concern of the Negroes and the poor alone."[47]

Among poor whites, the group's most extensive program was the North Nashville Project. As integrated groups of volunteers worked in black neighborhoods elsewhere in the city as part of the project, white activists Ronda Stilley, a SSOC staff member and co-leader of the North Nashville Project, Don Boner of Nashville, and John Hill of the University of Tennessee, tried to organize residents of the poor white neighborhood of Cheatham around such issues as deteriorating housing conditions, the lack of municipal services, and the paucity of job opportunities. By living in the neighborhood and spending time with its residents, the activists hoped to inspire them to create issue-oriented neighborhood organizations to attack the community's problems. Importantly, the students believed that creating these groups would be the first step toward uniting Cheatham whites with the black residents of the other neighborhoods in the North Nashville Project. The work of neighborhood organizations, the students believed, though focused on particularly local concerns, would reveal that "poor whites suffer in a manner similar to poor Negroes," and thus allow for white organizing to give way to interracial activism.[48]

The students' commitment to working with non-student whites did not lead to the creation of numerous community organizing campaigns. The North Nashville Project was the focal point of SSOC's effort to organize in lower-class white communities, and by the end of 1967, nearly all of the whites had dropped away from the project.[49] SSOC's inability to develop successful white community programs stemmed from the students' unwillingness to work in white neighborhoods. Unfamiliar with their surroundings, certain that the locals resented their presence in their neighborhoods, and unsure even how to reach out to lower-class whites, most SSOC students sought to avoid the white community altogether. That few of the students were willing to assume the burdens of working in the white community was not unexpected to Anne Braden. She had warned the group in her spring 1966 memo that community organizing required perseverance and patience—"I am not saying it will be easy. We will fail many times." Braden understood that few white activists would make the sacrifices such work required. As she later reflected, "I guess when you say why they didn't, it's because it's always hard to do. It's easier to work with people you know than to go out and ring doorbells of people you don't. . . . I think there's always been a problem to get white people to do that."[50]

Reorganizing

At the same time that SSOC students were struggling to devise effective programs for connecting with non-student whites, they also reorganized the group in an effort to encourage activist white students to take a proprietary interest in the organization, to embrace SSOC as "their" group. To achieve this goal, SSOC leaders transformed it from a staff organization to a membership group in mid-1966.

The group began to discuss reorganization in late 1965. In November, Howard Romaine argued that restructuring would help campus activists "feel they are part of what is going on in the SSOC office in Nashville." Although SSOC had succeeded in putting local activists in contact with one another, "southern students," Sue Thrasher opined in early 1966, "are probably getting tired of 'communicating' with each other. We need to find more concrete things to offer them—and we need to do that now." As SSOC members, activist students would have direct relations with the group. Instead of joining a campus group that functioned as a SSOC affiliate, a rather ambiguous, ill-defined relationship, students could join SSOC itself and establish SSOC chapters on their campus.[51]

The group formalized its reorganization at a meeting in early June at Buckeye Cove, a campground located near Swannanoa, in the mountains of western North Carolina, which its owners, the Committee of Southern Churchmen, frequently allowed the SSOC activists to use for gatherings. At the meeting, the leaders also agreed that expanding the group's publication service would help it to cultivate a membership base. As regular dues-paying members, students would receive subscriptions to the *New South Student* and the sporadically published *Worklist Mailing*. While the newsletter aimed for a more general readership than just the group's members, the *Worklist Mailing*, with brief pieces on organizational issues and recent developments within the group, was designed, the leaders explained, to "relate SSOC staff thinking to the scattered members." SSOC also decided at the meeting to publish more original and reprinted essays on movement-related topics, thereby requiring the group to hire additional staffers to coordinate this work. Jody Palmour, the Emory activist who had planned "Conversation Vietnam," was one such person, as he first joined the SSOC staff to work on the literature program. Another was Earl Wilson, whose earlier participation in the Virginia Summer Project led to his involvement in SSOC. Wilson, more than anyone else in the organization, was responsible for the smooth operation of the publication service, as he ran the mimeograph machine and oversaw the production process. Finally, at the conclusion of the meeting, the activists adopted a formal constitution for SSOC, codifying the changes in the group's structure and reiterating the group's new raison d' être.[52]

SSOC's evolution into a membership organization fundamentally altered the way the group functioned. Most importantly, SSOC's new structure gave local activists a greater role in shaping the group's activities, thereby ensuring that the organization initiated programs tailored to the particular needs and interests of campus groups. As Nan Grogan explained in a letter to a Tulane student curious about the group's recent changes, "the decision to become a membership organization is hopefully . . . a way to provide students with a democratic process for deciding directions in which SSOC shall move and work." Moreover, converting SSOC to membership status would allow white southern students to identify more easily with the organization. The benefits of membership in SSOC—joining chapters, paying dues, and receiving a membership card and SSOC publications—would give the group more of a presence in students' daily lives and, the group's leaders hoped, would inspire greater organizational loyalty from its supporters. SSOC no longer would be content to inspire the activism of young white southerners and then watch as these students joined other national organizations or only became involved in local activist groups. Reflecting this new attitude, Howard Romaine lambasted his friends at the University of Virginia in May 1966 upon learning that they were considering forming an SDS chapter on campus. Given the historically close ties between SSOC and Virginia activists, Romaine all but insisted that his colleagues remain committed to the southern group. As he told them, "This is a new thing in SSOC called organizational chauvinism."[53]

With SSOC now refashioned as a member-oriented, chapter-driven group, the activists believed that the organization's future was bright with possibility. But although reorganization infused the group's leaders with a new sense of zeal, made SSOC more responsive to campus activists, and encouraged these activists to identify with SSOC, it also created new organizational problems for the group. For example, the shift to membership legitimized the debate of organizational issues within the group, for it signaled that the activists believed that *how* the group was organized was not an inconsequential matter. SSOC, of course, needed to concern itself with organizational matters in order to ensure the group's smooth functioning, but these issues often took on a life of their own, consuming large amounts of the group's time and overshadowing the actual work of the organization. As SSOC's advisory board prophetically warned the students in the spring of 1966, the shift to membership status would confront SSOC with "the danger of becoming concerned primarily with organizational problems rather than effective programs."[54]

As SSOC prepared for the start of the 1966–67 school year, it did not resemble the group it had been just 18 months earlier. While the white activists remained committed to civil rights, they had grown interested in an array of issues only tangentially related to the struggle for black equality. As time passed, protesting the Vietnam War and agitating for university reform took their place alongside demonstrating for desegregation on SSOC's agenda. The previous year-and-a-half also had seen the rise and fall of SSOC's brief experiment in biracialism. By the spring of 1966, SSOC had dropped its efforts to attract black students to the group and pledged to focus its future actions on developing white support for civil rights and other progressive causes. Additionally, SSOC had renewed its commitment to organizing non-student whites, both out of deference to SNCC's insistence that SSOC focus its work on the white community and because of the activists' genuine belief that reaching such whites was vital for bringing progressive change to the South. Organizationally, SSOC was different too, as it had made itself over from a leadership-driven staff group into a membership organization oriented toward campus activists.

Whether all of these changes would make SSOC a more effective advocate for progressive causes, however, remained to be seen. Over the next two years, the activists who steered the group from Nashville, along with the students who became the group's members and formed its chapters, pushed SSOC to take on new issues, develop more radical programs, and forge closer ties with national activist organizations, all of which both raised SSOC's profile in the region and exacerbated the internal difficulties that eventually helped to destroy the group.

Chapter 6
New Message, New Messengers

I had a theory that everybody was organizable, every single person.... Everybody had something they thought was wrong with the system. Everybody. They had noticed some injustice somewhere. And I thought you could find it if you talked to them long enough.
—Lyn Wells[1]

What alternative to electoral politics is posed to white middle-class students (and professors) who every year flock to the Stevensons, Kennedys, and McCarthys? ... The alternative is becoming ever more clear; build a movement based on the personal exploitation of middle-class students (and professors) and from that base attempt to relate to the needs of the black movement, the Chicano movement and the labor movement.
—Gene Guerrero[2]

In July 1966, a white high school student in Birmingham wrote the SSOC office with the news that more than 30 students at his school intended to start a progressive organization. The purpose of his letter was to ask SSOC why they should align themselves with SSOC instead of with the Students for a Democratic Society. The task of responding to the students fell to Ed Hamlett. Although Hamlett long had worried that SSOC lacked clearly defined goals and that it spent too much time organizing conferences and too little time initiating programs, by mid-1966 he was optimistic that the group's recent reconfiguration as a membership organization and reorientation toward southern whites had set it on the right path. In his letter to the student, Hamlett expressed his confidence about SSOC's future, stated that SSOC could send a campus traveler to work with the new group, and explained that because "we have just become a membership organization" the students could set up a SSOC chapter. He also discouraged them from affiliating with SDS since it "does not have the money or human resources to adequately service the South."[3]

Hamlett, though, was unable to persuade the students to join with SSOC. "We are ... affiliating with SDS," the Birmingham student replied. "But," he continued, " we are depending on SSOC a great deal. Could you send someone to help us organize? . . . Also we would like as much sample literature as possible." The student went on to condemn, for reasons unstated, former SSOC Auburn organizer Tom Millican as a "bastard" as well as to "hope and pray" that SSOC would organize a conference in Birmingham. The letter infuriated Hamlett, and he composed a scathing response. "Your letter griped the shit out of me," he seethed in September 1966. "You write to us for literature, you ask for SSOC help in organizing, you ask SSOC to have a conference in Birmingham, you called our former staff member a bastard, and you decide to affiliate with SDS." The last point troubled Hamlett the most. Although he conceded that "SSOC has not done the greatest job in the world . . . we haven't noticed anybody else clamoring to do the job. Simply stated, we need your help. . . . SDS doesn't need it. Southerners have got to do for themselves." The time had come, Hamlett insisted, for southern progressives to tell their friends from the North, " 'We're going to make it. Y'all can go home and fight.' "[4]

Hamlett's reaction reflected his belief that a revitalized SSOC was the most important activist group that progressive white southern students could join. The two years from mid-1966 to mid-1968 proved that Hamlett's optimism was not misplaced. Owing to the group's spring 1966 decision to focus exclusively on white organizing, SSOC now had a more clearly defined mission that, in turn, facilitated the development of a wide range of actions targeting southern whites, from draft counseling and antiwar protests to anti-poverty campaigns and labor organizing. This broad-based activist agenda attracted interested students to the group, and, starting with the University of Virginia, SSOC chapters began to emerge on campuses across the South.[5]

Equally important, SSOC's reorientation led to the rise of new leaders by 1967. In the hothouse, ever-changing atmosphere of the 1960s, these activists constituted a separate generation from those who founded SSOC in 1964. The boundary between the generations was fluid; some individuals involved in the group prior to 1967 successfully made the transition to leadership roles in later years, while other post-1966 leaders previously had participated in SSOC. Yet important differences in interests, strategies, and style distinguished SSOC's founders from those who succeeded them as leaders. The push for black equality was not a defining issue for SSOC's second generation, as it had been for the group's founders. Although the post-1966 leaders supported black civil rights, SNCC's turn to Black Power, SSOC's renewed interest in working with non-student whites in the South, the steady escalation of the war in Southeast Asia, and the pervasive though premature sense that the passage of federal civil rights legislation in 1964 and 1965 had dealt a decisive blow to segregation all combined to make the Vietnam War and white community organizing the top priority for the later activists. Compared to their predecessors, SSOC's second generation also adopted more radical tactics in pursuit of its agenda. From walking picket lines with striking workers to disrupting military recruiters' campus visits to orchestrating antiwar demonstrations across the South, second-generation activists demonstrated a willingness to initiate actions sure to thrust the group into the public eye and to garner publicity for the causes SSOC promoted.

While the emergence of new leadership and the group's reorganization and reorientation allowed Hamlett and the others to look to the future with hope, these developments simultaneously caused new difficulties and exacerbated old dilemmas. SSOC labored to strike a balance between serving the needs of local activist groups and developing programs that appealed to its members across the region. Additionally, SSOC's decision to commit itself to white organizing rekindled a dispute, dormant since 1964's White Folks Project, over which whites, students or non-students, were the group's primary constituency. This disagreement became more problematic in 1967 and 1968 because the continuation of the war, which was increasingly unpopular on southern campuses, directly competed with working-class organizing for primacy on SSOC's agenda, thus sowing confusion about SSOC's goals and purpose among the group's leaders and campus supporters alike. As a result, the group lurched between developing working-class organizing projects and devising campus antiwar actions, providing evidence for those who later would contend that the group lacked focus and consistency.

The new generation's radicalism also raised a series of difficult questions for the group. Should SSOC cultivate radical sentiment in the South or moderate its message and image in order to attract more broad-based support? Was it appropriate for the group to focus on issues that were national in scope or was its primary responsibility to concentrate on problems peculiar to the South? Could SSOC reconcile its growing propensity for radical action with its identity as an indigenous southern organization? Was SSOC becoming a radical student group that happened to be located in the South or was its growing radicalism connected to the organization's southern roots? Differences within SSOC over these issues created serious divisions that deepened and hardened over time and ultimately contributed to the group's collapse.

The Second Generation

As SSOC remade itself into a membership-based, white-focused organization, it experienced significant turnover in its leadership. Between the summer of 1966 and the spring of 1967, many of the individuals who had led SSOC since its inception either drastically reduced their role in the organization or left the group altogether. Sue Thrasher, the person most responsible for keeping SSOC afloat during its first two years, was one of the first to leave, departing in the spring of 1966 for a position at the

Institute for Policy Studies, the liberal think tank in Washington, D.C. By early 1967, Ed Hamlett also had left the group. SSOC's first chairman, Gene Guerrero, remained nominally on staff, though by the fall of 1966 he was working in North Carolina with the Textile Workers Union of America and thus was far removed from the daily workings of the Nashville office. Likewise, Anne and Howard Romaine distanced themselves from the Nashville headquarters. Howard Romaine had returned to his graduate studies at the University of Virginia and focused his activism on the local setting, while Anne Romaine channeled all of her energies into making the Southern Folk Festival an independent organization.[6]

Personal reasons prompted many first generation activists to move on or to redirect their efforts to new causes. An uneasiness with the incoming activists' penchant for radical rhetoric and tactics also informed their decision to disengage from SSOC. Sue Thrasher framed the differences between the two sets of activists by suggesting that the late 1960s activism of SSOC, in particular, and the movement, in general,

> was very stylistic and just sort of swaggering and being a militant and being this hot shot stuff, and not having to prove it very often. I mean, you could go on demonstrations and you could do this, that and the other, but it wasn't the same kind of situation that had been in the earlier sixties. And I don't mean to sort of sound like this sort of old fogey putting the young people down. I just think it was a different political situation that people were in. . . . You got thrown into things in the early sixties . . . that you didn't want to be thrown in and you changed as a result, and you didn't often get thrown into those same situations [in the late 1960s]. Even a little bit later, it wasn't quite the same.

Others agreed and doubted that the newer activists could hold the organization together. As Ronda Stilley, who in 1967 married Vanderbilt professor David Kotelchuck and moved to Ithaca, New York, later remarked, "I wasn't sorry to leave in June '67 because I could feel the centrifugal force pulling it apart."[7]

In the view of second generation SSOC activists, the founders were out of step with the times; they were timid, cautious, and too reluctant to initiate dramatic actions that would bring the group publicity. From her perspective as a rising leader in the group, Lyn Wells believed that the older activists "were all looking to get careers," though hastening to add, "I'm not knocking them; everybody has to make a living." Nonetheless, that they were now "looking for those kind of liberal institutional-based things that they could do with their lives" suggested to her that their commitment to the group had waned. Some of the recent arrivals in the group made no secret of their view that the sooner the founding generation left the group the better. At one particularly contentious staff meeting in March 1967, Stilley reported with outrage that Brian Heggen, one of the new staffers, ominously had warned Ed Hamlett and her that they "had better watch out, because in three or four months, there was going to be a power struggle and perhaps a coup and 'a change was gonna come' within the organization."[8]

The importance of civil rights to the newer activists became a particular source of contention between the generations. Many SSOC founders believed that few second-generation activists shared their commitment to black rights. This perception fueled their disenchantment with their successors. But their concerns were unfounded; the civil rights movement had inspired the activism of many newer SSOC activists. Indeed, prior to assuming an organizational role within SSOC, many second-generation leaders had participated in civil rights campaigns on their campuses. When SSOC's second generation arrived in Nashville, however, civil rights had ceased to dominate the group's agenda. While the second generation was enthusiastic about the group's new focus, SSOC's reorientation obscured the importance of civil rights in these activists' personal histories.

Like their predecessors, many of SSOC's second generation traced their support for black equality to childhood experiences or transformative events early in their collegiate careers. Shirley Newton, who served on the Nashville staff in 1967, and Mike Welch, SSOC's executive secretary in the group's final years, followed similar trajectories to involvement in the movement. Both were born and raised in Memphis, attended segregated Treadwell High School, and became history majors at Southwestern at Memphis before dropping out to work in the movement full time. For both, sympathy for the civil rights movement as high schoolers evolved into active participation in the movement while at Southwestern. The televised images of brutalized civil rights protesters in Birmingham and elsewhere deeply disturbed Newton, as did President Kennedy's assassination. At Southwestern, she took her first steps toward activism when liberal history professor John Hemphill, who she recalls once described himself as the "red dean of history," introduced her and other students to several SSOC activists. Intrigued by the organization, she grew close to several of its leaders and eventually moved to Nashville to join the staff.[9]

The Kennedy presidential campaign and the burgeoning civil rights movement also captured Welch's attention in high school. Among the youth at his Baptist church, he was the only one willing to speak in favor of the Massachusetts senator during a mock debate shortly before the 1960 election. Welch's interest in civil rights prompted him to consider volunteering for Mississippi's Freedom Summer project until he learned that the volunteers needed several hundred dollars to cover their expenses, money he did not have. Welch found other ways to demonstrate his support for black equality at Southwestern. In the spring of 1965, he attended a NAACP-organized demonstration to show support for the Selma marchers, and by early 1966 he had met Ed Hamlett and started to become involved in SSOC.[10]

Jody Palmour, one of the Emory student who organized "Conversation Vietnam" in 1965, was drawn to the civil rights movement long before he grew concerned about the war. Born and raised in the small community of Gainesville in the mountains of North Georgia, Palmour came from a well-off family with deep roots in the town. When he arrived at Emory in the fall of 1963, he quickly fell in with a group of older, serious, bohemian students who did not hesitate to advocate progressive causes. Contact with these students, combined with the increased violence directed at civil rights activists and the assassination of President Kennedy, propelled Palmour to action. Over the next two years, he participated in numerous local civil rights demonstrations, and by early 1966 he had left Atlanta to join SSOC's Nashville staff.[11]

In the second half of the 1960s, a small cadre of activist white students came together at Millsaps College, a small, Methodist, liberal arts school in Jackson, Mississippi. A commitment to black equality fueled these students' activism. Mike Cassell, whose uncle, a policeman in McComb, was accused by the NAACP of killing a black civil rights protester in 1961, first became involved in the movement in 1964 after a deacon at his church in Canton had repulsed him by giving him brass knuckles to use against protesters seeking to integrate the church's services. Cassell Carpenter, a wealthy banker's daughter who had a privileged upbringing at Dunleith, one of Natchez's antebellum mansions, was drawn to civil rights in 1965 after visiting Strike City, an encampment near Greenville of black cotton workers who had been evicted from their plantations after striking for higher wages. Everett Long, who grew up in towns in the Delta, joined Tougaloo College students in an interracial civil rights discussion group while he was a student at Mississippi College in Clinton, just outside Jackson. He also took part in the leafleting of segregated churches in the Jackson environs, where he encountered white students from Millsaps. The next year, after Mississippi College did not invite him to return, Long transferred to Millsaps.[12]

David Doggett spearheaded the group of Millsaps activists, prompting Mississippi State Sovereignty Commission Director Earle Johnston, Jr., to label him the "chief agitator at Millsaps." The son of a Methodist minister who served on the Millsaps Board of Trustees, Doggett spent his childhood in a string of towns in the Delta and northern part of the state. In early 1964, as a high school senior in Tupelo, several local black activists fired his interest in civil rights when they spoke at his church about the movement. At Millsaps, he came into contact with African American peers and developed friendships with several students from historically black Jackson State University. By mid-1966 he had made contact with SSOC and attended his first SSOC meeting. Shortly thereafter, in June 1966, he participated in his first demonstration, boldly joining civil rights activists as they marched through Greenwood, where his parents lived, after James Meredith had been gunned down as he attempted to walk through the state on his "March Against Fear." The following May, he and fellow Millsaps student Lee Makemson organized a march to City Hall by white civil rights supporters to protest the slaying of Benjamin Brown, a black student whom National Guardsmen killed during a protest on the Jackson State campus. To the surprise of downtown businessmen and afternoon shoppers, approximately 20 white Millsaps students, carrying signs with messages such as "Millsaps mourns the death of Benjamin Brown," invaded the city center. For Doggett, the march was testimony to the power inherent in white organizing. As he later explained, "It was one thing to work and demonstrate with northern students and blacks; it was many times more moving to do so with my own people." Over the course of the next two years, Doggett helped to make Millsaps a key outpost of SSOC activism.[13]

Two of SSOC's most prominent second generation leaders came to the group from the University of Virginia: David Nolan, who in 1967 became editor of the group's monthly publication, the *New South Student,* and Tom Gardner, SSOC chairman in 1967–68. Like others of the second generation, their activism pre-dated their association with SSOC; both were members of Students for Social Action at Virginia and had participated in the Virginia Students' Civil Rights Committee's summer project in Southside Virginia. Nolan grew up in the Bayside section of Queens, New York, where he was raised Catholic by his artist mother and newspaper reporter father. Unlike many of the native southerners who became active in SSOC, Nolan derived little meaning from his religious upbringing, as the formality of Catholic rituals and the fact that Mass was celebrated in Latin dampened his interest in the church. Politics, though, captivated him and led to an interest in civil rights, an interest that mushroomed at Bayside High School, where he took a class on "Problems in American Democracy" from an avowed Socialist and joined a student group that brought Norman Thomas and representatives of the NAACP to the school. Although Nolan had applied to the University of Virginia on a lark, his support of black equality made the school an appealing destination. "Going South at the time of the civil rights movement," he recalls, "I knew that I was going to 'major' in civil rights."[14]

Gardner, the son of a Navy dentist from Western Kentucky, was born in New Orleans and traveled around the South as his father transferred to new assignments. His family's frequent moves prevented Gardner from establishing strong ties in many of the communities in which he resided, thus, he later reflected, inoculating him "from the kinds of strong, community norms of segregation which many of my later peers in SSOC would grow up with."[15] Although he attended segregated schools in the South, the fact that the military had achieved a level of integration unheard of in the South enabled him to interact with the children of black personnel in base activities. His familiarity with black students helped to ease his transition to high school in New Jersey, where his family had moved following another transfer for his father. At a school with a significant black population, Gardner was drawn to the African American students because "they seemed more like southerners than white students . . . in terms of general demeanor [and] friendliness." By the time he was ready to depart for the University of Virginia, his family's military background had allowed him to form interracial bonds, thus sensitizing him to matters of racial discrimination and ensuring that he would not tolerate quietly the Jim Crow laws that governed life in the South.[16]

Earl Wilson had more direct ties than Gardner to the armed forces. The seventh of eleven children in a poor farming family in south-central Pennsylvania, Wilson had served in the Coast Guard from 1959 to 1964, stationed first in Norfolk, Virginia, and then in Puerto Rico. When his tour ended, he matriculated at segregated Old Dominion University in Norfolk. As a veteran in his mid-20s, Wilson shared little in common with his classmates. His refusal to dress in a coat and tie and his decision to live off campus further distanced him from the rest of the student body. So too did his willingness to express himself on any and all issues. As he later conceded, he had a "big mouth. I was always willing to say something in class. I was always willing to get my two cents in in every situation." Wilson soon gravitated toward the Emerson Forum, the Unitarian Club on campus, which was a refuge for other non-conformist students. There, he met Rives Foster, a veteran of the first summer of the VSCRC's Southside Project. Foster prodded Wilson to become actively involved in the cause, and in 1966 Wilson joined the Southside project for the second summer. Shortly thereafter, he left for Nashville to run SSOC's literature program and to serve as assistant editor of the *New South Student*.[17]

Lyn Wells was a full decade younger than Earl Wilson. But her fiery personality, passion for white organizing, and unrivaled ability to motivate young whites to support SSOC made her the most influential second-generation activist. She became SSOC's North Carolina campus traveler in 1967 at the age of 18, and her work in that capacity accounted for the surge in the number of SSOC chapters in the state. Her seminal 1968 essay, "American Women: Their Use and Abuse," played a key role in inspiring SSOC to champion the women's movement. That same year she was elected SSOC's program secretary, giving her considerable power to shape the group's agenda. Although neither a college student nor a native southerner, her strong interpersonal skills rendered irrelevant her differences from the people she worked with and organized. Wells's forceful personality also put her at the center of most controversies during SSOC's last two years. While she inspired fierce loyalty from her supporters, she earned the enmity of those who disagreed with her tactics or dissented from the direction she sought to take SSOC. However, nearly all agreed that her talents as an organizer heightened SSOC's profile around the South and vastly increased the number of students active in the group. In deference to her organizing skills, David Nolan lamented long after

SSOC's demise, "If we had had twelve Lyn Wellses, we would have been hell on wheels."[18]

Born in Lancaster, Pennsylvania, Wells was raised in the Maryland suburbs of Washington, D.C., by parents who were active in leftist politics; in the 1940s and 1950s, her father was a union organizer, and both parents were sympathetic to communist causes. Her parents did not inculcate her with their political views; initially she knew, and cared, very little about their background. In fact, she did not learn of her parents' politics until the Cuban Missile Crisis. With war seemingly imminent, her mother told her that the government might detain communists, just as it had detained Japanese Americans at the outset of World War II. "And your daddy and I were communists and they'll probably round us up," Wells remembers her mother explaining. "And if they do, call this one, and if they're gone call that one." Her mother's words shocked Wells. "You were communists?" she recalls asking her mother. "Oh, God, what else?"[19]

Gradually, though, Wells grew interested in political and social issues, particularly civil rights. In 1963, before the start of the ninth grade, she attended the March on Washington with her mother, and shortly thereafter she took part in her first protest, a local demonstration in the wake of the bombing of the Sixteenth Street Baptist Church in Birmingham. Soon, she began skipping school to participate in civil rights demonstrations. The murder of the civil rights activists during Freedom Summer prompted her to begin volunteering at the SNCC office in Washington at 15. She was very active in the group; she took part in a sit-in at the White House after the attack on civil rights marchers in Selma, organized a national high school Friends of SNCC conference, and worked regularly with such SNCC activists as Stokely Carmichael, Ivanhoe Donaldson, and Ralph Featherstone. By the end of the summer of 1965, though, Wells decided that her future in the movement was to organize among southern whites. In 1966, as she began to plot her move south, she came into contact with some of the white activists, including Sue Thrasher, who recently had relocated to Washington. Later that year the two of them, along with Jim True, a former University of Virginia student and VSCRC volunteer, organized a Washington conference on white organizing in the South. Impressed with her organizing abilities, SSOC asked Wells to become its North Carolina traveler. She accepted, and in February 1967, just after her eighteenth birthday, Wells moved to Greensboro to begin her work.[20]

Just as it had for SSOC's founders, the activism of second generation SSOC members created turmoil in their families. Sometimes it caused difficulties for their parents. Tom Gardner recalls that after he had traveled to Czechoslovakia to meet with North Vietnamese as part of a delegation of American antiwar activists, his father's commanding officer summoned him to a meeting where, after referring to his son's F.B.I. file, alleged that the Communist Party had paid his travel expenses and that he was a communist sympathizer. Gardner's father knew better; as he explained to his surprised supervisor, *he* had paid for his son's trip abroad. Mike Welch's activism also caught the attention of his mother's employer, who fired her from her job as a shoe salesperson shortly after he had helped to organize a protest of Southwestern students at a segregated restaurant across the street from her store. Given her fine sales record and the fact that her superiors knew that he was her son, it was obvious she had lost her job because of his organizing work. Her firing transformed her into a civil rights supporter. It "became a turning point in her life, that instead of getting mad at me she got mad at them," Welch recalls. "It was like once there was a crack in what her view of the world had been up to then, . . . suddenly she saw everything in a different way."[21]

More commonly, second-generation activists encountered stern parental opposition to their activities. David Doggett's parents considered his activism counterproductive and dangerous. "They thought you had to do everything behind the scenes slowly," he remembers. "And so here I was wanting to lead demonstrations." Nor did many parents share the activists' liberal views. Jody Palmour, for instance, recalls his father chiding him, "don't ever call a black person a 'guest,' " after he had used the term in reference to a black deliveryman. Cassell Carpenter, who in 1966 was crowned Queen of the Natchez Pilgrimage, an annual celebration during which Natchez's leading white families gave tours of their mansions and estates, saw her relationship with her parents deteriorate after her activism increased. For these students, activism frayed ties to their families, and some familial relations ruptured under the strain. Tom Gardner estimates that his involvement in progressive causes alienated him from his mother for at least a decade, and Shirley Newton notes that she lost contact with her family shortly after she moved into the SSOC house in Nashville, a move her parents reacted to "with shock, horror, and dismay."[22]

Some parents were troubled more by their children's demeanor than by their activism. David Nolan's parents believed their son had become rude, insolent, and generally unpleasant since he began to participate in activist causes. In a bitter exchange of letters in the spring of 1965, his father, while not objecting to his participation in the summer project in Southside Virginia, took issue with what he called Nolan's "intolerably know-it-all attitude on everything. It's not often that I volunteer advice," he continued, "but I'll give you one word: Change! Not for your Mother or for me. But for yourself. Unless you do, promptly, you'd better make your own arrangements for school or work in the fall." In his angry and sarcastic reply, Nolan wrote, "I was happy to learn that I had been disinherited. It was so nice of you to tell me. I shall assume that you were either drunk or insane when you wrote the last letter, but be that as it may, I shall not take such insults from anyone, regardless of his misbegotten familial relationships." While his letter is filled with the bluster and disrespect typical of a youth chafing under parental authority, the fact that Nolan was about to take part in a summer-long civil rights project suggests that he also was rebelling against social conventions and the status quo. And although his family professed acceptance of his commitment to social reform, his decision to drop out of college in order to work on the project full-time devastated them and clearly exceeded what they felt to be appropriate behavior. As Nolan's grandmother wrote in a futile attempt to persuade him to return to school, "You have done your humanitarian deed—as you wished to this summer—but remember, your first duty is to your parents. You have crushed them . . . by telling them you did not want to go back to college. . . . I am asking you, begging you, for the first time in my life, to return to college—not next year, but this year, this month, now."[23]

David Doggett's parents also clashed with their son about his attitude and behavior, particularly his personal lifestyle choices. As Doggett's activism increased, so too did his alienation from mainstream values and cultural forms. He frequented black nightclubs, drank liberally, experimented with drugs, and became sexually active. His mother and Methodist minister father strongly disapproved of such behavior, and though perhaps unaware of everything he was doing, they knew enough to know that he was involved in what they considered unsavory activities. Their concerns intensified when he was arrested for indecent exposure at a black nightclub in Jackson during his sophomore year at Millsaps. That the charges were trumped-up hardly assuaged their shock at his arrest.[24]

Doggett was not the only SSOC activist drawn to an alternative lifestyle. The countercultural revolution launched in the San Francisco Bay area, with its concomitant emphasis on spiritual transcendence and earthly pleasures, won numerous adherents among second generation SSOC activists. In the spring of 1967, SSOC activists were among the students at the universities of Florida and Virginia who organized Be-Ins, consciously emulating the Human Be-In in San Francisco's Golden Gate Park in January of that year. In 1968 in Richmond, SSOC activist and former Lynchburg College student Bruce Smith joined with several others to open a head shop, the "Liberated Area." The shop became a gathering point for Richmond radicals and hippies, and despite constant police harassment, the store remained open until 1970.[25] Activists with SSOC ties also worked on two of the most significant alternative southern newspapers of the era. The *Great Speckled Bird* was founded in Atlanta in late 1967 by, among others, former SSOC chairmen Howard Romaine and Gene Guerrero and Pacific Northwest expatriates Tom and Stephanie Coffin. The paper's first issue signaled that it would occupy radical ground in Atlanta. Its purpose, wrote Tom Coffin, was "to offer some alternative to what some call 'The American Way of Life.' . . . People, especially the young, are now tired of the pap-feeding, the absurd sloganeering, the lies, the bullshit. With the discovery that our plastic civilization is hollow and void, the 'turned on' seek meaning. Through involvement. Political Activism. Art. Drugs. Involvement." With its focus on cultural trends and its regular skewering of both the liberal Democratic mainstream and the staunchly conservative, segregationist business and political elites of Atlanta, the *Bird* nourished political radicals and social nonconformists in Atlanta and inspired the creation of underground papers elsewhere in the South.[26]

In Jackson, Mississippi, the *Kudzu,* named for the fast-growing vine that covered the state, started publishing "Subterranean News from the Heart of Old Dixie" in September 1968. Founded by SSOC activists and Millsaps students David Doggett and Everett Long, the *Kudzu* united the hip with the political, combining sharp attacks on the state's reactionary political leadership with celebrations of the counterculture.[27] The paper's focus on cultural and lifestyle issues reflected the founders' belief that, as Doggett wrote, "youth culture is the major happening of this decade, and will prove to be one of the major events of the last part of the twentieth century." And suggesting the breadth of *Kudzu*'s ambitions, he editorialized, "We do not seek a narrow, violent political revolution. We seek a much more profound revolution, a revolution of a whole culture." Such countercultural sentiments earned the paper the opposition of white Jackson. Throughout the *Kudzu*'s life, the Jackson police constantly harassed the staff, and city officials worked to prevent the paper's publication and curtail its distribution. In October 1968, for instance, 12 *Kudzu* staffers were arrested on chargers ranging from vagrancy to assaulting an officer for attempting to sell the paper outside a local high school. Two months later, police arrested Doggett, Long, and two other staff persons for attempting to sell the paper—which authorities branded "obscene literature"—at a high school and junior high school. Despite the harassment, the *Kudzu* managed to survive until 1972 as the primary expression of white progressive thought and cultural dissent in Jackson.[28]

The disapproval of mainstream whites did not distinguish the *Kudzu,* the *Great Speckled Bird,* and other southern alternative newspapers from their northern contemporaries. The intensity of the opposition they provoked did. These papers were perceived to be genuine threats to the social order. They attained such status not merely because they celebrated sex, drugs, and rock 'n' roll, but because they accepted, promoted, and glorified racial reform. To do so in the North was hardly noteworthy. In the South it was radical. Whether the papers retained a political focus and championed desegregation and civil rights legislation or were decidedly apolitical but featured discussions of black culture and interracial relationships, southern alternative newspapers made themselves targets of white reaction.[29]

If alternative newspapers were one means by which SSOC activists could express their radical political and cultural sentiments, protests and demonstrations were another. At the University of Florida, SSOC activists organized actions designed to reflect students' varied interests and concerns. In 1967, more than 300 people took part in a march for "Love, Sun, and Peace in Vietnam" under the leadership of SSOC's Alan Levin. And on election night 1968, the local SSOC chapter organized an anti-Nixon, anti-establishment demonstration-cum-party attended by hundreds of students in the Plazas of the Americas at the center of campus. Levin, who became SSOC's national vice-chairman in 1967, concedes he was slow to see the benefits of reaching out to more culturally focused radicals. In his view, they trivialized the causes he embraced by diverting attention to their outrageous behavior. "They'd go around painting their bodies green . . . with smiles on their faces and flowers and stuff like that," he remembers. "And I thought . . . we should be angry and we should be serious and we should be organizing and we should be demonstrating. . . . They kind of sometimes would just smile, and they would say, you know, 'lighten up.' And I didn't appreciate the message." His views of these people, though, became more accepting after the campus SSOC chapter was able to draw some of them into protest actions and after he began to experiment with LSD in 1968.[30]

Vietnam: Protests, The Draft, Peace Tours

The second generation's embrace of the counterculture deeply troubled many SSOC founders. To them, the newer activists appeared more interested in the quest for personal growth and gratification than the fight against the Vietnam War or the struggle for civil rights. Some founders additionally believed that the radical tactics of their successors signaled a descent into frivolousness and triviality and betrayed a sense of hopelessness for the future. But such views mistake style for substance. Second-generation activists believed that SSOC's reorientation toward southern whites, the apparent intractableness of the South's racial and economic problems, and, especially, the deepening of the crisis in Vietnam necessitated more provocative actions in order to focus public attention on the causes it promoted.

The Vietnam War was a focal point for the second generation's activism. The war had intensified by mid-1966. Hopes for a peaceful resolution to the conflict were dashed by President Johnson's late January order to resume bombing raids into North Vietnam after a month-long pause and his decision in late June authorizing the bombing of petroleum, oil, and lubricant storage depots in Hanoi and Haiphong. The new bombing campaigns stirred domestic opposition to the war. In January and February, Senate Foreign Relations Committee Chairman J. William Fulbright of Arkansas, a former Johnson ally, convened nationally televised hearings that challenged the Johnson administration's Vietnam policy. In the spring, opponents of the war around the nation observed the Second International Days of Protest, and demonstrations against the Dow Chemical Company, the leading manufacturer of the flaming petroleum jelly napalm, became common.[31]

SSOC's opposition to the war picked up mid-year as well. In May, the *New South Student* announced that the group had set itself the modest of goal of "organiz[ing] people to talk and learn about why we are involved in Vietnam and what the arguments are." Many newer SSOC activists sought to add an activist component to this educational role. The recent escalation of hostilities, they believed, necessitated that they become more outspoken in their opposition to the war and that they initiate or, at the very least, participate in highly visible, provocative antiwar activities. More veteran SSOC activists disagreed, arguing that while such actions momentarily might catapult SSOC into the limelight, they ultimately were counterproductive since they would taint SSOC as an extremist group.[32]

These tactical differences broke into the open when President Johnson visited Nashville in March 1967. Reasoning that his presence in the city afforded the group a unique opportunity to capture national attention, several SSOC activists plotted to disrupt his Ides of March speech at the statehouse with a dramatic protest. The proponents of the protest were a group of new SSOC staff members, including Janet Dewart, the lone remaining black staffer in the organization, and Shirley Newton, the manager of the Nashville office. The lead organizer was Brian Heggen, a native of San Jose, California, and the current editor of the *New South Student* who already had gained a measure of notoriety in Nashville for threatening to chain a dog to a tree on the Vanderbilt campus and set it afire to protest the use of napalm by American troops. After discarding a plan to have demonstrators throw burned rice, representing destroyed Vietnamese crops, at the president's motorcade, the activists led a protest in which the approximately 30 placard-carrying demonstrators paraded outside the Capitol while the president spoke inside. In the protest's most dramatic moment, Heggen threw himself in front of Johnson's departing limousine, which had to swerve to avoid hitting him. Secret service agents and police officers quickly swarmed Heggen, and as Newton and Dewart tried to join him, the police hauled all three away.[33]

The demonstration and arrests reverberated through SSOC. Activists who believed that the acceleration of the war necessitated the use of more strident and confrontational tactics considered the protest appropriate. SSOC chairman Steve Wise, who remained a student at Virginia, fully supported the demonstration, thereby implicitly giving it the organization's stamp of approval. Other activists objected that the protest was an incendiary action whose sole purpose was to inflame passions and create controversy. Ronda Stilley thought that the demonstration reflected newer SSOC members' proclivity for extremist actions that, unless checked, would reduce SSOC to irrelevancy. "I believe that the base they will build," she declared shortly after the protest, "will be one of fringe people who will have little constructive and possibly considerable destructive effect on the whole movement." Lyn Wells related that she was "pretty upset about the recent adventure in Nashville." As someone who always worked to fit in with her surroundings, Wells undoubtedly faulted the protesters for failing to insinuate themselves in the Nashville community. From her union organizer father she first had learned that "if you're trying to organize people . . . you should be as much like them as was palatable." In her view, the Nashville demonstrators had not learned that lesson, or else they never would have undertaken an action that so obviously violated community norms.[34]

Other antiwar actions undertaken by SSOC also highlighted the growing fissures in the group over the purpose and tactics of antiwar organizing. On the one hand, educational efforts designed to cultivate antiwar sentiment remained common. In 1967 students at Florida and Virginia organized teach-ins on the war, while 30 Atlanta activists distributed more than 10,000 antiwar pamphlets at an Easter Sunday sunrise service in a local sports stadium.[35] On the other hand, some SSOC activists participated in highly visible and provocative actions. In February, SSOC students at Millsaps, unfazed by the white community's staunch support for the war, organized a daring series of protests centered on Secretary of Defense Robert McNamara's February visit to Jackson to give a fundraising speech for Millsaps. In Atlanta shortly after the Easter demonstration, SSOC helped to organize antiwar activities as part of the National Mobilization Against the War protests.[36] In September, Tom Gardner represented SSOC at a meeting in Bratislava, Czechoslovakia, between American antiwar activists and representatives of the National Liberation Front and the North Vietnamese government. The following month, SSOC led a contingent of southern students to the massive "Confront the Warmakers" demonstration in Washington during which protesters suffered arrests and violence at the hands of army troops as they attempted to lay siege to the Pentagon.[37]

SSOC's opposition to the draft also reflected the activists' differing approaches to antiwar work. As some SSOC men discovered, figuring out how to express their opposition to the draft was a challenge that had deep personal consequences. As a result, many agonized over what to do. Would seeking conscientious objector status and agreeing to perform alternative service be an immoral and cowardly act since, though exempting them from military service, it did not challenge the legitimacy of the draft itself? If their conscientious objector applications were denied, should they submit themselves to the draft or continue to resist it and thereby risk jail? Alternatively, should they take the non-cooperation route from the outset and refuse to comply with the system altogether? If so, should they quietly go into hiding or emigrate to Canada, or was it more honest and forthright to make their stance public, say, by burning their draft card or preaching for others to resist the draft as well? Sorting through these options was not an easy task. "It was a tough, tough, tough issue," Gene Guerrero remembers of his struggle over whether or not to cooperate with the system. "It was very hard to figure out what to do about it."³⁸

Tom Gardner concluded that although he wanted to avoid a "confrontation" with his draft board over his status, he would not stay in school just to retain his student deferment since, he wrote in 1968, that "would have been total prostitution and humiliation." Consequently, when he received his 1-A draft card in 1966, he wrote his board that while he believed he could avoid service through established procedures, to do so would be to recognize the legitimacy of the system and to condemn to warfare those less fortunate than he. Later that year, though, he had a change of heart as a consequence of his arrest in Winston-Salem, North Carolina, shortly after he had witnessed a judge sentence two men to two years in prison for non-cooperation with the draft. His failure to stand when, unbeknownst to him, the judge reentered the courtroom after the sentencing, led to his arrest, swift conviction on contempt charges, and sentencing to 30 days in jail. The 16 days of the sentence that he served convinced Gardner that not cooperating with "the system" would be a futile gesture on his part. As he wrote in notes he kept while incarcerated, "I was thinking, in regards to noncooperation, that one of the valuable things would be to confront the Judge and other officials with the burden of having to imprison me, etc. After watching Judge Gordon throw two guys away for two years and imprison me, as I sit here in jail, I can't help but give more pensive thought to the possibility of a two- to five-year term. Judge Gordon was not affected in the slightest, nor were any of the spectators.... It's somewhat ridiculous to let them put me away for four damn years." Thus he decided that he could be of most use to the movement by organizing against the draft and the war outside prison.³⁹

Alan Levin's journey through the selective service system was more complicated than most people's. One of the most outspoken activists in Gainesville, Levin initially was granted conscientious objector status. In November 1966, however, he was reclassified as draft eligible after he distributed antiwar literature at the draft physical he was required to take. His change in status left him in a quandary: "Do you go to jail? Do you go to Canada? Do you go underground? . . . I didn't know what was going to happen." Ultimately, Levin did not let his opposition to the draft prevent him from seeking and accepting a deferment because he was about to become a father. He made clear to his draft board, though, that he still opposed the war and the draft. He recalls writing the board a letter to the effect of, " 'Well, I don't believe in your authority and, I think you're a bunch of fascist militarist pigs, but I want you to know this fact of my life. And if you want to do anything with this, that's fine. I have a pregnant wife.' " When the draft board offered him a deferment, he was overjoyed. And while it was not a deferment of conscience, "I didn't say," he notes, ' "I'm not going to accept this.' " He simply was happy to be re-classified and to have the issue permanently resolved.[40]

Unlike Gardner or Levin, Bob Dewart made the difficult decision to resist the draft by leaving the country. Dewart had fought the possibility of conscription for a long time. In 1966, SDS's *New Left Notes* printed a letter he had sent to the draft board in his hometown of Erie, Pennsylvania:

> For some time now I have been receiving unsolicited mail from your office. You have sent me questionnaires that request information of such a personal nature as to be classified indecent.... Most recently you have pushed me to the limit of endurance by expecting me to prove my conscientiousness or face the consequences of indentured service in a dehumanizing machine of mass murder operating under the euphemism of the Department of Defense.... As I cannot consider myself a member of your despicable little club, I have destroyed those silly membership cards which you sent me.

In early 1967, Dewart disappointedly resigned as *New South Student* editor and returned to Erie, where he spent the better part of the year arguing with the draft board about his status. When it became clear that the board would not re-classify him, Dewart left the country for Canada, forever separating him from the group. As he wrote Tom Gardner shortly after his arrival in Toronto, "I regret having lost the sense of involvement that I had for the few short months while working for SSOC."[41]

SSOC developed a variety of programs to oppose the draft. Some embodied the educational orientation that some of the activists preferred. For instance, SSOC members held draft counseling sessions at colleges and universities throughout the region. In Nashville, local SSOC activists were among the founders of the Draft Resistance Union of Metropolitan Nashville, a group that frequently leafleted against the draft as well as provided legal aid for those who sought to resist the system. At the University of Virginia, the SSOC chapter organized a November 1967 demonstration at the local draft board's office by 13 students from Lane High School in Charlottesville, while SSOC activists at the University of Florida drew attention to the draft by holding a teach-in to publicize the plight of Lavon Gentry, who had been arrested the previous summer for posting "Bust the Draft" posters on university buildings.[42]

Other anti-draft work expressed some activists' growing desire for provocative action. At an October 1967 SSOC-organized demonstration at the induction center in Gainesville, Brian Heggen and Michael Meiselman, SSOC's South Florida campus traveler, were arrested for trying to block the departure of the bus carrying recruits to the local induction center.[43] Two months later, SSOC activists took part in a demonstration at the Raleigh induction center in support of George

Vlasits and Joseph "Buddy" Tieger, both of whom had decided to resist the draft after failing to persuade their draft boards that they deserved conscientious objector deferments.[44] That same day, SSOC organized a protest in Atlanta in support of Gene Guerrero's refusal to submit to the draft. While 75 protesters picketed outside the induction center, Guerrero refused induction inside. Despite the tension of the moment, Guerrero was able to find some humor in the proceedings because, he recalls, "the major in charge of the induction center at the time was a Major Herrera.... He was a Hispanic. And he was really pissed off that the first person he'd come across who was doing this was a Hispanic." The major's anger with Guerrero was indicative of the military's response to his action, for within a year he had been indicted, tried, convicted, and handed a five-year prison term. Not until the Supreme Court ruled that one could be a conscientious objector without having a traditional orientation to religion or formal ties to religious organization was his conviction overturned and the case dismissed.[45]

SSOC leaders hoped to capitalize on the growth of antiwar sentiment among white southern students through the Peace Tour programs. The project was the brainchild of David Nolan and Tom Gardner. Nolan and Gardner envisioned the Peace Tour as a caravan of SSOC activists who would spend several weeks touring campuses in one particular southern state. During their travels, the activists would encourage and help students on campuses bereft of activism to organize antiwar protests, join demonstrations on campuses where students already had begun to speak out against the war, and everywhere participate in teach-ins and give lectures on U.S. foreign policy, in general, and the Vietnam War, in particular. Importantly, the Peace Tour potentially could bridge the tactical divisions in the group by combining educational programs with direct action. And that the Peace Tour would visit many isolated southern campuses that previously had been immune to movement activities made the program especially provocative. The Peace Tour program was SSOC's longest-lasting and most far-reaching antiwar project. Between February 1967 and December 1968 the Peace Tour traveled to dozens of predominantly white colleges and universities in six southern states.[46]

Nolan and Gardner began organizing the first tour in late 1966 before either had joined SSOC's staff. At the time, Nolan was living reclusively in Charlottesville while Gardner worked on civil rights issues for the National Student Association's Southern Project.[47] Gardner, though, was anxious to turn his attention to Vietnam, and he persuaded Nolan to work with him to develop an antiwar project. Gardner started by writing to SSOC contacts throughout the South to gauge their interest in hosting the tour on their campus. Florida quickly emerged as the likely destination for the first tour because, initially, Phil Mullins of the SSOC group in Tallahassee and Alan Levin at the University of Florida were the only ones to express enthusiasm for the project. Throughout the winter, Gardner and Nolan assembled a list of nine college campuses and two high schools to visit. In addition to taking part in local actions and meeting with antiwar students at each stop, they planned to give a public address. Nolan would speak on U.S. policy in Vietnam, Gardner would focus on American foreign policy "and its relation to the centralization of power in the U.S.," and Nancy Hodes, a Southern Conference Educational Fund worker in Nashville who had spent part of her childhood in China and whom Gardner had recruited to the project, would talk about the Cultural Revolution and "the myth of Chinese aggression." In order to devote his full attention to the project, Gardner officially resigned his position with the NSA and joined the SSOC staff. With their itinerary set, in late February Gardner, Nolan, and Hodes made the long drive from Nashville to Tallahassee, the Peace Tour's first stop.[48]

Throughout the five-week tour, the three activists, the *Southern Patriot* reported, engaged in the "constructive 'stirring up' " of the issue of the war on campuses throughout the state. Their primary means for reaching students were their encounters as they manned literature tables on campus and the public address they made at each stop, an event that typically drew an audience of several dozen people. They also met with smaller groups of students whenever possible. Occasionally, sympathetic professors invited them to speak to their classes, and frequently the activists counseled draft eligible men on how the selective service system worked and what options they had for avoiding conscription. To generate publicity for the tour and to bolster local antiwar efforts, Gardner, Nolan, and Hodes participated in several events associated with the "Florida Days of Judgment on the War in Vietnam," a statewide series of antiwar activities. In Gainesville, they joined the march for "Love, Sun, and Peace in Vietnam," and the following day they took part in a peace vigil in Tampa. The trio attracted media attention wherever they traveled. In Miami, then-radio personality Larry King interviewed them on his program, while a Sarasota television station covered their visit to the New College. And in Tampa, where University of South Florida administrators originally had banned the activists from campus, the threat of television coverage forced them to relent and allow the Peace Tour to stop at the school.[49]

The South Florida administrators' reaction to the Peace Tour was not atypical. Across the state, opposition to the Peace Tour was common among students and school officials alike. Nancy Hodes recalls "being the object of incredible hostility most of the time . . . going in front of crowds who were ready to rip us apart." During their public sessions, antagonistic students frequently fired hostile questions at them and showered them with insults. Others created a scene at the literature tables, denouncing the peace workers and threatening to tear up their brochures. Sometimes this opposition assumed more dangerous forms. In Miami, a bomb threat disrupted one meeting, and in Gainesville, renegade students set fire to the Conestoga wagon local activists had decorated with Peace Tour posters to publicize the event. School administrators frequently tried to keep the activists off their campuses, fearful that these "card-carrying dropouts," as one college dean referred to them, wanted to initiate a student revolt. However, school officials unwittingly brought the activists more attention and served to make academic freedom and free speech, rather than the war itself, the central issue. As a result, students who may not have shared the activists' politics or position on the war but who were outraged by administrators' attempts to silence them, agreed that they should be allowed to speak on campus.[50]

Nowhere was this more clear than at the North Campus branch of Miami-Dade Junior College, where officials denied the three travelers permission to speak because they had rejected administration demands that they limit their talk to five minutes and refrain from distributing literature. On March 29, the activists openly defied the administration's ban and went to campus to distribute antiwar materials. After about 50 people had gathered, Gardner began to address them but was quickly forced to stop by two policemen. When Gardner refused to report to the dean's office, he was arrested, and the photo of riot-helmeted police officers dragging him away became a symbol of SSOC's resistance to the war and frequently was reproduced in the group's literature. Nolan and Hodes then tried to speak, and they, too, were arrested, as was South Florida campus traveler Michael Meiselman. The four were charged with disorderly conduct and resisting arrest, and they languished in jail for three days before they could raise bail. The arrests angered students at the school, particularly the fervently anti-communist and pro-war Cuban students, who found the situation all-too-reminiscent of the dictatorial actions of the authorities in their homeland. Nolan recalls one Cuban student telling him, " 'We saw this happening in Cuba. I never thought I would see it happen here.' " The arrests inspired several students to form a Students Rights Committee to talk about how to respond to the administration's actions. The arrests also drew the attention of the Miami media. Ultimately, the uproar the arrests created prompted the administration to reverse itself and to allow the activists to speak on campus. In an effort to save face, the administration required only that their talk be part of a bipartisan program with the war's proponents. Shortly after their release from jail, Gardner, Nolan, and Hodes made a triumphant return to campus where, in front of more than 250 people, they engaged in a three-hour debate with three members of the school's debate team.[51]

The events in Miami brought the Florida Peace Tour to a rousing conclusion. Gardner and Nolan believed that its success could be replicated elsewhere in the South, and they persuaded SSOC to support similar ventures in other states. These subsequent tours focused on schools that, like Miami-Dade Junior College, had seen little, if any, movement activity but that were not located in urban areas, such as James Madison and Sweet

Briar in Virginia, Wofford and Furman in South Carolina, Davidson and St. Andrews in North Carolina, and Milligan and Southwestern in Tennessee. The activists predicted that, as one flyer promoting the tours declared, their presence at these schools would "be of assistance in bring[ing] out those who are concerned about the war, but have not been active." Bruce Smith, a long-time SSOC activist who joined these tours, believed this was the greatest service the tours provided. "What we did," he reflected much later, "is we provided a focus of support and a way for people who were antiwar activists at these different little schools to get together."[52]

Smith's recollections, though, understate the difficult nature of the activists' mission. Randy Shannon, a recent addition to the SSOC staff in Nashville, came to recognize the enormity of the challenge facing the activists when he traveled on the North Carolina Peace Tour. "The need for really basic organizing, campus by campus, person to person, was brought home to me in a tiringly repetitious manner every time we came to a new school," he wrote shortly after the tour's conclusion. Not surprisingly, the peace activists frequently encountered apathetic, uninterested, and occasionally hostile students, faculty, and administrators. At Arkansas State University, the activists' engagement was canceled because of rumors that administrators intended to expel any student who attended their presentation. Elsewhere, the travelers faced more dangerous situations. At Erskine College in Due West, South Carolina, one student threatened the activists with a loaded .45 caliber handgun. Pointing the gun at Nolan, the student explained, Tom Gardner remembered two years later, " 'I just might kill you' " since " 'those damn gooks' " had killed his brother in Vietnam. Though Nolan managed to coax the student to put the gun away, the incident was a frightening experience for the activists.[53]

Their experience on the North Carolina Peace Tour testified to both the difficulties and the possibilities inherent in this campaign. At Appalachian State Teacher's College in Boone, students who Bruce Smith considered "right-wing fascists" and "super-patriot war freaks" ripped up the materials on their literature table and then surrounded and beat on their car as they attempted to drive off campus after the dean of students canceled their activities at the school. Undeterred, the Peace Tourers returned to the campus that evening in an effort to engage students in informal discussions about Vietnam and free speech issues. Gardner went to the student union by himself, and over the course of several hours had serious conversations with more than 50 students, several of whom were among those who earlier had driven them off campus. When some students tried to disrupt the gathering, others shouted them down and told to leave or to join the discussions. "So some of them would leave and some of them would stay," Gardner remembers. "And then another group would come in, and we'd go through this, you know, every half hour or so there was a new lynch mob that had to be quieted down." In addition to urging the students to educate themselves and to oppose the war, he used the arguments they offered in support of U.S. actions in Vietnam to highlight the importance of free speech issues. As he recalls, he would ask them, " 'what are these people [the South Vietnamese] fighting for? . . . You say they're fighting for freedom? What freedom?' [They'd say], 'freedom of speech.' 'Well, let's talk about freedom of speech for a minute. I want to talk about our policy in Vietnam and you want to hang me for it.' " Later, Bruce Smith and North Carolina campus traveler Lyn Wells met with almost two dozen students who wanted to form a study group, receive SSOC literature, and invite the Peace Tour to return to the school sometime in the future.[54]

Shortly after their tumultuous visit to Boone, the activists traveled to Belmont Abbey College in Belmont, where they found administrators and many students just as adamantly, though more peacefully, opposed to their visit. Given the tumult their presence at Appalachian State had created, school officials reneged on their pledge to allow the activists to spend all day on campus. Instead, they offered the Peace Tourers only one hour to visit, arguing that remaining for longer would put their safety at risk. The travelers objected to the revision of their schedule. More importantly, so did the president of student government, Ray Smith, who believed that the administration's actions had made free speech and academic freedom rather than the war the central issue. As he put it, "The issues have now changed. It is no longer anti-war opinion versus pro-war opinion. It is now a question of unnecessary restrictions on free speech." Thanks in part to Smith's comments, the administration agreed to hold a one-hour event on campus in which the students would listen to brief presentations from the tourers and then vote whether to allow them to return to campus for a full day. The SSOC travelers approached the event with trepidation, and justly so, as, initially, the vast majority of students booed and jeered them and held up signs commanding "Commies Go Home" and declaring "God Is A Marine." But after Gardner, Smith, and Hodes each gave a brief summary of the issues they wanted to talk about and stressed, along with Ray Smith, the significance of the First Amendment and the importance of the university as a place for the free exchange of ideas, the students voted to invite the Peace Tourers to spend an entire day at the school.[55]

The SSOC activists' willingness to engage students on the war on such seemingly hostile campuses as Appalachian State and Belmont Abbey was the Peace Tour's most notable achievement. The activists believed that students everywhere, when educated about the war, would recognize its folly and oppose it. For SSOC, the Peace Tours were the ideal vehicle for spreading its antiwar message. As David Nolan recounts, "the alternative would have been to raise $150,000 and get J. William Fulbright to go and speak at every campus in the state. . . . [T]his way we trained ourselves . . . and we went around and we did it ourselves."[56]

SSOC and Labor

Not everyone in SSOC shared Nolan's enthusiasm for the Peace Tour. By the spring of 1967, a substantial number of activists instead favored developing programs among non-student whites, a cause to which SSOC had committed itself at its April 1966 conference. They especially wanted SSOC to reach out to blue-collar whites on such issues as poverty, unemployment, and economic empowerment. These activists believed that making inroads into the working-class white communities could help make the much-dreamed-of interracial movement a reality. In their view, interracial cooperation in particular, and progressive reform in general, would remain elusive goals until SSOC connected with working-class whites. As one frustrated SSOC member asked near the end of 1966, "When is SSOC going to get at the job of organizing poor whites that must be done if the NS [New South] is to be a reality?"[57]

The question for SSOC was how to reach these whites. In 1966, the North Nashville Project was the only white community program SSOC implemented after it had shifted its focus to southern whites. The absence of additional white community projects from the group's agenda reflected the difficult nature of such work. Most whites in SSOC had no idea how to approach white communities as activists. And even if they had known how to work in these communities it is unlikely that they would have met with much success, for community organizing required an inordinate amount of time and an abundance of patience, neither of which they had. Bob Moses, who had seen SNCC struggle to transform itself into a community organization, explains that neither black nor white students had "enough understanding or patience" to be effective community organizers. Organizing in southern communities was slow and tedious work in which activists, rather than bringing about changes through their own actions, needed to persuade residents to become engaged in struggles for their own benefit. As young people, students typically wanted to make things happen themselves. "So it's hard," Moses later reflected, "getting them to see the power of helping other people get in motion."[58]

That white-community organizing put their physical well-being at risk also caused many SSOC members to lose their resolve for this work. The activists recognized that white neighborhoods were not a particularly safe environment for them. It was hardly unheard of for whites to respond with violence when faced with activists advocating black equality and interracial cooperation. While most students learned to shrug off the verbal taunts and threatening remarks, the instances when whites attacked students convinced both their victims and the larger community of young white activists that organizing in the white community was not just difficult, but potentially dangerous work.

SSOC's inability to engage working-class whites in their communities led the group to cultivate ties with labor unions in the South in an effort to appeal to them in the workplace. SSOC's desire to work closely with unions might seem curious given the virulent racism that characterized much of organized labor's attitude toward black southerners in the twentieth century. But the organization long had viewed unions as potential allies, and many activists believed that the interracial labor struggles in the southern past, despite their failures, provided SSOC with a model for interracial cooperation in the present. The Populists of the late nineteenth century, the Southern Tenant Farmers Union of the 1930s, and the Congress of Industrial Organization's campaigns of the 1930s and 1940s all taught SSOC that organized labor, by bringing blacks and whites together, could be a force for progressive change in the 1960s South.[59]

Because wages, job security, working conditions, and respect from supervisors were concerns of all southern workers, SSOC activists believed that working with unions would help to unite black and white workers. Once they illuminated the common problems that black and white workers faced, the activists were confident that whites would cast aside their lingering antipathy toward African Americans and willingly ally with their black peers. They also were optimistic that racially inclusive labor organizing would facilitate the development of biracial cooperation in the broader community. In the words of union official Peter Brandon, "Once the people organize, black and white together, and beat the 'boss man' in the shop, the idea will carry over into community issues." SSOC had a strategic reason for approaching whites through labor unions as well, namely, the belief that student involvement in recent union campaigns among minority workers in the South could serve as a model for similar work with white laborers in the region. In 1965 and early 1966, white students at both Duke and Emory mobilized in support of black service workers' demands for union recognition and collective bargaining rights. The Emory campaign ultimately failed to secure union representation for the workers. But at Duke, workers created Local 77 of the AFSCME, the government employees' union, and prodded the university to raise wages, establish a formal grievance procedure, and rehire workers fired for union activity.[60]

Gene Guerrero, who was a key supporter of the Emory employees' campaign, was the driving force behind SSOC's alliance with labor unions in the South. In April 1966, he helped organize a SSOC conference on students and labor at North Carolina College in Durham. The conference participants, who included student activists and union representatives, concluded that the two groups should develop projects that enabled students to work with unions in the South.[61] To that end, ten SSOC-recruited students traveled between Florida and Michigan as part of a joint SSOC-AFL-CIO-sponsored drive to organize migrant farm workers in the summer of 1966, a project that culminated in early 1967 with the formation of a United Packinghouse Workers of America local by farm laborers in Belle Glade, Florida.[62]

In fall 1966, Guerrero began working with organizers from the Textile Workers Union of America (TWUA) on a series of unionization drives at North Carolina textile mills. Over much of the next year, SSOC would play an active role in these campaigns. In the spring of 1967, nearly half of the organization's staff and hundreds of its rank-and-file members were working on the drives. The TWUA effort focused on eight textile mills operated by two national corporations: the New York-based National Spinning Company's mill in Whiteville, in the southeast corner of the state, and seven Cone Mills' plants centered on Greensboro, where the company was headquartered.[63] The substantive issues that drove each campaign were nearly identical. At both companies' plants, the union called for immediate wage increases and improved benefits, including a more generous pension plan for Cone workers and sick pay for employees at both mills. Another set of demands aimed to help the union institutionalize itself at the mills. Additionally, TWUA organizers pressed both companies to agree to binding arbitration of grievances, to rehire workers previously fired for their organizing activity, and to permit voluntary union (dues) check-off.[64]

Union organizers happily discovered that many mill employees not only were eager to air their grievances with management but that, contrary to the stereotype of the apathetic or anti-union southern worker, they were not opposed to joining the union. In just three days in August 1966, for example, 383 National Spinning workers, representing 68 percent of the mill's workforce, signed union cards indicating they wanted the TWUA as their bargaining agent. The growing number of younger workers in the mills proved critical to the TWUA organizers' initial success. At National Spinning, the average worker was a 25-year-old who never before had held a mill job and who thus, one organizer hypothesized, "lack[ed] the crippling heritage of the mill village." Without firsthand knowledge of the bruising and ultimately losing union battles undertaken by an earlier generation of textile hands, these workers brought to the mills an attitude of open defiance. In Gene Guerrero's words, they simply "weren't beaten down."[65] Significantly, some of these newer employees were African Americans. Although whites long had monopolized nearly all positions in the mills, the 1964 Civil Rights Act's prohibition against racial discrimination in employment led to the hiring of greater numbers of African Americans. The presence of black workers on the mill floor helped to remove race as a wedge issue that the company could wield in times of unrest. The appearance of black workers in the mills, one unionist explained, "took away the old weapon which had been used to destroy workers' unity before—the idea that, if black workers were hired, it would mean less jobs for whites." With a workforce that had become younger and more integrated, the TWUA organizers believed National Spinning and Cone were vulnerable to organizing campaigns.[66]

The labor drives heated up in early 1967, as both companies refused to engage in serious contract negotiations with the union. At the Cone plants, the union organized a series of short strikes in the winter and spring to demonstrate the workers' resolve and to disrupt production. At National Spinning, TWUA organizers initiated a May walkout that evolved into a four-month test of wills between workers and management.[67] SSOC activists became involved in the campaigns as the union prepared its actions against the companies. By January 1967, Guerrero and Nan Grogan, a SSOC staffer who had been one of the lead organizers of the VSCRC project in Southside Virginia, had married and moved to North Carolina to work on the campaigns.[68]

SSOC's work focused on recruiting students to the union cause. The TWUA drives gave the activists the chance to mobilize students who never before had participated in an activist campaign. A variety of factors drew these students to the TWUA, including their sense that union organizing was an appropriate local issue on which to work and the fact that textile work was in some of their families' background. Guerrero and Grogan visited college campuses across the state to promote the campaigns. Activists already based in North Carolina, including Harry Boyte at Duke and Chuck and Ann Schunoir at the University of North Carolina at Chapel Hill, also played an important role in galvanizing student support for the union. Of all the activists who helped build student support for the union, none was more successful than Lyn Wells, SSOC's North Carolina campus traveler. Wells's enthusiasm for her work and ability to relate to students at their level were key elements of her success.[69]

One of her first accomplishments upon arriving in Greensboro in February 1967 was to help develop a newsletter for activist students in order to keep them connected to one another and informed about the union campaign. Her work on the newsletter stemmed partially from the fact that she did not know how to drive and thus was confined to Greensboro, an odd predicament for the North Carolina campus traveler to face. She yearned, though, to begin visiting campuses, and, after prevailing upon Grogan to teach her to drive, began drumming up support for the textile workers and introducing SSOC to students at a wide range of colleges and universities, including the University of North Carolina at Greensboro, North Carolina Central University, Davidson, Queens, Livingstone, and Catawba colleges. Sometimes she had contacts at the schools she visited, and other times she searched for progressive students by examining the names on the signature cards in library books on the civil rights movement, Vietnam, and other politically relevant subjects. More often than not, Wells did not know anyone at the schools, and she simply set up a table on campus and began distributing union literature and talking to people about the campaigns. She proved to be remarkably persuasive; wherever she went she managed to cultivate support for the union among small groups of students. As testimony to her organizing skills,

Wells won support for the union not only from liberal-leaning students and those who hailed from working-class or mill families, but also from students whom most people assumed would oppose the organizing drives. Thanks to her work at Davidson, for instance, she persuaded Frank Goldsmith, the son of an executive with the Cross Cotton Mills in Marion, to participate in the drives, thereby scoring a public relations coup for SSOC and creating considerable stress in the Goldsmith household.[70]

Wells and the other activists motivated more than 300 students to participate in the Cone Mills campaign. In March, 300 workers and students took part in a pro-union march in Greensboro. Additionally, students walked picket lines at several Cone plants during a three day strike in February and a one-week walkout in April. In the February action, approximately 150 students joined Cone employees in picketing the seven targeted mills. At the White Oak plant, Guilford student and union supporter Richard Horne was "lightly hit" by a replacement worker's car, presumably as he tried to block it from entering the plant. And during a demonstration in Greensboro at the Proximity plant, the company's main office, police arrested two Davidson and two University of North Carolina students for trespassing when they inadvertently stepped over the "strike line" marking the picketing area.[71]

The organizers who stirred campus support for the Cone campaign also encouraged students to support the striking workers at the National Spinning Company mill in Whiteville. Grogan and Guerrero relocated to Whiteville to work on the drive, and the Schunoirs began working on the strike full time as well. In one instance, Chuck Schunoir, who was a particularly visible presence during the campaign, was one of two people arrested for trespassing at National Spinning's mill in Washington when he and 15 strikers and union staffers tried to distribute leaflets to workers outside the plant about the Whiteville strike. Subsequently, students from around the state, especially Chapel Hill, joined union pickets at the plant's gate. And to help draw public attention to the walkout, students from a number of schools, including Duke, the University of North Carolina, and the relatively far-away University of Virginia, participated in a mass march and demonstration in Wilmington at the conclusion of a five-day, fifty-mile "Whiteville-to-Wilmington" march by thirty strikers.[72]

The textile employees at both companies appreciated the students' support. Initially, though, many workers feared that the students would turn out to be long-haired, drug-abusing radicals who would alienate their supporters in the community and undermine the entire campaign. To counter such sentiments, the young organizers counseled their peers to obey all laws, eliminate profanities from their speech, dress nicely, and, in frequently repeated advice, shave all facial hair since, as one union organizer matter-of-factly explained to interested students, "beards make people think of things like LSD."[73] Gradually, the workers warmed to the students and made a concerted effort to stir student interest in the drives.

In April, workers from Cone joined with student activists to organize a "Conference on Textile Workers' Rights" in Greensboro in an effort to build support for the union. Students from Duke, Guilford, Livingstone, Davidson, Wake Forest, and the University of North Carolina campuses in Greensboro and Chapel Hill were among the more than 300 people to attend the meeting. And National Spinning workers traveled to Chapel Hill to encourage students and faculty on the North Carolina campus to support the union.[74]

For SSOC, the student-worker alliance was particularly heartening since it gave southern activists the chance to relate directly to working-class whites. Additionally, SSOC activists hoped to use their new relationship with white millworkers to encourage interracial unity in the mills. Interracialism, in fact, became a significant feature of both campaigns, as whites, with a readiness that surprised the activists, courted black involvement in the union. White enthusiasm for black participation in the campaigns did not necessarily result from a recognition that all workers, regardless of race, were engaged in a common struggle. More commonly, white workers recruited blacks to the union for a tactical reason, namely, the hope that the union would benefit from what they perceived to be African Americans' superior organizing skills, skills honed during the civil rights struggle. At the Cone plants, Gene Guerrero recollects that "whites were really eager to get black involvement 'cause blacks, as they saw it . . . had demonstrated that they knew how to get what they wanted because they had passed the civil rights bills." At the National Spinning plant, Nan Grogan discovered that whites were willing "to let blacks step forward and play a leading role . . . on the picket line . . . because they 'knew how to do it' from their being in civil rights stuff." And in a sign of their newfound respect for African Americans, white workers throughout the Cone mills, one unionist noted, even began "to use the polite word 'colored' " instead of the usual epithet to refer to their black co-workers.[75]

In addition to bringing blacks and whites together in the mills, the campaigns served to unite white workers and black college students in common cause. The SSOC activists, particularly Wells and Guerrero, included black as well as white campuses on their itineraries as they traveled the state promoting the union. The SSOC organizers believed that involving black students in the campaigns could play a crucial role in attracting black workers to the union. Moreover, they reasoned that forging an alliance between young blacks and working-class whites—groups that, in one black student's words, had been "life-long adversaries"—would be further testimony to the interracialism of the drives.[76] To the delight of the SSOC activists, many black students responded positively to their entreaties for union support. In Salisbury, black students at Livingstone College, which both Wells and Guerrero had visited, became active in the drive at the Cone Mills plant in the town. Prior to the week-long strike in late April, approximately 40 of the students helped with pre-strike organizing, and during the walkout many of them picketed the plant along side the workers. The black students' support left some white workers grateful but confused. Wells remembers that some of the white Salisbury workers had trouble making sense of the black students' presence at the plant. On one occasion, she recall several long-time women workers commenting, " 'that was real nice of them to come after we treated them like this all these years.' " Other white workers who welcomed the students' support had to work hard to encourage their peers in the plant and in the community to set aside their old animosities. In one revealing incident, Guerrero was at the union hall in Salisbury one day when

> this call came in from the college, from our students at Livingstone College, saying they couldn't come to be on the picket line because the president had called them in saying they were getting threats from the Ku Klux Klan so they couldn't participate any more. So I explained this to some of our strikers who were sitting there, and one of them gets up, picks up the phone, dials a number, "Sam, listen we just got this call. Y'all stop that now." That was it. So we call back [the college], and we got our students back. Just as simple as that.[77]

The biracial nature of the union struggle at the two textile chains, though gratifying to the SSOC activists, could not forestall the collapse of both campaigns in the summer of 1967. The TWUA bore much of the responsibility for the unsuccessful end to the drives. The union's refusal to back a protracted walkout at the Cone plants doomed that campaign, while its lack of financial support for the National Spinning strikers precipitated the collapse of their walkout. The failure of the drives distressed the SSOC activists, and they directed their anger at the union. Steve Wise summed up how many of them felt about the union when, shortly after the TWUA withdrew from the Cone plants, he bitterly wrote, "I am really beginning to think it a shitty union. Ten months of work on Cone shot down the drain. And we had organized a helluva lotta students around the Cone situation. . . . A really promising situation has been shot to hell." With a more loyal union ally, SSOC activists believed that the workers and students could have forced the companies to accede to their demands.[78]

Despite the unhappy conclusion to the campaigns, SSOC had shown that white students were capable of connecting with working whites. Indeed, its involvement in the union drives enabled SSOC to bring to life its commitment to white organizing. More than this, the union campaigns represented a moment of possibility, a point in time when it seemed conceivable that workers and students, black and white, could unite in common cause, thereby bringing lasting change to the workplace and taking the first step toward redressing the region's racial and economic problems. With greater student involvement and SSOC's continued commitment the activists believed future actions could succeed where these failed. Such sentiments did not sit well with the many in SSOC who resisted working with non-student whites. These students saw little to gain from such work. Few whites beyond the campus, they argued, showed any inclination to get involved in progressive causes. Moreover, they reasoned that since SSOC was an organization comprised primarily of students, the group stood the greatest chance of developing successful projects if it focused exclusively on students on southern campuses. They firmly believed that SSOC needed to concentrate on the issues that resonated with students, most notably the war.

SSOC and SLAM

The disagreement over the group's direction and constituency was the primary topic of discussion at SSOC's May 1967 Southwide conference at Buckeye Cove, the campground near Swannanoa, North Carolina. With a stated purpose of determining the appropriate role for southern activists in the national New Left, the meeting was an opportunity to debate SSOC's future. Adding to its importance was the fact that conference organizers invited representatives from a wide range of progressive and radical groups, including SNCC, SDS, NSA, the Progressive Labor Party, the DuBois Club, and the Young Socialist Alliance, to participate in the meeting.[79]

The activists arrived at Buckeye Cove armed with numerous proposals for reshaping the organization. Tom Gardner and David Nolan presented the most far-reaching one. Provocatively titled "Toward a Southern Student Organizing Committee," it outlined an aggressive program for making SSOC more relevant on southern campuses. Nolan and Gardner recommended that the *New South Student* be made into a slicker, more professional-looking publication that offered plenty of useful information for campus activists. They called for more action on university reform issues, identifying compulsory R.O.T.C., in loco parentis policies, and speaker bans as issues ripe for agitation. They also encouraged SSOC to expand its literature, campus traveler, and speakers programs in an effort to attract new supporters and improve service to its current members. They focused on organizational issues as well, insisting, for instance, that SSOC take greater care in hiring staffers. Too often in the past, they noted, the group had "hired people at random to organize," given them specific assignments, such as founding "an Afro American library in Baton Rouge" or building "a Texas student movement," and then "never heard from any of them again."[80]

Gardner and Nolan also weighed in on the brewing dispute over the group's focus and constituency. Not surprisingly, the two Peace Tour veterans advocated that SSOC make antiwar organizing its top priority.

While they believed that reaching out to non-student whites was a worthwhile endeavor and agreed that the group should continue to seek student support for labor actions, they argued that such activities should constitute only a small percentage of SSOC's total work. On a practical level, Gardner and Nolan noted, "we do not have the finances to support every 'good thing' that happens in the South." More significantly, they insisted that students were the group's main constituency and, thus, that working-class organizing should always take a back seat to organizing efforts on the predominantly white campuses of the South. As they pointedly noted in the proposal, "our work is with the campus, not with labor. . . . It is valuable for us to relate students to the sufferings of some of the workers in the South, but we should not turn into a southern labor organizing committee."[81]

Some of the assembled activists dissented from the proposal's recommendation that SSOC deemphasize its work with labor and blue-collar whites. Among these conference participants were a number of activists from Atlanta who believed that this was precisely the type of work white activists needed to do in the future. They doubted, however, that SSOC ever would make working-class organizing a priority. In their view, student organizing always had, and always would, dominate the group's agenda. Consequently, prior to the Buckeye Cove meeting, the Atlanta-based activists came together to create the Southern Labor Action Movement—SLAM—a student group that would focus exclusively on organizing among white workers.

Built on the premise that, as Alan Levin recollects, student activists "really needed to be working with working-class people and organizing a radical working-class movement," SLAM was the brainchild of Sam Shirah, the veteran white activist who, as one of SNCC's white field secretaries, played an instrumental role in SSOC's founding. Throughout his long activist career, Shirah maintained a deep interest in organizing among working-class whites. He, along with Ed Hamlett, was the driving force behind SSOC's 1964 White Folks Project in Mississippi, the group's first attempt at white community organizing. More recently, Shirah had joined the staff of the International Ladies' Garment Workers Union (ILGWU), and, from his base in Atlanta, begun organizing southern support for the union. But Shirah quickly became disenchanted with the union. Nelson Blackstock, a neighbor of Shirah's in Atlanta and an activist with sporadic ties to SSOC, recalls that Shirah felt the union spent too much time trying to persuade management of "the benefits of getting the union label on their garments" and not enough time talking to the workers about their concerns and explaining to them the advantages of unionizing. Hoping to work more closely with southern laborers, he began talking with friends, colleagues, and acquaintances about creating a worker-oriented activist organization. By the early spring of 1967, Shirah and a handful of others had founded SLAM.[82]

That Shirah was able to interest people in his ideas about SLAM was testimony to his powers of persuasion and the force of his personality. Like Lyn Wells, Shirah attracted the support of young white southerners wherever he went and regardless of the particular organization—SNCC, SSOC, or SLAM—he advocated. As everyone who knew him noted, Shirah had an unmistakable magnetism, a sense of presence, that captured people's attention and drew them toward him. Loud, foul-mouthed, and an increasingly heavy drinker, Shirah typically was the center of attention, entertaining and transfixing people with cutting remarks—delivered in his deep drawl—and his guitar playing while simultaneously winning converts to his positions by preaching the virtues of working-class organizing and the need for radical action. The fact that others considered him, in Levin's words, "a real working-class person" and not an earnest, naïve college student or a theory-obsessed intellectual, gave him instant credibility as an organizer. As Bo Lozoff, a colleague of Shirah's in SLAM, simply put it, "Sam was the genuine article." Consequently, many people readily deferred to his judgment on the question of how to reach working-class whites or, at the very least, wanted him on their side as they attempted to do so. Undoubtedly, such thinking contributed to the ILGWU's decision to hire Shirah. Lozoff's brother, Mike, who also worked with Shirah in SLAM, believes that Shirah's allure to the ILGWU would have been irresistible. "If you were a Jewish union executive from Manhattan and you're thinking of the South," Lozoff later speculated, "you're thinking, 'shit, I've got to go organize those hillbillies.' And then you met Sam, you'd hire him in a second. He would be the one to do it. You wouldn't want to send [someone named] Sol Weinstein. You'd send Sam Shirah."[83]

Shirah's talk about launching a labor-oriented student group stirred the interest of several Atlanta activists. Jody Palmour was one. After serving on SSOC's Nashville staff for most of 1966, he returned to Atlanta near the end of the year to help found an at-large SSOC chapter in the city. He was joined in that effort by Harlon Joye, a recent arrival from New York who had moved with his wife to Atlanta with the intention of conducting research for a study of SSOC. Having traveled in radical circles in New York, Joye sought to remain politically active as well, and within a short time he became acquainted with Palmour and then met Shirah, leading to his involvement in SLAM.[84] While Palmour and Joye had no experience with southern workers, the same was not true of Tommy Martin and Bo and Mike Lozoff, three individuals who played key roles in the SLAM. Martin, an Arkansan, and the Lozoffs, who were from Miami, previously had participated in an organizing drive among migrant farmworkers—Martin and Mike Lozoff, in fact, took part in the joint SSOC-AFL-CIO migrant organizing campaign in the summer of 1966. They also spent six months working on a unionizing drive among black farmworkers in the south-central Florida community of Belle Glade, which first had gained national notoriety in the Edward R. Murrow documentary *Harvest of Shame*. Bo Lozoff joined the effort too. As a student at the University of Florida, he founded the Farm Workers Support Committee and was instrumental in bringing students from Gainesville to Belle Glade to work on the campaign. Eventually, the Lozoffs and Martin relocated to Atlanta, where they quickly came into contact with Shirah and joined him, Palmour, and Joye in creating SLAM. They then traveled to SSOC's spring conference in search of funding and the organization's endorsement of their new group.[85]

SSOC's introduction to SLAM was the prospectus the new group presented at the meeting. The proposal outlined a dual role for students in their work with SLAM. First, they would serve an educational function, ensuring that workers understood their legal rights, highlighting the power employers wielded in the workplace, and educating working whites about labor's radical past. SLAM believed that by exposing workers to the history of earlier organizing campaigns in the South, they would encourage them to see current organizing efforts as part of a long tradition of labor radicalism, thereby heightening worker militancy. Additionally, SLAM planned for students to promote and participate in unionizing drives and strikes; they would encourage workers to walk off their jobs to press their demands, work to build support for walkouts at nearby college campuses, and join workers in walking picket lines. In SLAM's view, student involvement in labor actions would help to break down the barriers between white student activists and working-class whites. "Rather than seeing 'workers' and 'students' as air-tight compartments," the group's prospectus remarked, "SLAM views them as a continuum with students becoming economic and political organizers after they graduate and many workers become increasingly competent students of the problems of their choice." With its focus on white workers and its willingness to throw itself completely into labor work, then, SLAM distinguished itself from SSOC. To Shirah and his confederates, SLAM would function, in Mike Lozoff's phrasing, as "a SSOC with balls." SLAM would be the fulfillment of Shirah's desire to build an organization geared exclusively toward engaging southern workers. To help SLAM carry out its work, the proposal requested that SSOC provide the new group with $2,600 in seed money spread out over six months.[86]

Gardner, Nolan, and other activists who sought to steer the group toward more antiwar and campus-based activities opposed supporting SLAM. They considered it a naïve, ill-advised attempt to reach white workers, and they doubted Shirah's organizing skills and commitment to the new group. By 1967, most who knew him, both friend and foe, believed Shirah to be one of the legion of walking wounded among movement veterans. Like countless black and white civil rights activists, the stress, tension, and constant threat of violence had taken a toll on Shirah, making him less reliable as a peer and less effective as an organizer. Though SLAM's opponents sympathized with Shirah on a human level, they were extremely reluctant to offer financial support to a group in which he not only was involved but which he had created. And it was obvious to all that Shirah exerted near total control over the group. SLAM, after all, Bo Lozoff later wryly remarked, was merely "Sam with an 'L.'" Shirah's detractors believed that to invest SSOC's scarce resources in SLAM would be to throw good money after bad.[87]

As the discussion of the SLAM proposal concluded, it was apparent that the criticisms of Shirah would not prevent SSOC from backing the new organization. Many of the conference attendees believed supporting SLAM was a means by which SSOC more fully could bring to life its commitment to working-class organizing. And given that this debate took place in May 1967, the high point of SSOC's involvement in the North Carolina textile drives, enthusiasm for labor-oriented work was running high within in the group. The misgivings that some of SLAM's opponents expressed about working with Shirah also failed to move many of the students who only recently had become active in the group. This especially frustrated his critics. To them, he was a loose cannon who no longer could be trusted. More likely, he was a troubled individual with numerous personal problems who nonetheless was a sincere and steadfast proponent of working with non-student whites. Newer SSOC members, perhaps meeting Shirah for the first time at the Buckeye Cove gathering, undoubtedly found him a refreshing presence and considered him an honest voice of white working-class discontent. Though Shirah's detractors might roll their eyes and exchange knowing here-we-go-again glances at Shirah's antics at the meeting—at one point, Nolan recalls, Shirah refused to elaborate on the SLAM proposal because he said "you might tell George Meany and ruin the whole thing"—younger activists were just as likely to find his actions entertaining and not entirely unreasonable. At the end of the meeting, nearly 60 percent of the conference attendees approved the SLAM proposal and voted to fund it at the level of $100 per week for six months, thus giving the group the $2,600 it had requested from SSOC.[88]

The vote in favor of supporting SLAM notwithstanding, the May Buckeye Cove meeting did not resolve questions regarding SSOC's direction and constituency. The conference revealed that although working-class organizing had assumed a new importance in SSOC, the activists remained committed to cultivating support in the South for a wide range of progressive causes. The meeting confirmed that antiwar work would remain an important focus of the group's work. The activists, for instance, approved a key component of the Gardner-Nolan proposal by voting to organize an expanded Peace Tour program in the 1967–68 school year. Additionally, the composition of the new leadership cohort elected at the end of the meeting reflected the group's multiple interests. On the one hand, the election of Peace Tour advocate Tom Gardner as chairman, Alan Levin, a leading antiwar activist at the University of Florida, as vice-chairman, and David Nolan as *New South Student* editor ensured that Vietnam would remain near the top of the group's agenda. On the other hand, four of the nine activists elected to the SSOC executive committee—Jody Palmour, Sam Shirah, Bo Lozoff, and Mike Lozoff—were associated with SLAM. The Buckeye Cove conference highlighted SSOC's hesitancy to define itself as a narrow, single-issue organization and its continued desire to work with a broad range of white southerners on a multiplicity of issues.[89]

The decision to support SLAM had a sharp and immediate impact on SSOC. In one of his first memos to SSOC members as chairman, Gardner complained that funding SLAM would exacerbate SSOC's financial problems since it committed to SLAM more than half of the $5,000 SSOC had to sustain its programs into the fall months. In anticipation of the expected financial crisis, Gardner proposed that the group reduce office expenses and cut salaries, hoping such measures would enable SSOC to meet its obligations to SLAM as well as maintain the other components of its program. Given SSOC's many interests, Alan Levin noted in the SDS weekly, *New Left Notes*, "the problem of limited resources bites hard."[90]

The SLAM–SSOC relationship did not endure beyond the summer of 1967, and SLAM itself disintegrated soon thereafter. Although SLAM played an important support role in a wildcat strike by 400 white women at a Levi-Strauss plant in the remote Appalachian town of Blue Ridge, Georgia, this was the only project it initiated in its brief life. Moreover, SLAM had no apparent organizational structure—it had no formal leaders and held meetings only irregularly. Rancorous infighting developed among the activists as well. Shirah was at the center of the turmoil, as he became convinced that the Lozoff brothers were trying to oust him from the group. By August, SSOC had seen enough. Convinced that SLAM was not an effective means for reaching working whites, the SSOC executive committee decided to stop funding SLAM and to break all ties with the group.[91]

Without regular cash infusions from SSOC, SLAM stopped functioning in any meaningful way. Most of the activists associated with it drifted away or channeled their energies into new projects. Those who remained in the group presided over a descent into extremism, as SLAM was reduced to plotting violent acts of desperation to bring attention to itself. With horror and astonishment, Bo Lozoff remembers that he, Shirah, and a few others actually devised a plan for bombing the Atlanta federal building. Invoking the violent turn that some in SDS later took, Lozoff later concluded that SLAM had metamorphosed into "a vehicle for being able to be more violent, more like a Weatherman part of SSOC, rather than a wholesome reaching out to white laborers." With few supporters, no money, and an extremist agenda, SLAM quickly crumbled.[92]

The SLAM episode vividly dramatized the lack of consensus in SSOC in 1967 and 1968 about the group's direction and goals. SLAM's collapse hardly discredited white organizing work among the activists. To the proponents of such work, SLAM's demise suggested the messenger, not the message, was flawed. The campus-oriented SSOC activists, of course, could not disagree more. In their view, SLAM's ineffectiveness testified to the folly of trying to reorient SSOC toward non-students. The differences over SSOC's constituency and mission prevented the group from crafting a consistent program of activities. The group's work with both students and non-students, on both the war and union campaigns, reflected the growing divisions in the group. As 1968 progressed, this divide became increasingly difficult to bridge, as new developments further strained relations among the activists and made consensus even more difficult to achieve.

1. David Kotelchuck ducks a punch from an employee of Herschel's Tic Toc Restaurant during a demonstration against the restaurant's continued segregation, December 1962. Photo by Bill Preston, *Nashville Tennessean*.

2. Vanderbilt graduate student Ron Parker and Scarritt College student Mary Pless prepare to distribute handbills calling for a boycott of the Campus Grill to protest its refusal to serve nonwhites, November 1963. Photo by Jack Corn, *Nashville Tennessean*.

3. Founding SSOC members at the first Continuations Committee meeting in Atlanta, April 1964 (seated on floor, left to right): Dan Harmeling, Harry Boyte, Sue Thrasher, Cathy Cade, Marjorie Henderson; (top row, left to right) Sam Shirah, Jerry Gainey, Roy Money, Gene Guerrero, Ed Hamlett, Jim Williams, John Shively, Bob Potter, Bob Richardson, Marion Barry, Jr. Photo by Gunter's Studio. Whi-24492, Carl and Anne Braden Papers, Wisconsin Historical Society.

4. Robert Pardun, left, and Judy Schiffer at the Freedom Summer training session at Western College for Women in Oxford, Ohio, June 1964. Photo by Steve Schapiro.

5. SNCC's John Lewis and SSOC's Archie Allen posing outside SNCC's Atlanta office, July 1964. Photo courtesy of Archie Allen.

6. Norman Thomas, the dean of American socialists, during his 1964 visit to the University of Virginia, where he had spoken on the invitation of the Jefferson Chapter of the Virginia Council on Human Relations. Pictured with him are Virginia students (front, left to right) Bill Leary, Richard Muller (rear, left to right), David Messner, and Carey Stronach. Photo by David Nolan. Whi-24919, David Nolan Papers, Wisconsin Historical Society.

7. Virginia Students' Civil Rights Committee staff members (left to right) Howard Romaine, Duke Edwards, David Lubs, and Ben Montgomery at the office in Blackstone, Virginia, ca. July 1965. Photo by Anne Braden. Whi-24921, David Nolan Papers, Wisconsin Historical Society.

8. Antiwar protesters on a Ft. Lauderdale-Miami march during the "Florida Days of Judgment," March 1967. Photo © Tom Gardner.

9. SSOC's David Nolan, center, holds a "Stop the War" sign during a peace vigil at the Tampa, Florida, Post Office during the "Florida Days of Judgment," March 1967. Photo © Tom Gardner.

10. From left to right, Millsaps students Mike Gwinn, Doug Rogers, Carolyn Davis, and Sue Barnes, at a march in downtown Jackson to protest the killing of Jackson State student Benjamin Brown by National Guardsmen, May 1967. Courtesy of Mississippi Department of Archives & History.

11. David Doggett, center, at a march in downtown Jackson to protest the killing of Jackson State student Benjamin Brown by National Guardsmen, May 1967. Courtesy of Mississippi Department of Archives & History.

12. SSOC Chairman Tom Gardner during his arrest for attempting to speak at Miami-Dade Junior College during the Florida Peace Tour, March 1967. *Southern Patriot,* April 1967. Photo by John Massey.

13. University of Florida activists (clockwise, from top right) Alan Levin, Judith Brown, Tom Sharpless, Ed Freeman, and Bob Fierstein, 1968. Photo © Ray Fisher.

14. The Kudzu staff, including David Doggett, top left, Doug Rogers, top right, and, standing left to right at the bottom, Mike Cassel, Cassel Carpenter, Everett Long, and Peggy Stone, at the Ruins of Windsor, Claiborne County, Mississippi, November 1968. *Kudzu*, 1:9 (9 December 1968). Photo by Ernie Fado.

15. Founding members of the Furman University SSOC chapter (left to right): Chuck Evans, June Manning, Dennis Calvin, Dee Savage, David Eicher, Chris Pyron, George Johnson, Jack Sullivan, Joseph Vaughn, Joel Flowers. *Bonhomie,* Vol. 68 (Greenville, S.C.: Furman University, 1968), 177. Courtesy of Special Collections, Furman University.

16. A light-hearted moment for the 1969 Furman SSOC chapter. *Bonhomie,* Vol. 69 (Greenville, S.C.: Furman University, 1969), 280. Courtesy of Special Collections, Furman University.

17. Atlanta SSOC activists at a joint SSOC-SDS rally against racism on the grounds of the state capital, February 1969. Photo by Tom Coffin.

18. The SSOC contingent in Sherman Square during the counterinaugural demonstrations on the eve of Richard Nixon's swearing-in, January 1969. *Phoenix* 1:5 (ca. February 1969). In the author's possession.

19. The main building at the Mt. Beulah Conference Center near Edwards, Mississippi, site of the final SSOC meeting in June 1969. In the author's possession.

20. Celebrants at the Southern Student Organizing Committee Thirtieth Reunion and Conference, Charlottesville, Virginia, April 1994. Photo © Daniel Grogan.

Chapter 7
Short-Term Gains, Long-Term Costs

> Sororities attempt, and often succeed, to make antebellum idiots of their members. The dorm honchos try to match the sororities and produce solid middle-class dishwashers, childcare experts, grooming fanatics, and fake politicos. In the classroom, the brighter coeds learn how to restrain their threatening (intelligent) critiques. . . . Even in the so-called radical groups, she is ignored, exploited, or coopted. Everywhere she plays a foot-shuffling, head-scratching, second-class game.
> —Judith Brown[1]

> One thing the radical movement at Furman showed me was that you can do something for yourself in this world. You can create a community of people who are not alienated and who share common values in a very intense way. In the very beginning . . . just about three people were involved, but there were a lot of people around who shared these values. Once the spark was ignited, it grew very rapidly. By 1968 we had a lot of support.
> —Jack Sullivan[2]

In 1968 and 1969, developments within SSOC strengthened the group in the short term while simultaneously undermining its long-term prospects. SSOC's growing enthusiasm for the burgeoning women's movement and its strong support for locally driven activism, though increasing the group's visibility in the South and heightening its status among progressive activists, intensified long-standing internal problems. That SSOC became a champion of women's liberation won it the accolades of southern feminists and made it more relevant to the daily lives of collegiate women in the South. But SSOC's embrace of the feminist movement caused tensions in the group, since it occurred neither quickly nor, initially, with everyone in the group's full support; until its last year, many men resisted incorporating the feminist agenda into SSOC's program. SSOC's support of the women's movement also revealed serious differences among SSOC women over the tactics, goals, and overall purpose of women's liberation. Some of the questions that divided them were: How radical should their actions be in support of the feminist cause?; Were men the cause of women's subordination or did blame more properly lie with an economic system that encouraged the exploitation and subjugation of women?; Was the primary goal to achieve parity for women in public places or to liberate them in their personal lives? Women's rights, thus, like the Vietnam War, labor organizing, and white community work, created discord among the activists, thereby adding to the centrifugal forces slowly pulling SSOC apart.

SSOC's efforts to cultivate local activism in the South proved to be just as troubling. Undoubtedly, its promotion of campus- and community-level action brought the group new supporters and helped to spark the surge of student activism that occurred in 1968 and 1969. This activism transformed some predominantly white campuses into battlegrounds while bringing movement activity to numerous schools where student protest activity was rare. Yet the fact that the group did not control the form or content of this activism meant that it supported a wide array of programs and projects, from those organized around national or regional issues, such as civil rights and Vietnam, to many that were peculiar to a specific locality, such as tenure disputes and speaker bans. The variety of issues that animated these efforts made it difficult, if not impossible, for SSOC's Nashville-based leaders to stay abreast of the often rapidly changing situation far from the group's headquarters. As a result, many campus activists came to see SSOC as a remote and distant organization with little knowledge of their specific concerns. To them, SSOC frequently was nothing more than a group that published an interesting newsletter rather than an organization to which they felt connected. And while these students often formed SSOC chapters, they did so not because they identified with the group but because adopting the formal organizational structure of a "chapter" permitted them to seek recognition as an official student group, a designation that typically gave them access to campus facilities crucial to their organizing efforts, such as meeting rooms, and which at some schools entitled them to official funding. In an effort to bring the organization closer to the local activists, SSOC restructured itself in 1968, the second time it had done so in two years. But this failed to fortify most students' commitment to SSOC and only strengthened the impression among them that the group was not particularly important to their work. Within SSOC, this inability to make the group more vital to local activists sharpened the disagreement over SSOC's ultimate purpose and mission and created further divisions among the activists.

Women's Liberation and SSOC

The emergence of women's issues within SSOC in 1968 and 1969 was reminiscent of their earlier rise to prominence in the Student Nonviolent

Coordinating Committee and the Students for a Democratic Society. As in both of those organizations, SSOC women were the first and the most frequent advocates of feminist issues within the group. Most SSOC men, like their counterparts in SNCC and SDS, initially were skeptical of the relevance of women's rights to the group's work and, therefore, resisted devoting any of the organization's time or resources to it. Additionally, just as women's experiences in SNCC and SDS had radicalized them on the question of women's rights, so, too, did involvement in SSOC transform its female members into advocates of the women's movement. The subtle and not-so-subtle instances of discrimination women activists encountered in SSOC as well as their growing awareness of their second-class status, an awareness heightened by their participation in the movement for black equality, facilitated the development of a feminist consciousness among these women.[3]

Feminist sentiments did not surface in SSOC as early as they did in SNCC and SDS because SSOC women initially did not have a sense of women's subordinate position in society. Moreover, prior to 1968, women's issues were easily obscured by the other problems and concerns on which SSOC focused. Expressing a view common among SSOC women, Ronda Kotelchuck recollects that she had "no great consciousness [about] men [and] women" when she first became active in the group. It was not until much later that women's issues came into focus for her and that, as she puts it, "my feminist instincts came into full blossom." Though a few SSOC women more readily were aware of women's issues, their nascent feminism typically did not translate into social activism. Some of them were reluctant to encourage SSOC to devote time to women's issues because, like the men in the group, they were preoccupied with building opposition to the war, working to end segregation, and trying to reach out to non-student white southerners. Others had yet to find the language to articulate their concerns. Sue Thrasher had vague notions that her gender worked against her in the group since her male peers did not accord her as much respect as she thought she deserved. But she dismissed these sentiments as the concerns of a middle-class woman and not someone working in the trenches of the civil rights movement.[4]

The sexual imbalance in the group also slowed SSOC's turn to women's issues. Men dominated the organization throughout its life. Fewer than one-fourth of the white students who gathered for SSOC's founding meeting were women, and by 1968 approximately three of every four SSOC members was a man.[5] More women were not actively involved in the group at least partially because men significantly outnumbered women on southern campuses, the recruiting ground for the organization.[6] Equally important, gender-specific considerations discouraged female students from associating with activist organizations such as SSOC. Women students could not help but be aware of the masculinized nature of life on college campuses. "In the virile world of Southern universities," historian Clarence Mohr notes, "white college women encountered a range of less than subtle cues concerning their perceived status as ornamental husband-seekers." From football to fraternities, R.O.T.C. to student government, college life revolved around men. To gain acceptance in this world, women needed to conform to the expectation that they were in college for the social life or to pursue a "Mrs." degree. Women who took their studies "too seriously" or became politically active risked being excluded from the mainstream social life on campus.[7]

Male domination thus made SSOC more, rather than less, like the campuses from which women came to the group. But SSOC women's minority status did not render them invisible. Sue Thrasher, Janet Dewart, and Shirley Newton held staff positions that gave them responsibility for the operation of the Nashville office, while Lyn Wells won appointment as SSOC Program Secretary. Beyond the Nashville office, women such as Cathy Cade and Kathy Barrett in New Orleans and Judith Brown in Gainesville were leading activists in local organizing campaigns. All of these women proved to be outspoken in their support of progressive causes and unhesitant about participating in actions that might expose them to danger or threaten to make them outcasts on their campuses and in their families. This is not surprising since the group attracted students who were willing to speak and act out for what they believed. For SSOC women, as for the men in the group, fears about the consequences of their activism dissipated as their involvement in SSOC deepened.

The presence in the South of other activist women also helped SSOC women to find their voice in the organization. The college students of SSOC stood in awe of these women. "They were fully self-actualized," Anne Romaine reverentially explained years later. "They could do whatever they wanted as far as I could tell. And they were breaking new ground." The students took inspiration from the words and actions of black women such as Ruby Doris Smith Robinson, one of the key leaders in SNCC, and Fannie Lou Hamer, the Mississippi sharecropper who gained national prominence for testifying at the 1964 Democratic National Convention about being jailed, beaten, and evicted for trying to register to vote. That these women had endured constant harassment and terrible violence for their efforts made a lasting impression on many SSOC women; the black women's actual suffering made the white youths' fears about the potential consequences of their own activism seem trivial in comparison and served to strengthen their commitment to the movement. Even more important to them was the example set by other young white women who were working in the southern civil rights movement. Connie Curry, Dorothy Burlage, Casey Hayden, Jane Stembridge, Dorothy Miller, and the other white women involved in the early days of SNCC created a model for how white women could contribute to the movement for black equality as well as demonstrated how working for the freedom of others could sensitize one to her own subjugated status.[8]

No white activist in the South was more influential among SSOC women than Anne Braden of Southern Conference Educational Fund. SSOC women valued Braden not merely because she was a "SSOC elder" who offered the group important advice but because she was a role model for them.[9] Through her close contact with black civil rights activists and her work as editor of the SCEF newsletter, the *Southern Patriot,* Braden demonstrated that women could become effective activists in the southern movement. Braden also helped to raise the younger women's awareness of how gender shaped their lives, for she balanced her work in the movement with the raising of her three children. More explicitly, she constantly encouraged SSOC and other movement women to confront the inequality they faced in their personal lives. Braden herself was openly critical of men in the movement who treated women as second-class citizens or who viewed women as lesser able than men. In 1963, for instance, Bob Zellner, whom she had helped to become SNCC's first white field secretary, angered her by opposing her suggestion that Dorothy Burlage succeed him because he felt a man needed to hold the position. "I didn't argue with him," she wrote Burlage, "but just told him that when he got ready to fight for the rights of the whole human race to let me know."[10]

The presence of these other female activists in the South provided SSOC women with encouragement and helped to raise their awareness of women's issues. SSOC women took notice as Casey Hayden, Mary King, and other white women in SNCC began to question gender and sex roles in that organization. Just as importantly, involvement in the civil rights movement brought to the surface concerns among SSOC women about gender discrimination and male chauvinism. On the most basic level, working for black rights highlighted for SSOC women their own status as second-class citizens within both their organization and the broader society. On a deeper level, they recognized that in the South racism and sexism were intertwined, that the region's ossified racial order preserved women's secondary status. Many came to understand that the myths that deemed southern white women fragile, pure, and inherently unsuited for public affairs drew their strength from the segregationist system. Thus, smashing the latter would go a long way toward destroying the former.[11]

For most SSOC women, their experiences within the organization, more than anything else, inspired them to take an active interest in women's issues. The gendered division of labor within the group was a particular source of unease. Men dominated the posts that represented SSOC to the public. Women, on the other hand, primarily assumed administrative positions that revolved around the mundane tasks of running the SSOC office. While some women did serve as campus travelers, and though Lyn Wells emerged as the single most important individual in the group in its last years, they were exceptions. SSOC women were not unaware of the division between men's and women's work. Sue Thrasher, who as SSOC's first executive secretary almost single-handedly kept the group afloat in its early years, was deeply ambivalent about her role in the organization. She took pride in the fact that her work "was holding the office together," and she clearly recognizes that she was "*the* main contact person inside the office." Nonetheless, she could not help but feel that the less desirable tasks, the boring and unglamorous work that none of the men wanted to do, had fallen to her. As she puts it, "there's no getting around that sending out newsletters is shit work, answering letters and stuff is shit work."[12]

So, too, were the domestic chores SSOC women often found themselves saddled with in SSOC offices and communal houses. Many women resented being stuck with such work. And that this was occurring within an organization devoted to equal rights only heightened their awareness of the contradiction between their personal lives and their activist goals. Increasingly, they complained about how the men treated them, even—or especially—when the primary targets of their anger were their boyfriends, lovers, and husbands. Nancy Lewis, who was married and had a small child while she was one of the leaders of the SSOC chapter at the University of Florida in the late 1960s, remembers growing resentful of her husband's lack of involvement in caring for their son or helping around the house. On a typical day, she recalls, "I . . . had classes from eight until noon, back to back. And I would get up, take Eric to the babysitter, where I had to pay a babysitter while Wes could sleep in. . . . Then I'd come home at noon, he'd be getting up, and his cronies would all be coming over. And then he expected me to fix him lunch. And then they'd disappear down to the local bar to hang out and talk politics and be cool." As time passed, it became impossible for her to reconcile her political radicalism with her subordinate status in the home. She began to talk to other women who shared similar feelings, and eventually she left her husband.[13]

Female SSOC members' growing discomfort with women's work made gender relations in the group more difficult and less harmonious, especially after 1967. SSOC men did little to smooth relations with their female colleagues, for whether intentionally or unwittingly, out of ignorance or insensitivity, they generally demonstrated only slight, if any, interest in women's issues, and they persisted in acting in ways that many women considered inappropriate and sexist. Consequently, words and actions that men considered to be innocuous took on more pernicious meanings when seen from women's perspective. SSOC men and women also interpreted sexual relations within the group differently. All agreed that such interactions created tension. But men tended to see such conflict and unhappiness as a natural by-product of a situation in which, as David Nolan puts it, "Everybody's living together in small quarters, pairing-off, and then playing musical beds and pairing-off again." Some women, however, blamed men for the consequences of these couplings since, in their view, the men were the ones who initiated the bed-hopping.

Sue Thrasher left no doubt where she believed responsibility for these liaisons rested when she later remarked that "the boys would sort of sleep around with different women and that would create different frictions at different times."[14]

Male SSOC members' initial lack of enthusiasm for women's issues further strained gender relations in the group. Gradually, however, SSOC men came to accept the legitimacy of women's concerns and to agree that SSOC was an appropriate vehicle for making the case for women's rights. This did not happen quickly; most of the men did not readily appreciate women's concerns. And many only slowly shed the sexist views that they long had embraced unquestioningly. Indeed, such views had encouraged SSOC women to focus on gender issues in the first place. As Tom Gardner accurately noted in 1970, SSOC women's feminism resulted "not from SSOC's advanced level of countering male chauvinism, but more from its level of male chauvinism."[15] Nonetheless, by early 1969, the men in the group had recognized women's liberation as an important element of its progressive agenda. Fittingly, it was a man, Sheldon Vanauken, an iconoclastic professor at Lynchburg College in Virginia who had ties to SSOC activists in the state, who helped to sensitize SSOC men to women's issues. In his awkwardly titled "Freedom for Movement Girls—Now," a 1969 essay that SSOC distributed widely, Vanauken argued that "The parallels between sexism and racism are sharp and clear." Women, he warned movement men, will no longer tolerate discriminatory treatment. "It's like the black awakening a few years earlier," he wrote, only that "Women have been second-class citizens far longer than the blacks. And now girls are awakening to the fact that the sexist myth is dead, just as the racist myth is dead—and things will never be the same again."[16]

SSOC men's growing support for women's issues did not resolve the tension in the group over women's concerns, for serious differences existed among the female activists about the meaning and purpose of a movement for women's liberation. These disagreements mirrored the disputes that divided women activists elsewhere in the nation during the early phase of the women's liberation movement. Historian Alice Echols usefully has termed this a conflict between "feminists" and "politicos." Feminists believed that male supremacy, more than political structures or economic systems, was responsible for women's oppression. Consequently, they argued that women's liberation was not merely another component of the broader Movement but rather a goal that women needed to pursue apart from the male-dominated movement groups. SDSer Anne Koedt, a founder of New York Radical Feminists, expressed this point of view in 1968 when she pointedly argued that women should "begin to expose and eliminate the causes of our oppression as women . . . rather than storming the Pentagon as women or protest[ing] the Democratic Convention as women." Politicos, on the other hand, advocated close ties between women liberationists and the larger Movement, and thus favored remaining connected to pre-existing organizations. In their view, the capitalist system and not male chauvinism was the great enemy. Many politicos shared Chicago SDS activist Evelyn Goldfield's sentiment that a "women's movement which confines itself to issues which only affect women can't be radical."[17]

In the South, Gainesville activists Judith Brown (Judy Benninger, before she married and took her husband's name) and Beverly Jones offered the earliest articulation of the feminist position in their widely circulated 1968 paper "Toward a Female Liberation Movement." Both women were involved in SSOC and other activist groups at the University of Florida, Brown as a student—she had helped to lead the first desegregation protests at businesses near the campus—and Jones as the wife of activist professor Marshall Jones. Their paper identified male chauvinism, and, more radically, men themselves, as the chief impediment to women's equality. "In the life of each woman, the most immediate oppressor . . . is 'the man.' Even if we prefer to view him as merely a pawn in the game, he's still the foreman on the big plantation of maleville," they wrote. Importantly, they stressed that though women may fervently support efforts to end the war or ensure black civil rights, they needed to recognize that women's liberation was a prerequisite for other progressive reforms because, "There cannot be real restructuring of this society until the relationship between the sexes are restructured." Women, therefore, had "to stop fighting for the 'movement' and start fighting primarily for the liberation of women." By organizing themselves separately from movement men, Brown and Jones suggested that women would have the freedom to discover and to talk about their common plight as victims of male domination. As a result, Brown and Jones hoped more women would advocate the restructuring of home and family relations in an effort to achieve greater equality between the sexes.[18]

Southern politicos vigorously dissented from Brown's and Jones's vision of women's liberation. In December 1968, Lyn Wells wrote "American Women: Their Use and Abuse," an essay that served as a retort to the Florida women's recently published paper. Wells did not oppose the idea of women organizing among themselves, but she disagreed with feminists over the purpose of such efforts. Wells believed consciousness raising that focused on women's personal lives served no useful purpose. At the first such gathering she attended, a meeting of six or seven women in Durham, she was appalled and embarrassed when it "denigrated into this session about their husbands being assholes, and then [women] started talking about their sex life in front of everybody. . . . It was all just really alien to me." Likewise, she rejected the idea that women should organize apart from men to wage battle for their own equality. Instead, she believed organizing women as women could be a valuable tool for awakening them to their common status within the capitalist system. As Wells later put it, she opposed the notion that "women had to liberate themselves from men. . . . I thought women had to liberate themselves from the system."[19]

In the view of politicos such as Wells, all of women's problems, from their political powerlessness to their secondary status in the household, stemmed from the subordinate role ascribed to them in the capitalist order. "American females, from the time we stepped off the ships at Jamestown, Virginia, have been USED," Wells wrote. "We have been economically exploited for profits and we have been used to make the system run most effectively for the good of those who profit. Many of our personal identity problems and our lack of political power are symptoms that can be traced to the economic usage of our sex."[20] The remedy lay not in blaming men for women's problems but in working with men to change the system. "As women, we must be concerned about our own destinies. We must begin to build a mass movement that holds our self-interest as primary; but the *goal* our activity is directed towards, as radicals, must be for *all humanity*. Our goals should be: a society in which people are no longer exploited by any force; a society that is a democracy in which all human beings have a voice in the decisions which affect their lives."[21]

The competing visions of the women's movement espoused by Wells and Jones and Brown helped to shape SSOC's most visible action in support of women's liberation, the "Women, Students and the Movement" conference attended by 180 white southern women in Atlanta in February 1969. Organized by Lyn Wells, Ann Johnson, a SSOC activist in North Carolina, and Maggie Heggen, who worked for SSOC in Nashville, the conference agenda reflected both the feminists' and the politicos' interests and views, thereby fulfilling organizers' hope that "many of the current positions on 'women's liberation' can be aired at the meeting." Not surprisingly, the session Wells led on "Southern Women's Struggle for Her Own and Others [sic] Freedom" focused on women abolitionists, Populists, and labor leaders, thereby linking women's struggle to the wider struggle of the dispossessed and the lower classes. Other women at the meeting shared her belief that true equality for women was impossible under capitalism since the system rewarded the exploitation of women. Sessions on "Women in the South: A Class Comparison" and the "Role of the Woman as a Commodity" spoke directly to the concerns of politicos like Wells in their emphasis on capitalism's exploitation of women. For the feminists at the conference, panels on the "Southern Sex Myth," "Coed Rights Organizing," and "How the Southern College Molds Women's Roles" were especially significant because they illuminated the many ways in which male superiority operated to oppress women. To them, capitalism was not the problem, but rather gender conventions that condoned the treatment of women as anti-intellectual sex objects capable only of raising a family and functioning in the wider world as consumers.[22]

In the South as elsewhere in the nation, the feminist view of women's liberation was ascendant among activist women. Partly this resulted from their experience with sexist movement men who, by dismissing their concerns as trivial, made clear that they would be unable to achieve equality in New Left organizations. Equally important was that many women, particularly first-time activists in the South, were drawn to the women's movement by the blatant discrimination and obvious double-standards that existed in their everyday lives. Just as it was easier for novice civil rights activists to recognize the unjustness and unfairness of segregation than to become aware of the ways in which the American political and economic system helped to perpetuate Jim Crow, so too was it easier for southern women to denounce their unequal treatment than to connect their second-class status to the country's economic arrangements. Driven by their feminist views, these women helped to spread women's liberation activity across the South in 1968 and 1969.

Working through SSOC chapters, chapters of radical national organizations, such as the chapter of Women's International Conspiracy from Hell—WITCH—that emerged at the University of Virginia, and local women's groups, such as Gainesville Women's Liberation, the first women's liberation group to appear in the South, feminists made male chauvinism, male superiority, and the unequal treatment of women the target of their work. Some groups also sought to improve women's economic standing by supporting "equal pay for equal work" measures or by pushing employers to raise wages in occupations traditionally dominated by women. In early 1969, for example, a women's groups in Nashville sought to help win pay raises for secretaries who earned the minimum wage at two insurance firms that were among the city's major employers. Other organizations worked to undermine gender stereotypes. In North Carolina, women's groups from Durham and Chapel Hill joined forces for a protest at the March 1969 Miss Durham Pageant, where they chanted "we shall not be used" and, "to the tune of 'the farmer in the dell,'" reported the radical newspaper the *Guardian*, "sang verses on this order: 'the woman trades her brain for a stopped-up drain / Hi-ho the derry-o the housewife loses all.'"[23]

No issue received more attention from these organizations than the many special rules that governed the daily lives of female students at southern colleges and universities. In loco parentis policies that set curfew hours and dress codes for women were the focal point of numerous demonstrations and protests on campuses throughout the region. Such actions resonated with a broad cross-section of campus communities—liberals and conservatives, freshmen and seniors, women and men. Undoubtedly, notions of self-interest infused the drive to do away with women's rules. But while women, of course, were anxious to have more freedoms, and men, no doubt, were enthusiastic about the prospect of their dates being able to stay out longer in the evenings, the existence of separate regulations for women violated many students' sense of fairness and equality as well as struck them as a wholly inappropriate way to treat adults. As a result, anti-women's rules actions occurred throughout the region in late 1968 and early 1969. At the University of Arkansas in Fayetteville, women students leafleted and organized mass meetings to push for changes in the rules that shaped their lives on campus. A campaign against women's policies also got underway at Clemson University when several female students created a women's group on campus. Similarly, the SSOC chapters in South Carolina initiated a drive to eliminate curfews for women on campuses in the state, and 1,500 male and female students at Vanderbilt University participated in a demonstration calling for the abolition of social rules for women.[24]

The challenge for the female opponents of women's rules was to convince men on campus that the dispute was fundamentally about the unfair and unequal treatment of women. While many men agreed, on some campuses female activists discovered that their male allies viewed the struggle against women's rules through the lens of university reform, a perspective that enabled them to place their own interests, as much as women's, at the center of the debate. At the University of North Carolina at Chapel Hill, for instance, the male-dominated SSOC chapter played an important role in the campaign that successfully forced school officials to scrap the rules that governed visitation between the sexes in on-campus residences. In late 1968 the group manned literature tables, organized a rally, and collected more than 3,900 signatures on a petition demanding that women be allowed in men's dormitory rooms. But the SSOC men did not perceive the issue to involve a challenge to gender conventions. Speaking of male undergraduates, Sam Austel, the SSOC leader on campus, declared, "It is our position that students have the right to decide if and when they want visitation in their rooms." Likewise, the editors of the student newspaper, the *Daily Tar Heel*, seemed to be thinking only of campus men when they explained that "the petition on coed visitation can only be taken by the Chancellor . . . as a sign that students want the opportunity to run their own lives, to decide such things as whether coeds may enter their rooms."[25]

Though men's and women's understanding of the issues at stake in challenges to women's rules did not always coincide, their efforts often successfully pressured administrators to revise policies. At some institutions, such as North Carolina, officials responded quickly to student demands. At many others, though, administrators resisted student calls for the abolition of gender-specific restrictions. But these school officials failed to comprehend that by refusing to bow to pressure to revise or eliminate women's rules they unwittingly helped to strengthen student opposition to the policies and to inspire students who never before had considered participating in protest actions to join the effort to overturn women's rules.

Such a situation developed at the University of Georgia in the spring of 1968. Early in the semester, approximately 25 men and women founded the Movement for Co-Ed Equality, a group that advocated the lifting of all rules that treated women students differently from men. One of its leaders was David Simpson, who later became a campus traveler in Georgia for SSOC and played an important role in the group during its last year. A tenth-generation Georgian who hailed from the university town of Athens, Simpson spent three years in the Navy before returning to Athens in 1966 with a wife and one-year-old child to attend the University of Georgia on the G.I. Bill. Having absorbed the values of the segregated world of white Athens, he had opposed the civil rights movement as a youth. As a 16-year-old, he once counter-demonstrated a desegregation protest at the Varsity Restaurant, dressing in white and parading opposite the civil rights protesters with a sign asking "Don't whites have rights?" Simpson's Navy experience, however, altered his views, as the personal relations he developed with black seamen persuaded him to support civil rights. Simpson returned to Athens as an advocate of black equality, and he began to take part in local desegregation efforts. He also focused his energies on the university's vast web of student regulations. As a non-traditional student, he found the school's in loco parentis policies particularly odious. "I was twenty-one-years-old, I was married, I was a veteran, I had a kid, and the rules seemed really juvenile to me," he remembers. "It might not have seemed that way to an eighteen-year-old who just came from home, . . . but I think it made a difference to me, being in the position I was, to see the absurdity of those rules."[26]

The issue of special regulations for women, ranging from a policy that required women to live in dorms until they turned 21 to a rule forcing female dorm residents to abide by an 11:30 P.M. curfew, angered Simpson and numerous students on the Athens campus. In the Movement for Co-Ed Equality's most significant action, the new group led more than 500 students in a march through the April rain to the administration building to protest women's rules. After presenting administrators with a petition demanding equal treatment for women students, 300 of the marchers started a sit-in in the building. Although the protesters voluntarily left the building after two days in an attempt to facilitate negotiations, school officials failed to take advantage of the students' overtures. Instead, they responded by trying to crush the movement, refusing to negotiate with the students and obtaining a temporary restraining order that barred campus demonstrations from blocking building entrances and prohibited them during class time. This prompted the Movement for Co-Ed Equality to organize a march mourning the death of free speech on campus and to send three students to disrupt the inauguration of the university's new president, Fred C. Davison, by raising a banner that read "The Emperor Wears No Clothes" at the ceremony. The administration retaliated by beginning disciplinary proceedings against the leaders of both marches, a process that culminated with the handing down of a one-year suspension to David Simpson. But the administration's reaction did not have the desired effect, for instead of dampening interest in the issue of women's rules it further enflamed students and helped to draw more people to the cause. When classes resumed in the fall, so too did protests, and in January 1969 the administration took the first steps toward abolishing women's rules by doing away with women's hours.[27]

The movement to end women's rules followed a similar trajectory at the University of Tennessee. There, school officials' refusal in late 1968 to lift curfew hours for over-21 female dorm residents, as requested by the campus organization Associated Women Students (AWS), sparked a broad-based campaign against such policies when classes restarted after winter break. The AWS kicked off the drive by passing out buttons that bore the picture of Lucy from the *Peanuts* comic strip in a cage with a ball-and-chain around her foot in order to "symbolize the locking up of UT women at night," explained one AWS member. The AWS was not alone in calling for an end to women's rules on the Knoxville campus. The administration's intransigence on the issue prompted such typically establishment institutions as student government and the school newspaper, the *Daily Beacon*, to become outspoken proponents of gender equality. Just as importantly, the university's defiant stance benefited opponents of women's rules by delivering them the support of students for whom women's hours seemingly was an irrelevant issue, such as commuters and some men. While these students may have been drawn to the issue less out of an interest in gender equality than out of concern that the administration had trampled students' rights and acted in an authoritarian manner, they nonetheless threw their support behind the AWS's effort. That the student body had rallied to the cause became evident in early 1969. In the first week of February, 50 women defied curfew rules by staging a brief post-midnight walkout from their dormitory, Clement Hall. The following evening, as an estimated 2,000 male students cheered in support, 1,000 women left their dorms for a midnight march-cum-rally against women's hours. With another rally planned for the next night, the walkouts forced the administration to act, and the next day it announced the immediate implementation of a no-hours policy for all women over 21 who received parental approval. Though a curfew remained in effect for younger women and of-age women still needed parental consent to escape it, activists considered it an important first step toward gender equality on campus. As a joint student government-AWS statement declared, "we have met the issue of equality for women—and we have won."[28]

Connecting the Local to the Regional

SSOC's desire to work with young southerners on the issues that mattered most to them underlay its support of local women's rights campaigns since one of the group's tenets was to reach out to people "where they were at" in the hope that their current concerns would become a stepping stone toward greater involvement in progressive causes. This perspective inspired the group to support local student activism on a variety of issues, for doing so, the group reasoned, would help to radicalize the students involved and impel them to take an active interest in other subjects. SSOC, therefore, was quick to align itself with the local activism that developed in 1968 and early 1969 in the South. As elsewhere in the country, the turbulence of these years—the shock of the Tet Offensive, the assassinations of Martin Luther King and Robert Kennedy, the riots that engulfed the nation's cities after King's murder, the mayhem of the Chicago Democratic Convention, and the growing notoriety of the Black Panther Party—fueled the wave of student activism that washed over the South. Both campuses that never before had witnessed protests and demonstrations and those for which student activism had been a constant feature of daily life throughout the 1960s experienced tumult and upheaval in 1968 and 1969.

Student antiwar activism intensified in the wake of the Tet Offensive in January 1968. Antiwar demonstrations took place at a host of southern schools, including Lynchburg College, Hollins College, and the universities of Arkansas, Alabama, Virginia, and Tennessee.[29] In March 1968, Richmond SSOC activists picketed the hotel where Vice President Hubert Humphrey was speaking to the Young Democrats' national convention, while in May the joint SSOC-SDS chapter at the University of Georgia led 60 activists who had gathered at the university for SSOC's annual spring conference in picketing a campus speech by Secretary of State Dean Rusk. Several months later, the SSOC chapter at the University of South Carolina organized an antiwar march in Columbia of more than 100 students and active duty G.I.s from Fort Jackson, the large army base located in the city.[30]

In addition to Vietnam, labor issues continued to capture white students' attention in 1968 and 1969. SSOC activists in Mississippi, Tennessee, and West Virginia participated in strikes by manufacturing workers in their states, joining them on picket lines and distributing literature on their behalf on college campuses. SSOC students across the region also organized demonstrations and picketing at grocery stores in support of the United Farmworkers Organizing Committee's call for a boycott of California table grapes to support striking workers.[31] Students took a particular interest in labor disputes on southern campuses that pitted non-professional university workers, the overwhelming majority of whom were black and female, against white, male-dominated boards of trustees and administrations over wages and unionizing rights. At the University of Virginia, activist students held a teach-in in October 1968 to demand, among other things, raises for the cooks, janitors, groundskeepers, and other workers who ensured the university's smooth functioning. At Duke University and the University of North Carolina at Chapel Hill, students organized support for striking non-academic employees. Beginning in April 1968, in the wake of Martin Luther King's assassination, Duke students held demonstrations and vigils, including a two-day vigil in which 250 students occupied President Douglas Knight's home. The following year, students in Chapel Hill, led by the Black Student Movement and the campus SSOC chapter, supported the workers by picketing dining halls, boycotting classes, and holding rallies, including one that attracted 2,000 students and another—at which Joan Baez performed—that drew 3,000 people. At both schools the students' support for the workers helped to spur negotiations that eventually yielded a settlement granting the workers many of their demands, including higher wages and a shorter work week.[32]

Civil rights remained of deep interest to white southern student activists. In 1968 and 1969, they took aim at the racism that continued to infect their campuses, and they worked to create more opportunities for African Americans at their schools. One common goal was to pressure administrative leaders and goad admissions officers to recruit and to admit more African Americans. At the University of Virginia, these efforts paid dividends, as the university hired its first black admissions officer in early 1969 and established a summer preparatory program for admitted black students.[33] White activists also tried to make their schools more welcoming to black students. In 1969, the Clemson SSOC chapter joined with the African American student group in condemning and refusing to stand for the playing of "Dixie" at campus events. Elsewhere, sympathetic white students organized support for black students' demands for black studies courses and academic departments. Duke, the University of Georgia, Furman University, and the University of Virginia, were among the southern schools to develop black studies curricula in response to interracial students protests.[34]

University reform issues, ranging from academic freedom to in loco parentis, emerged as a central concern of white southern students in 1968 and 1969 on some campuses, even overshadowing Vietnam, labor organizing, and civil rights. The continued black student-led drive for racial equality and equal opportunity on campuses helped to stimulate their white peers' activist impulses on these issues, for the fact that black students were working to change the system highlighted for white students their own passivity. As a flyer issued by the Duke SSOC group elaborated,

> [T]he blacks have started to take their freedom, to cease being niggers—and now they have more rights than whites! Because they organized and fought, they have some power over their education destiny—and we have none! We are still niggers! ... WE MUST DEMAND THE SAME RIGHTS THE BLACK STUDENTS HAVE! ... WE MUST HAVE THE RIGHT TO HELP DETERMINE

OUR CURRICULUM, TO HELP DETERMINE WHO TEACHES US, TO HELP DETERMINE THE POSITION OUR UNIVERSITY TAKES IN WORLD AFFAIRS AND IN THE COMMUNITY!

Undoubtedly, the white activists' casual use of racial slurs in reference to themselves was unseemly and offensive to many people, black and white. Nonetheless, such language conveyed the increasingly commonplace sentiment among them that white students were treated as second-class citizens by college and university administrations. Like black students, the activists believed, whites needed to speak and act out in order to force school administrators to deal with their grievances.[35]

The promotion and protection of freedom of speech on campus was the single most important university reform issue to white students on many southern campuses in the late 1960s. White activists were acutely sensitive to infringements upon their constitutionally protected rights of free speech since they depended upon this guarantee to ensure that their views and opinions, which often were unpopular, would be heard on campuses. Such concerns were not new in the late sixties. In early 1966, at the University of Florida, for instance, activists Alan Levin and Lucien Cross drew attention to the issue by distributing a raunchy college humor magazine and an antiwar periodical, a deliberate violation of university policy banning from campus materials published by non-college organizations. The administration's decision to place the two activists on probation for violating official policy created a furor on campus and compelled school officials to devise a new policy that guaranteed students' free speech rights, including the right to picket, pass petitions, and distribute free literature on campus.[36]

By 1968, student interest in free speech had spiked in response to new efforts by university officials to erode students' free speech rights. Not infrequently, local civil rights and antiwar actions took on a free speech component as a result of administrative attempts to curtail or prevent protest activity from occurring at all. In early 1968, Duke students transformed what had begun as protests against the presence on campus of recruiters for the Dow Chemical Company, the main supplier of the flesh-burning petroleum jelly, napalm, into a demonstration that also took aim at university policy disallowing picketing on school grounds. Later that year, an antiwar rally at the University of Alabama metamorphosed into a free speech protest thanks to an administrative edict threatening to ban from school students who spoke at the rally. At East Tennessee State University, agitation against in loco parentis policies evolved into a fight over students' free speech rights when the university suspended eight students for distributing a leaflet on campus calling for university reform. And at the University of South Carolina, students married free speech concerns to their opposition to the war during the late 1968 joint student-G.I. antiwar march through Columbia. The march moved through the city's streets to the state Capitol building and then to the home of university president Thomas F. Jones, where the marchers protested his refusal to allow them to rally at the Horseshoe, the main quadrangle on the South Carolina campus.[37]

For some newly created SSOC chapters, freedom of speech was the obvious first issue around which to organize owing to their difficulty in winning university recognition. Typical was the situation at the University of Tennessee, where school officials cited the involvement of non-students in the group, the vagueness of the organization's goals, and its similarities to SDS—meaning that it was a radical, violence-prone group—as the basis for refusing to recognize the chapter. To SSOC activists, these rationales were merely attempts to obscure university officials' disregard for their free speech rights. In response, the new groups initiated actions to publicize their case. In March 1969 at Florida State University, students occupied the administration building for two days and started a vigil to protest school officials' refusal to designate the SSOC chapter an official student group. The same month, more than 200 students held a week-long protest in Tigert Hall, the administration building at the University of Florida, to protest the president of the university's unusual decision to reverse a student-faculty committee's recommendation to make the SSOC chapter an approved student organization.[38]

Beyond the core group of activists on southern campuses, free speech issues appealed to a broad range of students because virtually everyone could conceive of ways in which restrictions on free speech could affect them personally. Consequently, efforts by school officials to regulate campus speech served to unite students in opposition. In particular, students sharply attacked policies that prohibited certain individuals from speaking on campuses. Dubbed speaker bans by students, the policies denied campus speaking privileges to anyone school authorities believed would lace their talks with obscenities, promote anti-American ideas, or advocate violence. Although administrators designed such policies to protect their campuses from what they considered incendiary speech, students resented that officials reserved for themselves the power to determine who student groups could and could not invite to speak on campus. Moreover, the fact that students often considered restrictions on free speech to be yet another example of administrators treating them as less than responsible adults heightened opposition to such measures.[39]

In 1968 at the University of Tennessee, a campaign to liberalize the speaker policy drew support throughout the student body, picking up steam when the university's chancellor refused to allow black activist Dick Gregory to speak on campus since, he declared, Gregory's "appearance would be an outrage and an insult to many citizens of this state." For similar reasons, he also nixed a scheduled appearance by Timothy Leary, and he followed that by rejecting pleas for a more open speaker policy on the grounds that it would undermine his authority on campus.[40] Far from resolving the issue, the chancellor's actions sent shockwaves through the campus community and prompted groups as different as the local SSOC chapter, the Black Student Union, student government, the Faculty Senate, the Interfraternity Council, and the university's chapter of the AAUP to call for a new policy. In editorials in the *Daily Beacon*, resolutions passed by student government, and protest rallies on campus, including one that drew more than 1,000 people despite persistent rain showers, students and faculty continuously pressured the administration to adopt a more open speaker policy. The university's unresponsiveness to this demand helped to radicalize UT students, for it encouraged them to consider this another example of university officials trying to suppress their rights. It was not a coincidence that many of the students who attended the large demonstration in the rain reconvened that same evening to participate in or support the midnight walkout by female dormitory residents to protest women's hours. And activists on campus recognized that the administration's intransigence redounded to their benefit. "It has been to the advantage of UT-SSOC," remarked campus SSOC organizer Don Armour, "that the administration has acted in such an indecisive manner in regard to an open speaker policy by playing bureaucratic games with inalienable rights guaranteed by the Constitution." Finally, at the end of the 1969 school year, student agitation combined with a federal judge's ruling that the speaker policy was too broadly cast to pass constitutional muster, forced the administration to implement a new policy that removed its veto over campus speakers and gave student groups the freedom to invite anyone they wished to speak on campus.[41]

A speaker ban also provoked widespread student opposition at Furman University, a small, Baptist school in Greenville, South Carolina, that had been home to one of SSOC's most vibrant chapters since 1967. The SSOC activists on campus played the key role in making the university's restrictive speaker policy the dominant issue on campus. In early 1968, they openly challenged the speaker policy by inviting Black Power advocate George Ware to give a talk on campus, an invitation that the administration approved because it did not know anything about Ware. School officials thus were outraged by Ware's remarks, which included his condoning black violence to defend black communities. Correctly believing that SSOC intentionally had invited a speaker who would violate the speaker policy, the school's president placed SSOC on two-weeks probation.[42]

SSOC drew more attention to the speaker ban during the 1968–1969 school year. First, the administration nixed SSOC's attempt to bring Chicago Eight defendant Rennie Davis to campus. The ruling outraged students, including many who either never had heard of Davis or who disagreed with his politics. It also prompted organizations as diverse as the Pershing Rifles, Baptist Student Union, College Republicans, and Young

Democrats to form an unlikely alliance to propel Ronald McKinney, an advocate of revising the speaker policy, to a narrow victory in the race for student government president. Then in spring 1969, SSOC provoked a crisis by issuing a speaking invitation to SDS firebrand Bernadine Dohrn, an invitation that the administration promptly vetoed. Regardless of their political views, many in the campus community were angry with the administration's action. To them, freedom of speech was at stake. Steve Compton, the SSOC member who had issued the invitation to Dohrn, remembers, "It was a very clear case of the administration violating what seemed to be, to a lot of people, . . . a very basic right." In response, 20 faculty members and more than 600 students signed a petition calling for revision of the policy. More significantly, 150 students attended a free speech rally organized by SSOC and other speaker ban opponents in May in front of the administration building, the first student protest demonstration to take place on the university's usually serene grounds. The broad opposition that the speaker policy inspired was evident at the rally, as among those to mount the platform to denounce the policy or to promote freedom of speech were Ronald McKinney, SSOC founder Jack Sullivan, history professors Bill Lavery and A. V. Huff, and student senator Kurt Stakeman. A delegation of students and their faculty supporters then went to the president's office to present the petition. Bill Lavery remembers with amusement encountering "terror-struck secretaries who thought the next thing they were going to do was trash the building." But violence was far from the students' minds. The protest had been a quiet, moderate, and respectful event. If the rally lacked the radical edge of a protest at Columbia, Berkeley, or even Florida or Virginia, it nonetheless "marked a first," one student publication concluded, for never before had Furman students "stood idealistically in favor of an open and relevant education." In January 1970, in response to student complaints, the university adopted a new speaker policy that ceded greater control of the speaker selection process to students.[43]

Furman SSOC's opposition to the speaker ban was indicative of the type of local issues that animated the group. By focusing on local concerns, the activists believed it could be an agent of change on the campus and in the wider community. Thus, by working to build opposition to the draft, the group demonstrated, in Jack Sullivan's words, that "the Vietnam War was a local issue" since nothing revealed the local and personal consequences of the war in starker terms than the prospect of military service. Likewise, they focused their civil rights work on events close to home. For instance, in response to the fatal police shooting of black students at South Carolina State University, Furman SSOC students demonstrated outside the Federal Building in Greenville.[44]

Furman SSOC's success depended not just on its ability to activate students on issues of local importance but on its rejection of confrontational tactics. As Greg Wooten later explained, "how we protested things was geared toward not just simply being defiant. . . . [W]e wanted to positively influence people as well." Unlike SSOC chapters elsewhere, the Furman group aspired to mainstream status, for the activists understood that any trace of radicalism or extremism in their goals or tactics would doom its chances of attracting moderate and liberal students. The chapter's mainstream aspirations inspired it to seek and win approval as an official student organization, a designation that gave the group access to campus facilities for its meetings and projects and that enabled it to draw student activities' monies, providing the perpetually cash-strapped organization with much-needed funds. While few other SSOC groups pursued and received financial support from their institution or could be found joyfully posed for pictures in the school yearbook, the Furman activists believed they had to become part of the mainstream on campus if they were going to have a chance to reach many students.[45]

In this respect, Furman SSOC was no different from other chapters in the region, for all campus SSOC groups had to be sensitive to the local context in which they operated lest they risk alienating potential supporters by promoting unpopular causes and programs. While some attempted to smooth their radical edges or to portray themselves as centrist organizations, many chapters remained locally relevant by focusing considerable attention on community issues. Dwelling on such matters, however, often meant that the campus groups existed in a sort of vacuum; they happily "did their own thing," unconcerned or uninterested in what other chapters were doing or in what was happening in SSOC's national office. To these activists, SSOC's value was that it enabled them to do this work; it was the vehicle by which they could build a student movement on their campuses. Their decision to work with SSOC, then, was often strategic, based as much on tactical considerations as on a deep attraction to the group's mission and programs. At the University of North Carolina, many of the students involved with the SSOC chapter had no sense that their group was connected to a larger entity. What mattered to them was that SSOC at Chapel Hill was a means for organizing students. Scott Bradley, one of the founders of the group, recalled that SSOC's role was as "kind of a figurehead, a name. . . . [I]t wasn't a thing that people were particularly concerned about membership . . . it was more of an organizational tool . . . a central something for organizing." Furman SSOC members, like those on other campuses, almost never traveled to the Nashville office or to the region-wide conferences, and they rarely paid attention to the disputes among the group's leaders. As far as the Furman activists were concerned, *they* were SSOC; it was as if they were unrelated to the Nashville-based organization.[46]

The Furman activists' lack of commitment to SSOC was not unusual. SSOC's leaders were well aware that many of the group's rank-and-file members did not have a strong attachment to the organization, and they responded by instituting a series of changes in mid-1968 designed to decentralize the organization. The leadership hoped these changes would make SSOC more responsive to the needs of activists interested primarily in local affairs. The drafting of a new constitution was the most significant step the group took toward decentralizing. The constitution altered the group's leadership structure, as it did away with the chairman's office and made the more locally friendly sounding position of program secretary the top office in the group. More significantly, the constitution called for the creation of state caucuses to act as the key decision-making bodies within the group. Comprised of representatives from SSOC chapters in the state, the caucuses would have responsibility for devising statewide programs and actions. Interposed between the local chapters and the national office, the hope was that they would be able to respond quickly to developments within the state. As for the main office, SSOC Chairman Tom Gardner wrote the membership that with the group's new decentralized structure it would "serve as a collective information pool, fundraising and publication center and otherwise service center for state offices and travelers who would service the chapters and other campuses in their state."[47]

SSOC's restructuring, however, was ineffective as a means of moving the group closer to the local organizations. The state caucuses envisioned by the new constitution did not flourish. Few of these state-wide organizations were created in 1968 and 1969, and where they did emerge, they failed to bridge the gap between SSOC and the campus groups. Instead, the caucuses made the parent group seem more remote and insignificant, for they encouraged the campus-based activists to view the situation in their locality and state as predominant. From the perspective of local activists, though SSOC was a friendly and supportive organization, their campus groups were of far greater relevance. Moreover, since SSOC now believed that responding to the activities and programs initiated at the local level was its top priority, the group appeared to many local activists to lack focus. Beyond a general desire to assist local movements, many found it difficult to discern SSOC's goals and aims. Ultimately, SSOC's organizational changes further weakened its already frayed connection to local activist groups in the region and raised questions among them about SSOC's purpose and mission. Most local activists simply did not have enough invested in SSOC to come to the group's rescue when its survival was at stake, as it was in the spring of 1969.[48]

Chapter 8
Falling Apart: The Dissolution of SSOC

> Sending the troops in for a great moral cause and emerging with a colony primed for corporate exploitation and cultural domination is all too familiar in American history. In such respects southerners could feel some empathy with Third World countries. . . . Was there much difference between General Westmoreland's defoliation of the Vietnamese jungle with agent orange and General Sherman's burnt earth swath from Atlanta to the sea?
>
> —David Doggett[1]

> I still think what I've been saying for 25 years—that SSOC made two mistakes. The first was to organize in the first place. The second was to disband when it did.
>
> —Anne Braden[2]

"I feel like a doctor whose entire life savings are all tied up in shares of R. J. Reynolds and whose patient has just died of lung cancer after smoking three packs of Camels a day for forty-five years," wrote Ed Hamlett shortly after SSOC voted to disband in June 1969. His wry remark betrayed his deep ambivalence about the group's dissolution. As a founder of the organization, Hamlett, like many other early SSOC activists, was saddened by the group's collapse. But he recognized that continuing internal discord over SSOC's mission, agenda, and tactics had become a serious distraction for the group and undermined its ability to create effective programs or adequately support its members across the region. Nonetheless, Hamlett adamantly believed that forces beyond the group's control ultimately were responsible for SSOC's death. In particular, he singled out for blame the leading student organization of the day. SSOC, he insisted, "would not have been dissolved without the extremely active interference in its internal affairs by national Students for a Democratic Society (SDS) staff and other factions within SDS."[3]

Hamlett had good reason to accuse SDS of killing SSOC. In 1969, riven by its own internal divisions, SDS openly waged fratricidal war on SSOC, as the factions contending for power in the group believed that destroying SSOC and winning the allegiance of its members, especially its leaders, would enable them to prevail in the fight to control SDS. However, SDS was not primarily responsible for SSOC's fate; its actions were not the sole—or even the most important—cause of SSOC's dissolution. More significantly, SSOC's own internal problems precipitated the group's downfall. Chief among these was the group's lack of analytical focus. Throughout its existence, SSOC gave primacy to action over analysis; as the group's leaders never tired of pointing out, SSOC was an organization of doers, not talkers. Additionally, although a vision of a non-racist, non-militaristic, and more just society animated SSOC's work from its first days, the group did not have a clear theoretical perspective or a coherent prescription for progressive change. Eventually, its failure to develop ideological moorings or an analytical framework to guide and make sense of its actions heightened tension within the organization and drove a wedge between SSOC and SDS.

Much of the trouble stemmed from the fact that the group's vision of a New South encouraged the activists to consider the South as distinctive. Because the South was inherently different from the North, SSOC activists reasoned, the region's ills required peculiarly southern solutions. But southern distinctiveness was not a broadly conceived ideology. While it was a useful frame of reference for the young whites as they confronted the uniquely southern system of racial segregation, it was of limited value when SSOC faced issues that were national or international in nature, such as university reform, the Vietnam War, or labor organizing. By 1969, southern distinctiveness had become the newest and most serious issue to cleave the group. Moreover, it served to alienate SDS from SSOC, increasing its disenchantment with the group and fueling its offensive against it. A confluence of factors thus had brought SSOC to the brink of collapse by June 1969. It is understandable that in the immediate aftermath of the break-up, Ed Hamlett blamed SDS for SSOC's destruction since it openly advocated this result. Yet SDS's attack only accelerated a process that SSOC had begun itself by its inability to reach consensus about its goals and strategies or to project a coherent image. As the group degenerated into rival factions, it became vulnerable to outside meddling as well as weakened its own members' ties to the group, many of whom concluded that the organization was not worth saving.

Southern Distinctiveness

The notion that the South was a distinct region of the country had a long history in SSOC. That the group named itself the *Southern* Student Organizing Committee, adopted as its symbol an image of joined black and white hands superimposed on the Confederate battleflag, and self-consciously tied its founding statement, "We'll Take Our Stand," to the proudly southern manifesto of the Nashville Agrarians, all revealed that an awareness of southern difference had infused the group since its founding. The differences that mattered most to SSOC were those between the North and the South. In the late 1960s, SSOC activists gave special emphasis to the enduring economic differences between the regions. This issue captured the activists' attention not simply because the South lagged behind the rest of the nation by almost every economic measure but because they were convinced that the North was responsible for the South's economic troubles. In their view, the North intentionally had underdeveloped the region, exploiting it for cheap labor and raw materials. The South, they concluded, was a land of debilitating poverty because this served the needs of rapacious northern industrialists. The region's extractive industries, from coal to timber, provided raw materials for northern industries, while the prevailing wage structure and weakness of organized labor encouraged northern enterprises to relocate their operations to the South. As Tom Gardner seethed in 1967, "Northern bosses have extracted much while contributing almost nothing to the development of the South."[4]

SSOC considered the South's economic relationship with the North to be a product of the region's unique history. More generally, the activists found explanations for southern distinctiveness in the region's history. Drawing on the work of C. Vann Woodward, Kenneth Stampp, F. Ray Marshall, Henry Caudill, James Agee, V. O. Key, and others, the group learned that poverty, corruption, defeat, domination, and occupation defined the southern past and were the basis for the region's separateness in the present. As Woodward wrote, it was because of its history "that the South remains the most distinctive region of the country." SSOC activists were mindful that southern distinctiveness rested on such a sordid, unprogressive past. They were well aware that, to a large extent, southern history was the story of secession and civil war, endemic poverty and economic subjugation to the North, bitter white racism and abominable Jim Crow laws. When thinking of southern distinctiveness, Ed Hamlett wrote in the *New South Student,* "The Lerleens, Lesters, Sheltons, and Eastlands leap to the mind; lynchings, poverty, pellagra, rape, states rights—the list is seeming[ly] interminable." Yet despite the many shameful episodes of southern history, the activists believed that there were elements in the southern past in which white southerners could take pride. Finding such historical episodes and actions was an important task because, as Tom Gardner later noted, "we had to come to grips with being white southerners." They had to prove to themselves and others that "white southerner" was not shorthand for "benighted racist." SSOC, therefore, worked to locate positive elements in the southern past that would serve the group's contemporary needs. In Mike Welch's words, SSOC sought "historical encouragement" for its progressive activism.[5]

The activists found much in the historical record that they could use. Unsurprisingly, they focused particular attention on the dissenters and radicals of the past in an effort to place SSOC's activism within a tradition of indigenous southern protest and thereby make it seem less foreign to white southerners. Given the long shadows cast by slavery, secession, and civil war, few southern whites, including those already involved with the SSOC, knew much about this history. To bring it to light, the group commissioned a series of essays by prominent scholars and activists on radical moments and individuals in southern history for a new feature in the *New South Student* on "The Roots of Southern Radicalism." Beginning in late 1967, the magazine published essays by Herbert Aptheker on slave revolts, Norman Pollock on Populism, Myles Horton on the Highlander Center, H. L. Mitchell on the Southern Tenant Farmers' Union, Don West on abolitionists, Clarence Jordan on Koinonia, and Anne Braden on the Southern Conference for Human Welfare. SSOC also highlighted the lesser-known aspects of southern history at its conferences. For instance at SSOC's women's conference in early 1969, one session, "Rebels in the Past," focused on radical southern women in history. Shortly thereafter, SSOC sponsored an entire conference on radical southern history in Atlanta. With sessions on such topics as Populism, Reconstruction, and "Radical Politics in the Thirties," the conference was SSOC's most significant effort to educate its constituency about the region's radical past. "Radical southern history is our history," the group exclaimed in the history conference's program notes. "We are that struggle alive today—fighting the same oppression and many of the same problems. With an analysis of earlier struggles, we can develope [*sic*] a better understanding of what we're about."[6]

SSOC's interest in identifying its ancestors in radicalism did not blind the group to the distasteful parts of the southern past or prompt it to offer a whitewashed version of the region's history. Indeed, SSOC readily acknowledged that southern history offered southern whites much of which to be ashamed. But as an organization attempting to transform the region, some in SSOC believed that even the worst parts of the southern past had some redeeming value to white southern activists in the 1960s. Such an understanding of the past accounted for the growing prominence of Confederate symbols and rhetoric within SSOC during its last years. The most overt manifestation of this newfound respect for the Confederate past was the executive committee's 8–6 vote in 1967 to revive SSOC's old hands-over-flag symbol, which the group had discarded in late 1964 when it sought to transform itself into a biracial group. At the time of SSOC's founding, the symbol's supporters had interpreted it as a rejection of the Confederacy and all that it stood for; the linked black and white hands served to negate the racism and white supremacy implicit in the flag. When the emblem reappeared as the group's official symbol in 1967, this interpretation was joined by one that did not demonize the Confederacy as the source of the South's problems but rather suggested that one could work for civil rights and other progressive causes without having to denounce the actions of one's ancestors. "You didn't have to deny your identity and your heritage and your birthright to take a good stand on civil rights," Lyn Wells later reflected. For Wells and many others, the clasped hands did not erase the flag; rather, the flag stood on equal footing with the interlocked hands, making clear that the progressive goals of the present could be fused with the positive elements of the Confederacy—the bravery, loyalty, and devotion of the southern whites who took up arms to fight, not for slavery, but for their families and homes.[7]

SSOC also increasingly used the rhetoric of the Confederacy. In particular, talk of rebellion and secession crept into group's publications and the activists' conversations and speeches. SSOC invoked these historically portentous terms both to connect with their forebears and to symbolize the drastic, all-transforming changes they envisioned for the South. Although the activists did not actually contemplate a day when the South once again would secede or rebel, they knew that their ironic use of these terms would give added weight to their cause. As Tom Gardner recollects, "we started promoting kind of a positive view of being rebellious, a 'rebel' of a different kind, a rebel against racism, a rebel against the war." Ed Hamlett was one who fully appreciated the power of these terms. "We must make our own cultural, social, political, and economic revolution," Hamlett declared in 1967. "I am a secessionist—I urge moral and spiritual secession."[8]

Secession was a particularly useful concept for conveying SSOC's opposition to the Vietnam War. In the words of a SSOC-affiliated group in North Carolina, southern students should secede from "the oppression that our country is practicing in Vietnam by SECEDING from the SSS [selective service system] or . . . pressuring school administrations to SECEDE from the military-industrial complex by dropping ROTC, Defense contracts, and selling their stock in war industries." Additionally, SSOC used the term to connect modern-day antiwar activists to the southerners who opposed the Confederacy. Representing a small fraction of the South's population, these dissenters, separated by 100 years, both endured hostility, ostracism, and occasional violence from the majority for their refusal to support a war they saw as wrong, unjust, and amoral. "If isolated resisters can at least have the feeling that they are backed up by a proud and defiant history which is Southern and which is theirs," Tom Gardner wrote in 1967, "it may be easier for them to speak out and actively resist the masters of war." Consequently, the group made frequent reference to anti-Confederates, war resisters, and the men of the Unionist stronghold of Winston County, Alabama, who "declared their right to secede from the state," SSOC reminded its members in an effort to show that secession had "a radical tradition in the South."[9]

Secession was the basis for SSOC's contribution to the Ten Days of Resistance, a series of local antiwar protests and demonstrations across the country and around the world during the last ten days of April 1968. SSOC, which promoted the project in the South, decided that it would have more success by emphasizing a peculiarly southern perspective on the war than by calling attention to the fact that southerners would be part of a vast national and international protest. In February 1968, Gene Guerrero urged that "SSOC should send out a call saying we are part of the protest but because we are from the South, we . . . could call for one day of southern secession. It would give the South cohesiveness and a chance to bring out a Southern theme." SSOC expanded this idea into the Southern Days of Secession, a call for southerners to "secede" from the war. SSOC made its approach clear in "We Secede," its flyer promoting the project. "As young Southerners," the flyer declared, "we announce that we hereby *SECEDE* from the oppression our country is practicing: THE WAR AGAINST THE VIETNAMESE THE RACISM AND EXPLOITATION OF THE POOR AT HOME AND THE SELECTIVE SERVICE SYSTEM." "By seceding from our country's oppression, both nationally and locally," the flyer explained, "we reaffirm our determination to work together to create a new South, free from racism and exploitation and a world in which people are free to determine their own destinies."[10]

The flyer's reference to the "racism and exploitation of the poor at home" suggested that SSOC believed secession helped to illuminate the connection between the Vietnam War and the poverty and racism that permeated the South. Moreover, its criticisms of the war revealed that some in the group had begun to adopt a class perspective of the conflict. The parallels between the Vietnam and Civil Wars encouraged this line of thinking. To these SSOC activists, both southerners and Vietnamese had been victimized by northern aggression, and both were economically subservient to the North's industrial interests. As Jody Palmour argued in 1967, "The South has been militarily, economically, and politically dominated by what are organically the same forces interested in the perpetuation of the Vietnam War." The SSOC activists who turned to class analysis also did so as a result of their reading of southern history, for they learned that the southerners they identified as their ancestors in radicalism were radicals not just on matters of race but of class as well. They discovered that the union organizers, Populists, utopians, and abolitionists they admired frequently had advocated that poor and working-class southerners unite in common cause across racial lines since they shared many of the same problems.[11]

Because race and class were so tightly intertwined in the South's history, these activists believed that their challenge was to show how calls for white unity had been, and still were, a tool used by elite whites to protect their class position. Tom Gardner urged southern activists to recognize the class divisions in their society in a speech he delivered at the counterinaugural protest in Washington, D.C., on the eve of Richard

Nixon's swearing-in. After the SSOC contingent, marching under a "Southern Liberation Front" banner, had led rebel charges and let loose with rebel yells during the large protest march, they gathered, symbolically, in Sherman Square to listen to Gardner tell them that they could learn a thing or two about appealing to class interests from Alabama governor and recent presidential candidate George Wallace. Although they despised Wallace, Gardner counseled his peers to take note of the fact that Wallace's talk of the oppression of working whites by big business clearly had resonated with white southerners. But what made his views so nefarious, Gardner contended, was that his attacks on business were couched more in the language of suspicion of outsiders than in the rhetoric of class oppression. "What Wallace and other Biggies throughout our history have accomplished by pointing to the outside oppressor is the obliteration of class distinctions IN THE SOUTH," Gardner declared. And they had used race as the primary means to accomplish this end. "Against the Populists, the Southern labor movement, and now, on campus or in the plant, the Bosses have race-baited, red-baited and called for southern whites . . . to join together in one happy family to resist the influences of 'outside agitators.' That is the Solid South crap we're up against." Gardner argued that a truly radical southern movement would grow and prosper only by highlighting class differences. "We must put an end BOTH to Yankee corporate exploitation of the South and to the power of the Southern 'overseers' that continue to profit from the resources and the people of the South. . . . We must learn to . . . build movements that will sharpen, not ignore, the class distinctions in America. And finally, if the South is to be revolutionized, it will be Southerners who will accomplish it."[12]

In fact, Gardner and other like-minded SSOC activists shared the conviction that a revolutionized South could serve as an example for the rest of the nation of how to create an economically equitable, integrated society. In their view the South had a unique opportunity to assume this leading role because of its history. "The South has experienced a special history, both political and economic," Lyn Wells announced in 1969, "which can give us special abilities in the creation of [an] anti-capitalist movement." The formation of such a movement could make the South a model for the poor and displaced throughout the nation. As Ed Hamlett recalled saying at the time, " 'we get our shit together and, hey, maybe people in South Dakota and North Dakota might want to join up with us later.' You know, we were aware there was Mississippi . . . but South Dakota was in the top five poorest states too. So it wasn't just completely a southern thing, but it was a class thing." Thus, southerners' experience with economic subjugation and racial domination could be transformed from a liability into an opportunity.[13]

Most SSOC activists did not adopt class as a frame of analysis. Similarly, many worried that the group placed too much emphasis on the South's distinctiveness. The use of Confederate rhetoric and symbolism was a particular point of contention within the organization. On one level, some members believed that frequent invocations of the Confederate past wrongly implied that the South faced similar problems and had similar goals in the 1960s and the 1860s. This was especially problematic when the subject was the Vietnam War. As several activists argued, it was misleading for SSOC to raise the specter of secession since doing so negated the central reality of the war, namely, that it was a national and international issue. To argue that southerners could remove themselves in some way from the war or that the war, at root, was an act of northern aggression, diminished southerners' status in the country and implied that they were lesser Americans than people elsewhere. As one anti-secessionist emphatically put it, "We're part of this damn country as much as people in the North are."[14]

On another level, these members believed that association with the Confederacy was morally reprehensible. The group's readoption of the hands-over-flag symbol especially repulsed them as well as many who considered themselves SSOC friends and supporters. To these people, the flag was a symbol of race hatred, not one that conveyed a sense of pride in one's past. Such was the power of the flag that the clasped hands, though in the foreground, were overwhelmed by the Confederate emblem. Don West, the radical poet, political activist, and co-founder of the Highlander Folk School, tried to convey this to the group in a letter in early 1969. "There is nothing wrong with black and white hands clasped," he wrote. "But you should know that the Confederate flag NEVER symbolized anything good for the poor white South, nor for the Negro. That flag, like the Nazi swastika, symbolized a brutal power which kept millions in chattel slavery and millions more in poverty. . . . So how in the hell can you say that flag symbolizes anything good for the southern masses? If you're going to be 'rebels,' get another flag!"[15]

That only a minority of activists shared West's sentiments was evident from SSOC's reaction when Brett Bursey, its South Carolina campus traveler, burned a Confederate flag during a protest at the University of South Carolina in Columbia in February 1969, an act for which he later was arrested and charged with violating a state law barring desecration of the flag. To Bursey, the flag was repugnant. In a state where civil rights activists chafed at the sight of the battleflag flying over the capitol, Bursey could not support the use of the very same flag as part of SSOC's official regalia. From his perspective, the SSOC symbol was "a gimmick that . . . didn't work."[16] Most SSOC activists, though, were outraged by the flag burning, and throughout the spring they sharply criticized Bursey for his actions. Former SSOC chairman Steve Wise argued that destroying the flag was an ineffective way to try to organize white students since it sought to use guilt to motivate them. "White people cannot be organized to make a revolution by constantly telling them that they are guilty of oppressing blacks for 350 years," Wise insisted. "If you are full of guilt and hate for yourself, your land, your people, and your history, you will not be able to fight that which oppresses you." Other critics correctly perceived that the flag burning complicated SSOC's task of building support for progressive causes by damaging its reputation with whites who took pride in the flag. On the South Carolina campus, members of the conservative student organization Young Americans for Freedom (YAF) used the incident to try to tar SSOC as an extremist group hostile to southern whites. In a flyer distributed shortly after the protest, YAF charged that "These left wing Revolutionaries raise Viet Cong flags freely . . . and burn Confederate flags as a symbol of their intent to destroy a heritage which stands in the way of their new collectivists [sic] society." Then, perhaps unknowingly echoing SSOC's founding manifesto, YAF declared, "We take our stand . . . that the suppression of our creed in being proud of our heritage and displaying the symbols of that heritage is a denial of our rights under the Constitution."[17]

More revealing than the practical objections were those that were grounded in principle. At a staff meeting in Atlanta shortly after the incident, Bursey's infuriated opponents nearly forced his removal as South Carolina traveler on the grounds that he had defiled a sacred symbol of the South and, as he later recalled their words, "a legitimate working-class symbol of rebellion." What these activists failed to perceive, however, was that their hostile reaction to Bursey had the potential to do as much harm to the group as the flag burning itself. The passion and anger with which his detractors condemned him for destroying what they referred to as "our flag" easily could have raised concerns about SSOC's commitment to a progressive agenda with liberal-leaning students who had no special love for the flag. What type of organization would work itself into a paroxysm of anger over the destruction of the preeminent symbol of the Confederacy? Why would a putatively pro-civil rights group rush to defend the flag rather than applaud Bursey for reducing it to ashes? As Tom Gardner later reflected, SSOC "should have been unmistakenly foursquare against both the racism of the old and New South and all of its symbols and in total solidarity with black protests against racism. Bursey was right and SSOC was confused."[18]

Southern distinctiveness was thus a double-edged sword for SSOC. Its use of the battleflag, talk of secession, and emphasis on southern difference could attract students to the group but also cause some progressive whites to question the group's commitment to black equality. At one point, Jack Sullivan and the SSOC chapter he led at Furman University wrote the Nashville office to inquire if they could allow interested black students to join the group. That the chapter felt it had to ask this question stunned SSOC's leaders. "It was assumed," Tom Gardner remembers, that "we were anti-racists. We knew our history. We'd come out of the civil rights movement." But this was not obvious to students like those at Furman who only recently had associated with the group. To the Furman students, all that was clear in 1968 and 1969 was that SSOC was a white organization focused on the war, university reform, labor organizing, and, only occasionally, civil rights. SSOC "just forgot that all along the way we had to be more explicit about the anti-racist nature of our history and our work," Gardner concedes. "SSOC had not gone far enough in distinguishing its Southern particularisms from the racist traditions it sought to overcome."[19]

The turmoil and confusion southern distinctiveness caused within SSOC, though serious, paled in comparison to the damage it did to SSOC's relationship with SDS. SSOC's new emphasis on the uniqueness of the South led to the rupturing of ties between the two groups in 1969. From the beginning of their relationship, SDS leaders had worried that SSOC was too focused on the South and too invested in the notion of southern difference. By 1969, SDS had become increasingly global in its outlook, making SSOC's southern distinctiveness seem not only outdated but counterproductive to the building of a radical movement. Few of the northern activists saw the need for a regionally focused group of white southerners when the most important problems were national or international in scope and when the ultimate goal was to foment a revolution of working-class people the world over. SSOC shared SDS's discomfort with their relationship. From the outset, SSOC activists, though welcoming the legitimacy that ties to SDS conferred on the group, were concerned that some day SDS might abandon SSOC and seek to establish a southern wing. These fears intensified over time, fueled by what the southerners perceived to be SDS's condescending, lefter-than-thou attitude toward its southern allies and the northern group's growing organizational presence in the South.

The roots of the groups' future discord reached back to 1964, when they first allied. Despite SDS leaders' skepticism about southern distinctiveness, they invited SSOC to become a fraternal organization, reasoning that SSOC could give SDS traction in the South. Before 1964, SDS had few contacts in the region. In fact, its interest in the South had waxed and waned over time. As early as 1962 it had explored and rejected the possibility of working in the South because of doubts that southern students would be attracted to a northern organization. "It is hard as hell to get good Southern people active," complained Robb Burlage, a native Texan and one of SDS's founders. "It is all the more harder to get good Southern people trusting foreigners."[20] SSOC's founding revived SDS's interest in the South and inspired hope that the new group would serve as a conduit for bringing it southern members. SDS therefore began providing the group with literature to distribute on southern campuses and jointly funded SSOC's first campus traveler. But SDS found that even with SSOC's assistance it made little headway in the region. By 1965, only a handful of SDS chapters existed in the region, most notably at the

University of Texas and the University of North Carolina. It also encountered difficulty because SSOC developed a more proprietary attitude toward the South, especially after it reorganized as a membership group in 1966. By that point, the groups had reached an informal agreement in which SDS conceded the South to SSOC and pledged not to organize in the region.[21]

This arrangement, however, did not endure. Before the end of 1966, SDS chapters began sprouting in the region, a trend that continued through the rest of the decade. Several combined SDS–SSOC chapters also appeared in the South, some of which were joint creations, while nearly all of the others originally had been SSOC chapters. By 1969, the University of Miami, Clemson University, Florida State University, and the University of Alabama were among the schools to host SDS chapters, while Duke and the universities of Florida and Georgia were all home to SDS–SSOC groups.[22] SDS's growing strength in the South rankled SSOC. By allowing the formation of SDS chapters in the region, SSOC believed SDS had broken its agreement not to organize in the South. The southern activists also believed SDS's southward expansion indicated a lack of respect for SSOC. As Ed Hamlett bluntly noted in the SDS publication *New Left Notes,* "SSOC is not SDS-South." The words and actions of individual SDSers encouraged such sentiments. At the May 1967 Southwide conference at Buckeye Cove, for instance, SDS's Mike James offered a motion that called for SSOC to cease functioning as an independent organization and instead merge with SDS. The motion was defeated soundly, but it heightened SSOC activists' suspicions of SDS's intentions.[23]

In increasingly harsh tones, SSOC activists reminded SDS not only that it was independent, but that as an indigenous southern organization, it was the group best suited to work in the South. In a 1968 letter to SDS staffers Cathy Archibald and Les Coleman, Tom Gardner, expressing a frustration many of the southern activists shared, charged that SDS's work in the South suggested that non-southern activists were trying to tell southerners what they needed. "When Southern student-type activists . . . have been trying for four years to build their own movement, distinctly Southern, you can appreciate the feelings of those several hundred people when a few people . . . decide that its time for them to junk what they've been trying to build and get sucked into something national . . . from which they can't get as much servicing, organizing or just plain understanding." Indeed, nothing angered—or continued to anger—SSOC activists more than SDS's presumption that it knew what needed to be done in the South. "For some reason those people in SDS thought that they were more radical than us," Lyn Wells bitterly remarked 25 years later. "And it just rubbed everybody the wrong damn way. And then we got to thinking about it. . . . What is it that's more radical about them? What is it that they had that was so grand that they needed to deliver it to us in person like missionaries? It was arrogant, antagonistic, and bullshit."²⁴

SSOC, however, overlooked the fact that SDS had not forced itself on southern students. Rather, on southern campuses where SDS had a presence, students had chosen to join the group. Some were drawn to it explicitly by its radicalism. SDS's developing radical critique of the United States, its theoretical coherence, and its increasing willingness to employ provocative tactics, resonated with these students. To them, SSOC was too liberal in its outlook and too timid in its tactics. In the view of a University of Texas activist who had ties to both groups, SSOC simply seemed to have more in common with Young Democrats than with SDS. Other southern SDS supporters may have found SSOC's southern distinctiveness distasteful. SSOC's emphasis on southern difference, not to mention its prominent use of Confederate rhetoric and imagery, would have alienated southern students who believed SSOC's regional orientation was ineffective for conceptualizing and responding to the problems that were national and international in nature.²⁵

SDS's decision to hold its March 1969 National Council meeting in Austin—its first major gathering in the South since a conference in Chapel Hill in 1962—convinced many in SSOC that SDS wanted to compete with it for support on the South's predominantly white campuses. Although Austin was an SDS stronghold and SSOC had few supporters in Texas, the southerners believed the meeting was a prelude to SDS's raiding the South. Whether it meant to or not, SDS certainly gave this impression in its announcement of the meeting. "This will be the first N.C. [National Council meeting] to be held in the South and hopefully will contribute to the creation of a new political dynamic in the South," Bernadine Dohrn, SDS's inter-organizational secretary wrote the membership. In a response notable for its angry tone, Steve Wise, a former SSOC chairman, excoriated Dohrn for her ignorance of the region. "As a matter of fact," he sneered, "there has been considerable movement in the South since February 1, 1960," the day the sit-ins began. He then took SDS to task for its condescending attitude toward SSOC. "We have enough problems down here without having Movement people contribute to the feelings of cultural inferiority which American culture perpetuates upon every Southerner. And we are goddamn tired of having Yankee voyeurs coming South to remark on how politically sophisticated they are and how politically naive Southerners are." Wise concluded with a menacing warning: "If such fools do not learn to hold their tongues, they will find themselves shipped home in wooden boxes."²⁶

Wise's bravado aside, he and his colleagues were correct to surmise that SDS envisioned the Austin meeting as the start of a new push to recruit in the South. The brewing factional struggle within SDS underlay this effort. The key divide in 1969 was between the supporters of the Revolutionary Youth Movement (RYM) and the advocates of the

Progressive Labor Party (PL). The numerically superior RYM faction argued that SDS should seek alliances with Third World insurgents and black revolutionaries in America, positions long supported by many SDSers. Where RYM broke with the past was in its insistence that SDS shift its focus from students to the working class, for it had adopted the Marxist view that class conflict was central to revolutionary change and that workers would be the ones to lead the Revolution. Students, by themselves, RYM explained in "Toward a Revolutionary Youth Movement," the *New Left Notes* essay that outlined its views, were not capable of initiating the revolution to "bring about the downfall of capitalism, the system which is at the root of man's oppression." The essay was written by Michael Klonsky, a 25-year-old "red diaper baby" from California who won election in 1968 as SDS National Secretary, the top leadership position in the group. Klonsky, who would be a pivotal figure in SSOC's demise, made clear in the essay that RYM had given traditional Marxism a slight twist: not all workers, but rather working-class youths, held the key to revolutionary change, for their class status gave them personal experience with capitalism's contradictions while their youth ensured that they would have the courage and attitude necessary to spark the Revolution. Student activists still had an important role to play in the movement, RYM's adherents believed, but their job now was to help activate working-class youths around their class status and to persuade students to ally with their working-class peers, thereby creating a revolutionary force of young people. SDS's task, then, was to unite all young people in a "class-conscious anti-capitalist movement," the essay explained. "The nature of our struggle is such that it necessitates an organization that is made up of youth and not just students, and that these youths become class conscious. . . . We must . . . build SDS into a youth movement that is revolutionary."[27]

The Progressive Labor Party, founded in 1962 as a Maoist splinter group from the Communist Party of the United States, challenged RYM for control of SDS. PL, as it was called, was a highly disciplined organization that stressed the primacy of the class struggle and urged the overthrow of the liberal-capitalist system by a working class-led socialist revolution. PL believed that students formed an important support group for the workers who would lead the revolution. In 1967, it began advocating a worker-student alliance through which students would help to radicalize the working class and then follow workers into the Revolution. PL urged students to support striking workers, back the demands of nonprofessional employees on their campuses, try to draw workers into the antiwar movement, and, most importantly, take jobs in factories and other industrial settings in order to gain an intimate familiarity with workers and their experiences and problems.[28]

PL's conception of the relationship between students and workers differed in a subtle but important respect from RYM's. Unlike RYM, which insisted that workers and students would play equally important roles in the coming Revolution, PL believed that students always would be the junior partner in the worker-student alliance. In PL's view, RYM simply placed too much emphasis on youth and not enough on the working class. PL's differences with RYM and the rest of the student movement on this issue accounted for its opposition to street demonstrations and its condemnation of drug experimentation, casual sex, and the hippie subculture. Street protests, marijuana smoking, long hair—all of these things, PL argued, alienated working-class Americans, making it harder for students to reach them, impeding their radicalization, and delaying the onset of the Revolution.[29]

PL first began working within SDS in 1965, and by 1968 it was a powerful force in the organization. PL's strength never was clearer than during SDS conferences and meetings, when the party managed to tie up the gatherings in long, protracted discussions of theoretical issues, block passage of proposals and resolutions made by non-PL members, and even win support for some of its own. Much of its success rested on the fact that despite having the allegiance of just a small minority of SDS members, PL was a disciplined group within a decentralized, consensus-seeking, factionalized organization. PL's ideological rigidity also helped it to win support from the newer students in SDS who often were impressed by PL members' revolutionary ardor and what long-time Texas SDSer Jeff Shero called their "comprehensive ideological structure with which they could interpret the world." PL, in other words, had an answer for everything, and that answer was always the same—build a working-class revolution. RYM's emergence in late 1968, with its focus on the working class and its call for the uniting of students and working-class youths, was testimony to PL's growing strength in the group and a clear effort to steer students away from PL.[30]

With each faction believing it possessed the right formula for bringing about the Revolution and certain that it knew who would be in the Revolution's vanguard, their battle for organizational control of SDS became all-consuming. The group slowly degenerated into a caricature of its former self in which the organizational struggle took primacy over all other matters, resulting in the group's eventual implosion. But not before it helped to destroy SSOC.[31]

The SDS Attack

In the spring of 1969, the southern movement emerged as the new battleground for the fight for control of SDS. At the SDS National Council meeting in Austin in March, SSOC became hopelessly entangled in SDS's factional war. SSOC, of course, would have preferred to remain detached from the turmoil in SDS. The southern activists understood they had nothing to gain from the dispute, and, outside the Nashville office, few of the activists understood or had an interest in the theoretical issues that were at the heart of the conflict. The SDS factions, however, believed that gaining power in SSOC, or failing that, destroying it, would give them leverage in the struggle for control of SDS.

Progressive Labor moved against SSOC first. This came as no surprise to SSOC, for PL had been on an anti-SSOC crusade since at least 1966. PL's two most vocal critics of SSOC were Ed Clark and Fred Lacey, both of whom operated from within the New Orleans Movement for a Democratic Society, the PL stronghold in the South. Clark and Lacey worked to destabilize SSOC by encouraging SDS to organize in the South and by attempting to promote PL ideology within SSOC. Clark even made an ill-fated run for the SSOC chairmanship in 1967, a race he lost badly to Tom Gardner. Unable to seize control of SSOC, PL settled on a strategy of trying to destroy the group. By forcing SSOC to dismantle, PL hoped both to deprive RYM of support in the South and to rid the movement of an organization whose politics and tactics it despised. Thus, despite Clark's loss, PL was encouraged by the formation of southern SDS chapters, believing, correctly, that they would be useful allies in attacking SSOC from within SDS. The Austin meeting was the denouement of this plan, as Clark and Lacey led representatives from these new chapters in presenting a resolution calling for SDS to break ties with SSOC.[32]

The resolution, entitled "Build SDS in the South," embodied Progressive Labor's criticisms of SSOC, which clustered around two issues: SSOC's "bourgeois liberalism" and its "Southern exceptionalism." PL argued that at its core SSOC was a hopelessly liberal organization. The resolution mocked SSOC for believing that promoting liberalism among white southern students could be the first step in their radicalization, and it scored the group for accepting financial support from what it charged were "Kennedy controlled" liberal foundations. In short, SSOC represented "one of the ruling class's main efforts to build its kind of student movement in the South." The resolution also attacked SSOC's emphasis on southern distinctiveness. On the level of symbolism, the resolution berated SSOC for its use of the Confederate battleflag, a symbol that is "offensive to all blacks and to anyone opposed to racism." Substantively, the resolution condemned southern distinctiveness for its implication that southerners were a "nationally oppressed people" who needed to be liberated from their northern tormentors. What SSOC failed to comprehend, PL charged, was that class rather than region was the main fault line in American society. Southern distinctiveness was dangerous because it suggested that the southern working class was inherently different from the northern working class and that the two groups thus had different interests. As the resolution explained, "SSOC's theoretical position on the South as a 'colony' is . . . pernicious because it deflects the South and the Southerner from the class struggle. . . . The basic contradiction is between proletariat and capitalist, not Southern proletariat and Northern capitalist." The resolution concluded that SSOC's liberalism and regionalism disqualified the group from leading a radical movement in the South. "SSOC has tried to exist on the basis of being all things to all men: loyal servant to Ruling Class foundations, helpful ally to emerging bourgeois liberalism, radical leader to discontented and rebellious Southern white students and workers. We have examined SSOC and found it wanting." Consequently, PL and its allies in SDS urged the group to dissolve fraternal relations between the organizations and, as the title of the resolution suggested, to expand into the South.[33]

"Build SDS in the South," though a rhetorically powerful denunciation of SSOC, did not offer an accurate assessment of the group's past and reflected unrealistic views about the potential for radical change in the South. The resolution betrayed the anti-SSOC forces' ignorance of the fact that southern distinctiveness had been an important component of SSOC at its founding, and that the group had wrestled with the concept for years. The resolution, however, declared that its emergence in SSOC was a more recent development and one that was not well-thought out. It was the result, the resolution incorrectly explained, of "a simple-minded linear extension of the 1966 black power and black self-determination principle of organization along cultural and ethnic lines." Moreover, PL's advocacy of working-class revolution and a worker-student alliance was a path to irrelevancy on the southern campuses SSOC worked. SSOC's emphasis on southern distinctiveness and its willingness to reach out to southern students "where they were at" politically infuriated PL but represented a sharper analysis of what was possible in the South. PL may have had doctrinal purity and a rigorous analysis, but it did not have many members. SSOC did. And though it struggled to find ways to appeal to southern whites, it had more success than PL ever did.[34]

While the southern activists were not surprised that PL was pushing an anti-SSOC resolution, they were deeply troubled that SDS's governing body was set to pass judgment on SSOC. "We were totally threatened by the size and prestige of SDS," David Doggett recalls. "Most of us couldn't believe that SDS would care about SSOC." Some SSOC staff members suggested that SDS's views on SSOC were irrelevant since SSOC was and always had been independent. Their view, as Doggett characterized it, was "who cares about SDS? Let 'em pass what they want to. We'll go back home and do the same thing we've always done." But most staffers recognized that SDS's assessment of SSOC did matter, for they understood that nothing positive could result if the largest and most radical student organization in the country, a group with which SSOC had been connected since its founding, was to condemn the southern organization as unworthy of fraternal status.[35]

SSOC, therefore, quickly determined that it needed to fight the resolution, which had circulated among southern activists before the SDS National Council meeting. At an executive committee meeting in Atlanta in mid-March, two weeks prior to the Austin gathering, SSOC supporter Jim Gwin reported that the group resolved to "ask the Austin NC to accede to its organizational hegemony in the South." To prepare for the SDS meeting, the staff hurriedly drafted a series of position papers outlining the group's purpose and goals. These papers amounted to an impassioned defense of SSOC as it then existed, particularly its orientation toward students and its promotion of southern distinctiveness. Because students were the backbone of the movement for southern liberation, Lyn Wells explained in the paper she wrote, it was vital that SSOC work to radicalize them. Reflecting the class analysis that had begun to seep into the group's discussions, Wells made clear that SSOC was not organizing students in isolation but rather striving to forge ties between students and southern workers. A paper signed by 17 SSOC staffers that was based on Wells's essay stressed that "Our primary activity now is to organize students not in their narrow group interests but to [persuade them] to reject their class and race privileges and rejoin the militant Southern rebel working class." The papers also reiterated SSOC's view that the region's unique culture and history justified its emphasis on southern distinctiveness. Additionally, the activists argued that southern distinctiveness could inspire students to join the movement by highlighting the rebellious and revolutionary aspects of the southern past. "By reaffirming what is good in our cultural heritage," the multi-authored paper declared, "we hope to build a movement of students who will join a rebel army totally committed to the liberation of our people." The paper ended on a defiant note, insisting that SSOC was the best organization for working in the South and urging SDS not to break ties to the group. "If the NC condemns SSOC, consequences will follow which will split and hamper the Southern movement and impede the development of a national revolutionary movement. . . . Our existence is a fact; we are and have been the movement in the South. Our experiences are the experiences of the South."[36]

The position papers, however, did not form the basis of the activists' defense of SSOC at the SDS meeting. The night before the gathering, the ten SSOC staffers who had traveled to Austin held what David Doggett remembers was a chaotic meeting during which Lyn Wells and her close colleague David Simpson convinced the others that SSOC should try to appease their opponents by accepting the validity of some of their criticisms and by downplaying southern distinctiveness, the issue that most disturbed PL.[37] That two of SSOC's most outspoken members and ardent proponents of southern difference advocated that the group take a low-key, non-confrontational approach to the PL resolution was a consequence of a meeting between Wells and RYM's Mike Klonsky at the Alabama Conference on Radical Social Change in Tuscaloosa the week before Austin. After they gave speeches and debated one another before the assembled activists, they had a long discussion in which Klonsky convinced Wells of the legitimacy of the criticisms of SSOC articulated in "Build SDS in the South." Just as importantly, she agreed to support RYM in its fight with PL in exchange for Klonsky's pledge to oppose the PL resolution.[38]

Strategic considerations shaped both activists' decision. Klonsky's interest in SSOC and the South was contingent upon PL's interest in the group and the region. Although he subsequently revealed that he shared many of PL's criticisms of the group, he believed that shielding SSOC from PL's attack would prevent it from using SSOC or the South as a base from which to attack SDS. As he later remarked, the South " 'was too important to be left to Progressive Labor.' " Consequently, Klonsky traveled to southern campuses before Austin to build support for RYM among SSOC activists and to make clear that he and his allies opposed the PL resolution. From Wells's perspective, the decision to commit herself, and, by extension, SSOC, to supporting RYM was worthwhile if it made RYM SSOC's ally at the Austin meeting. She believed that RYM would be outspoken in its opposition to the PL resolution and protective of SSOC's right to exist. With RYM's backing, she believed SSOC could survive "Build SDS in the South."[39]

A more ominous result of the meeting for SSOC was that Wells had admitted that there were legitimate criticisms to be made of the group. Until the Alabama conference, Wells never had met Klonsky, and given that the Austin meeting was only a week away, she expected a tense, confrontational encounter with him. To her surprise, though, she found him to be a thoughtful and reasonable person with a number of provocative and appealing ideas. For example, she was intrigued by his argument that the South was not a nation but rather that blacks constituted a seperate nation in the South. While this challenged SSOC's southern exceptionalism as well as echoed the views of the communists of the 1930s, Wells later recalled that it "made more sense to me right then than 'the South as a nation' because, clearly, there are two cultures here . . . The South might have been a colony, but as a nation with a common culture, it didn't really stand up." Additionally, Klonsky's criticism that SSOC's lack of ideological development inhibited its ability to play a leading role in the revolutionary struggle resonated with Wells. And while Klonsky hinted that SSOC really was not needed, that a southern SDS should lead the movement in the former Confederacy, he conceded that the South was unique and, therefore, different tactics and strategies were required for organizing in the region. Having gone to Alabama expecting to tell Klonsky "to get the hell out" of the South, she left agreeing "to entertain the ideas[s]" he proposed and, more importantly for the short term, with assurances that RYM would oppose the PL resolution.[40]

Unfortunately for SSOC, the Wells-Klonsky understanding was unable to stem the anti-SSOC momentum building in SDS. The debate on SDS's ties to SSOC took place in Austin on March 30, when Phil Sanford of

Florida State SDS and Dick Reavis of Austin SDS joined PL's Ed Clark and Fred Lacey in bringing the PL resolution to the floor. Lacey led the attack, condemning SSOC for its liberalness and southern exceptionalism. Most National Council members responded positively to his presentation. Although some initially may have doubted Lacey's motives for attacking SSOC given his PL credentials, the fact that more than a dozen SDSers from Louisiana, Alabama, and Florida had signed their name to the resolution enhanced its credibility in these members' eyes. The resolution also benefited from most of the NC members' ignorance of SSOC. Because they knew little about the group, they readily accepted Lacey's harsh assessment of it. David Doggett was appalled to overhear SDSers from the north "educating" other northerners about SSOC when all they were doing was reiterating Lacey's attack. Though SSOC activists like Doggett felt that Lacey presented a distorted picture of the group, the burden was on them to disprove the charges.[41]

But SSOC did not help itself with its defense. Following through on the agreement with Klonsky, Wells and David Simpson made conciliatory remarks in which they acknowledged that SSOC had problems and accepted the validity of the criticisms of the group. They even denounced as racist SSOC's use of the Confederate battleflag, an astounding position for them to take since they had been leading defenders of the group's use of the flag. They closed their presentation by insisting that the group was righting itself and by moving that the NC postpone its decision on breaking ties with SSOC in order to give the group a chance to prove that it deserved SDS's support. Their comments neither pleased the SSOC representatives at the meeting who desired a forceful defense of the group—Doggett later derided their remarks as "wimpy apologetic speeches"—nor assuaged PL and its allies, who insisted that SSOC could not be redeemed and that postponing the decision was a stalling tactic designed to delay the inevitable. To make matters worse, two Communist Party (CP) members from New York followed Wells and Simpson to the floor to speak in SSOC's defense. Given SDS's animosity toward the Old Left, in general, and the CP, in particular, these were not people SSOC wanted speaking on its behalf. Yet somehow they managed to seize control of the floor, and they proceeded to commend SSOC for its southern distinctiveness and to argue that the group's liberalism was an effective tool for organizing in the South. Predictably, their remarks made it nearly impossible for SSOC to rally support at the meeting. In David Doggett's words, the CP's support of SSOC was "the kiss of death."[42]

Almost, but not quite. With a strong endorsement from Klonsky and RYM, SSOC would have had a fighting chance at the meeting. But Klonsky offered SSOC only tepid support. In fact, by the end of the meeting, Klonsky had turned against SSOC and embraced the PL resolution, a move that, more than anything else, ensured that SDS would sever its ties to SSOC. Initially, as he had pledged, Klonsky opposed the anti-SSOC resolution. Before the NC debated the resolution, Klonsky, accompanied by Wells, met with the southern SDS representatives who sponsored the resolution to urge them to give Wells, Simpson, and their allies more time to break with SSOC's liberal past. Specifically, he asked them to table their resolution until SDS's June meeting because the SSOC radicals intended to try to win control of the group at its upcoming spring conference. But the southern SDSers turned aside his request, arguing that SSOC's incipient "left wing" was not radical enough. With it becoming increasingly obvious that the PL resolution would sail through the NC, Klonsky withdrew his support for SSOC. By backing the resolution, Klonsky hoped to prevent PL from using his support of the group as a weapon against RYM. Shortly after Klonsky's move, the NC members voted overwhelmingly in favor of breaking fraternal bonds with SSOC.[43]

Many SSOC activists felt humiliated by SDS's renouncing of its relationship with the group, and they were outraged by the northern group's presumption that it knew what the South needed. These activists blamed the meeting's outcome less on PL for pushing the resolution than on what they considered Klonsky's dishonest, manipulative, and traitorous actions in Austin. Ironically, many in SSOC shared Ed Clark's view that Klonsky had revealed him to be a rank opportunist who, when it became clear that he could not defeat the resolution, "played the old operators game and decided now to 'oppose' SSOC." Though SSOC activists disparaged PL as extremists long bent on taking over SSOC, Klonsky was supposed to be their ally, someone who would rally his supporters against the resolution. That he did not infuriated them. Lyn Wells, who felt the sting of Klonsky's betrayal more keenly than others, later complained that Klonsky "Pissed me off really bad . . . [since] he left us out to dry." Tom Gardner expressed his displeasure with Klonsky in especially vituperative terms. "Your cop-baiting of SSOC is not very funny—nor are your pseudo-revolutionary sand-box 'politics,' " SSOC's former chairman wrote in "An Open Letter to Mike Klonsky (read 'God')," shortly after the meeting. "SDS . . . has reached a new low when their National Secretary has nothing better to do than attack one of the few organizations that represents . . . the real potential for building the Southern part of the revolutionary movement to which we aspire. Check out the enemy, Klonsky, he's not all that hard to find . . . perhaps we should direct our energies at him rather than each other."[44]

Some of the activists attributed SDS's condemnation of SSOC more to the northerners' condescending attitude toward the southern group than to Klonsky's personal deviousness. Staffer Alex Hurder expressed this view shortly after the meeting when he dismayingly wrote Gardner, "We were fucked over again, and I can't for the life of me figure out how it happened, except that we all have a tendency to respect people and take them at their word which makes us easily exploitable by any fast-talking cocksure Yankee. I don't suspect what happened in Austin could have been avoided. The Yankees had made up their minds and didn't consider it a Rebel's place to contradict them." By Hurder's reckoning, Klonsky was the most dangerous Yankee in SSOC's orbit, and he worried that some in SSOC still did not see him for the threat that he was. "Mike Klonsky is still traveling around the South, and I'm afraid there's a bunch of SSOC people that still haven't figured out that he's the tip of the phallus."[45]

If Klonsky had not earned SSOC's everlasting enmity for his actions in support of the resolution, he did for his whole-hearted embrace of the anti-SSOC position after the Austin meeting. Klonsky publicly articulated his opposition to SSOC in a lengthy, sharply worded article he penned for *New Left Notes*. The essay elaborated upon rather than just recycled the criticisms leveled against SSOC in Austin. Klonsky argued that SSOC had accepted money from foundations that were not merely tied to the Kennedy administration but that were "CIA conduits," not an easily dismissed charge in the wake of the 1967 scandal that exposed the CIA connections of the foundations that supported the nation's largest mainstream student organization, the National Student Association. He also bluntly reproached SSOC for its continued emphasis on students and its failure to recognize "that the working class is the primary agent of revolutionary change." Klonsky especially criticized SSOC's regional orientation and its emphasis on southern distinctiveness. He accused the group of offering an "ahistorical and racist" analysis of the South for wrongly asserting that the South was a northern colony and for inappropriately focusing on only white southerners. He agreed with SSOC that the South was different from the rest of the nation, but he argued that its differences were rooted in the fact that "the historical basis for a separate black nation lies in the South.... If there is any oppressed colony in the South,... then it is obviously a black colony." Thus, he argued, "If we try and organize white people as white people, or believe that there is such a thing as revolutionary white culture or white history we will find that no basis for revolutionary class unity exists."[46]

The article incensed SSOC's supporters, who believed his assessment of the group was based on inaccuracies, half-truths, and myths. They rejected the argument that SSOC's foundation supporters controlled the group's agenda and they vehemently denied his suggestion that SSOC had ties through these foundations to the CIA—a suggestion none-too-subtly made by the mocking diagram of two clasped hands, one labeled SSOC and the other CIA, superimposed over a dollar sign, which appeared at the top of his article. To attack SSOC for its foundation support, its backers believed, was unfair since they were the only organizations willing to make funds available to the group. Tom Gardner bluntly explained that "SSOC has never 'accepted' money from the power structure, we have actively sought it ... [since] I know of no place that capital originates in the U.S. except from the power structure." Gardner also dismissed the charge that SSOC was unacceptably liberal as a reflection of the sectarian struggle in SDS and not an accurate representation of SSOC's politics or ideology. In a sentiment shared by many SSOC supporters, David Nolan remembers concluding that the SDS leader simply "didn't know what he was talking about."[47]

Lyn Wells disagreed. Klonsky's criticisms struck her as serious, constructive, and not easily dismissed as a cynical ploy to bolster RYM's position in SDS. She had felt this way after her first meeting with Klonsky at the Alabama conference, and his betrayal of SSOC at the NC meeting did not alter her views. As she explained later, "I thought Klonsky's arguments were sound . . . even though he screwed us." But, as she was well aware, acceptance of the criticisms was incompatible with her continued support for SSOC. How could she justify allegiance to a group that she acknowledged to be a liberal tool of the ruling class and a group too narrowly focused on southern whites? In Austin, Wells had finessed this issue by insisting that SSOC was becoming more radical. In the weeks after the meeting, though, she rejected this view and began to argue that SSOC's problems were irreparable. Because she was the group's most respected and influential organizer, Wells's embrace of the SDS criticisms had devastating consequences. Once she joined the attack on SSOC and moved toward SDS, other staffers and members followed her, leading to the emergence of a RYM-aligned, pro-SDS faction within SSOC.[48]

In a series of working papers and articles, one of which, tellingly, appeared in *New Left Notes,* Wells and her closest allies in SSOC—David Simpson in Nashville, George Vlasits in North Carolina, and Jim Skillman in Atlanta—unsparingly detailed their newfound criticism of the group.[49] Echoing RYM, they argued that the primary task of the southern movement was to reach out to working-class youths, a mission SSOC had ignored in favor of working on campuses for "increased 'house slave' privileges (student power)," in Wells's words. They assailed SSOC as well for its lack of radical political analysis, contending that because single issues typically dominated its agenda, the group never was motivated to develop an all-encompassing radical analysis of American society. All that concerned SSOC, they charged, was how to achieve the immediate goals—desegregate the school, abolish in loco parentis policies, ban Dow recruiters from campus. As they put it in the *New Left Notes* essay, SSOC "made decisions about action on the basis of expediency, rather than on a long-range plan of direction."[50]

Like Klonsky, they attacked SSOC's emphasis on southern distinctiveness and cultivation of southern pride for undermining its ability to fight racism and to bring blacks and whites together in a common movement. Parroting the RYM leader, Simpson wrote that too often "we speak of southerners and southern culture when we meant white southerners and white southern culture." The results were mistakes such as the group's use of the battleflag. "The fact that black and white hands are superimposed upon the flag does not negate the fact that the 'rebel' flag remains the symbol of 300 years of murder and oppression for black people," Simpson averred. The dissidents also argued that by stressing regional difference SSOC falsely implied that the South's problems were unique. For activists who believed the working class held the key to the revolutionary future, this was SSOC's greatest sin and the primary reason for their disavowal of southern distinctiveness after having been staunch advocates of it. In a letter to SSOC members, Wells, Simpson, and Vlasits wrote that although "we had be[en] strong proponents [*sic*] of SSOC's 'southern consciousness' . . . [w]e began to realize how [it] had backfired, causing the development of regional chauvinism instead of class consciousness. . . . [I]n the South, as in the rest of the country, the problem we face is CAPITALISM. The basic struggle must also be the same, and that is for SOCIALISM." Because the South was more alike than different from the rest of the nation, they argued that SDS should replace SSOC as the movement organization in the South. While they acknowledged that in the South poverty was worse, workers were more exploited, and racism was more intense, they did not agree, Simpson explained, "that a separate white southern student organization is necessary for us to effectively address ourselves to these regional characteristics. In fact, we believe the . . . continued existence of SSOC encourages a liberal analysis of the conditions we face in the South and engenders the twin danger of splitting the movement along either regional or racial lines."[51]

Their attack on SSOC was passionate and broad-based, but it was neither an honest recounting of the group's history nor a blueprint for a southern movement of the future that could inspire much support. To castigate SSOC for its failure to make the working class or working-class youth the primary focus of its efforts was to ignore the White Folks Project of 1964, the textile organizing in North Carolina in 1967, and the many programs initiated by local SSOC groups through which the group sought to reach out to working-class whites or to bring students and workers together. They also were misguided to suggest that SSOC was a single-issue organization. This was a debatable but plausible characterization of SSOC prior to 1965, when the group focused almost exclusively on civil rights issues. By 1965, though, Vietnam, university reform, poverty, and labor organizing all had begun to fight for attention within SSOC. Finally, that these formerly ardent southern chauvinists now denounced southern distinctiveness not only angered and confused SSOC members but cost these activists much of their credibility in the group. Speaking of Wells and Simpson in particular, Bruce Smith remarked years later that "I really felt very much betrayed by the people who were condemning SSOC . . . because they were the main advocates of all that southern nationalism and then turned right around acting like it was a disease . . . and condemn[ing] everybody they had talked into it for being some kind of fools or devils." Why would anyone put faith in them or their views again? Were they not likely to renounce RYM and the working-class approach sometime in the future? While Wells and company were convinced that they were now on the right path, others were appalled that they had so readily switched their loyalties and, in effect, renounced their former selves.[52]

The emergence of an internal critique of SSOC that closely conformed to Klonsky's analysis of the group precipitated its factionalization. The defining moment was an explosive staff meeting on April 20, at the end of the group's conference on radical southern history in Atlanta. Wells, Simpson, and Vlasits, with the vocal backing of several prominent SDSers who were allowed to attend the meeting, articulated their criticisms of SSOC and urged the group to dissolve and for its members to join SDS. They were not just speaking for themselves, for an informal pro-SDS caucus rallied behind them at the meeting. Almost all of their backers were disaffected SSOC members from Florida, Georgia, Louisiana, and North Carolina, most of whom already had established a relationship with SDS. Their support for Wells, Simpson, and Vlasits gave added weight to their criticisms.[53]

Led by Mike Welch, the group's executive secretary, and staffer Randy Shannon, SSOC loyalists responded by defending SSOC's emphasis on southern distinctiveness, stressing the need for an indigenous southern movement group, and asserting SSOC's right to exist. The last two points, ironically, Wells's herself futilely had put forward at the Austin SDS meeting, indicating just how far she had moved in the month since that gathering. They particularly were irked that she had not consulted with anyone else on the staff regarding her evolving views about SSOC but simply had presented it to the staff as her conclusions about the group. As Mike Welch later complained, "if your concern is [the] movement and bringing other people in . . . you don't stake out your self-appointed high ground and declare that you've got the right position now. You discuss through, even if you think you've seen something nobody else's seen."[54] Her attack of SSOC, he believed, warranted her expulsion from the group. In his view, the last thing SSOC needed was someone from inside the group working to destroy it. "If your intentions are to go around and organize SDS in the South," he said to Wells at the meeting, "I think you should be fired." His motion sparked a heated debate, with Wells's detractors urging that she be dismissed from the group and her backers threatening to bolt the organization if the motion passed. Near the end of the meeting, Wells gave an impassioned speech in her defense that neatly traced her journey from teenaged civil rights activist to rebellious southern chauvinist to working-class revolutionary.

> I was elected to develop programs for SSOC. The legacy I was left with was from the New Left, and especially SNCC. I worked on the old

SNCC theory that an organizer is someone who sits in the back of the room, finds out what people want to do, and helps them do it. The do-your-own-thing philosophy. We've reached a lot of people. But it was wrong. I built on anti-Yankeeism. Even though foremost in my mind has been link-ups with the working class, development of a working class ideology in the student movement, I still used an individualist approach to organizing, and that's a bad position.... I believe in the revolution of working people, the liberation of black people, the liberation of people around the world. If you can explain how firing me deals with these questions, then do it.

After her appeal, the motion narrowly failed, 14–11. Nonetheless, shortly after the vote, Wells confirmed her estrangement from the group by resigning from the staff.[55]

The effort to oust Wells irreparably fractured the SSOC staff. It did not, though, divide neatly into pro- and anti-SDS factions. Although they generally blamed SDS for infecting SSOC with factionalism, a number of staffers intentionally avoided choosing sides within SSOC. In the view of Joe Bogle, who worked for SSOC in North Carolina, individuals like him "felt like the fight was against racism and the war in Vietnam and not against ourselves." Wanting nothing to do with ideological battles, they disengaged from active participation in SSOC or, like David Doggett, quit the group altogether. But the fact that activists who had been committed to the group did not rally to the side of the SSOC loyalists was commentary on their dissatisfaction with those SSOC activists. Some, including Doggett, refused to get involved because they believed that the anti-SDS faction had become a pro-Communist Party faction intent on transforming SSOC into a communist organization. Though it is unclear if this actually would have occurred if SSOC had survived, Mike Welch notes that he and others were moving toward the CP as SSOC began to disintegrate, eventually moving into the Party after the group's demise. Others' dissatisfaction with the anti-SDS faction stemmed from their disapproval of the way Welch and his allies ran the SSOC office. Although he vehemently opposed SDS's meddling in SSOC, Tom Gardner argued at the meeting against firing Wells. While his opposition to the move to purge her from the group partly was based on the fact that he recently had dated Wells and remained fond of her, it also resulted from his belief that Welch had created a culture within the SSOC office that was elitist and cliquish. His discomfort with Welch overshadowed his support for SSOC and kept him from actively supporting the anti-SDS faction, even though, he later reflected, firing Wells might have stabilized the group and ensured its survival.[56]

The non-aligned activists' ambivalence about the group's defenders benefited those who wished to see SSOC destroyed. After the motion to purge Wells from the group failed, and after some of the staff had left the meeting, the pro-SDS faction pushed through a motion declaring that the scheduled SSOC membership conference in June in Mississippi would not be restricted to SSOC members but would be open to anyone interested in the southern movement. Ostensibly, this was to ensure a free-ranging, thorough discussion of the future of the movement in the South. As the announcement of the meeting explained, the recent turmoil necessitated "a long informal meeting with other Movement people to discuss the state of the Movement in the South, its strategy, and its relation to other Movement groups around the country." But in reality, the decision to allow non-SSOC members to attend the conference ensured that SDS members from both the RYM and PL factions would be present and that they would join the anti-SSOC faction within the group in voicing the need for SSOC to dissolve.[57]

Mt. Beulah

The conference took place June 5–8 at the Mt. Beulah Conference Center near Edwards, Mississippi, a small town 25 miles west of Jackson. SSOC chose Mt. Beulah because of its significance to progressive activists in the state. Located on the grounds of a former black junior college, the conference center previously had served as the headquarters of the Mississippi Freedom Democratic Party, the central office for the National Council of Churches—sponsored Delta Ministry, and the base of operations for the Child Development Group of Mississippi, one of the leading Head Start programs in the country. Many of the people who converged on the conference center neither were affiliated with the group nor residents of the South. Representatives of the Southern Christian Leadership Conference, the Southern Conference Educational Fund—in the person of Carl Braden—SDS, PL, the DuBois Club, and the CP were among the approximately 100 people at the meeting. SDS was particularly well represented. PL had its supporters at the gathering, while RYM's Mike Klonsky and Mark Rudd, two of its most visible leaders, were in attendance. Determined to keep their factional adversary from winning control of the southern movement, RYM and PL supporters had come south to push for SSOC's dissolution and to capture the support of the organization's membership.[58]

Those who made the trip to Mississippi quickly discovered that the meeting was going to be a chaotic, poorly organized affair. One SSOC sympathizer noted with annoyance shortly afterward that "People as they arrived were confronted with a situation in which no one was in charge, no agenda or starting time had been established; in fact, no one knew exactly what was to take place or what was proposed, because there were no proposals." SSOC had intended to publish a special issue of the *Phoenix* prior to the meeting to lay out the various arguments and proposals regarding SSOC and the future of the southern movement, but printing problems prevented the group from ever distributing it. Further complicating matters was that several SSOC staffers from Nashville, including Mike Welch and Randy Shannon, experienced car problems that forced them to miss the whole first day of the meeting. Their delay ruined the staff's plan to resign en masse at the outset of the meeting as well as threw the anti-SDS faction into turmoil since the loyalists were unprepared to defend SSOC without its key leaders present. Thus, with several important staff members nowhere to be found and without a program or even general agenda to guide discussions, confusion reigned as the conference opened.[59]

Into this vacuum stepped SSOC's RYM-allied, pro-SDS forces. Led by Wells and Simpson and with the support of Klonsky, Rudd, and the other RYM activists at the meeting, they seized control of the proceedings for the first day-and-a-half. But they were unable to bring a sense of order to the conference. Despite their control of the agenda, the only substantive debate they initiated was over the "National Question," the SDSers' contention that the Black Belt constituted a separate colony for black people. This discussion did not make a favorable impression on SSOC's supporters, many of whom either were intimidated or alienated by its abstract nature. "I was mainly turned off by what I heard," Ed Hamlett later remarked. "Why?—Style, tone, self-righteousness, the lack of a sense of humor, language, their being impressed with what they saw as their own importance." Much of the rest of the time the pro-SDS people—that is, the RYM faction in SSOC and the RYM activists in attendance—met as a caucus to plot strategy, except for one open workshop they conducted to explain why they opposed SSOC's continued existence. But few of the conference participants stayed for the duration of the workshop, as the theoretical and sectarian discussion bored and confused them. Instead, most spent their time swimming in the center's pool or, observed CP member Mike Eisenscher, "just lounging about waiting for something to happen."[60]

Welch and the others from the SSOC office finally arrived on the second evening of the meeting, and the next afternoon, Saturday, June 7, the attendees at last took up the issue that had drawn them to Mt. Beulah—the future of the southern movement and the fate of SSOC. Anticipating that the pro-SDS faction would push for SSOC's dismantling, the group's defenders floated a hastily crafted proposal that would allow SSOC to continue, albeit in drastically altered form. Composed primarily by Jim Bains, the group's Alabama traveler, and members of the Nashville office, the proposal built upon the recently aired criticisms of the group in an effort to show that SSOC was remaking itself. Reflecting the concerns that SSOC did not devote sufficient attention to the working class, the proposal called on SSOC to turn completely away from students and campuses and toward the white community. Workers and non-student whites would be the exclusive focus of the group's work, a move that would be symbolized by the dropping of "Southern" from the revamped organization's name; SSOC would become SOC—the Southern Organizing Committee. The proposal also addressed the complaint that SSOC was politically adrift by vowing, in Mike Eisenscher's words, to "giv[e] it politics, that is ending it as a loose confederation of [local] groups with no set line." But the proposal failed to placate the group's detractors, who argued that SSOC was beyond salvation and that disbanding it was the only sensible course of action.[61]

The criticisms of SSOC fueled the debate at this crucial meeting, which lasted well into the night and then reconvened early the following morning. A position paper drafted by the pro-SDS faction became the focal point of discussions. Entitled "Resolution: To Dissolve SSOC" and distributed near midnight, the seven-page paper articulated the criticisms of SSOC that had accumulated since the Austin SDS meeting. The first half of the paper was devoted to an exposition of the theoretical beliefs that underlay their criticisms of SSOC. With the exception of the CP members at the meeting, who found this discussion crude and unsophisticated, there was broad agreement among the assembled activists with the Marxist and anti-imperialist thrust of this section of the paper. Regardless of one's position on SSOC, everyone agreed that capitalism was an exploitative, brutally oppressive system and concurred with the sentiment that future work required "building a movement around an anti-imperialist, anti-racist, working-class consciousness."[62]

No such agreement existed over the critique of SSOC that comprised the second part of the document. The resolution accused the group of being too liberal, attacked it for an inability "to relate successfully to the Black" struggle, and castigated it for its reliance on foundation funding, its regional orientation, and its failure to develop a class analysis. SSOC, it argued, shared the liberal's worries about economic "stratification and social mobility" when the fundamental issue was the "relationship to the means of production" in society. Without a class analysis, SSOC "failed to develop beyond liberal politics based around [the] perception of separate, single issues;" that is, the absence of a radical class analysis prevented SSOC from recognizing the interconnectedness of civil rights, university reform, the war, and the numerous other causes it promoted. The critique also excoriated the group for its cultivation of liberal funding sources, though it did not repeat the never-proven contention that SSOC had accepted CIA-tainted funds. Instead, it argued that SSOC's reliance on foundation monies prevented it from forging stronger ties with its membership since the rank-and-file was not essential to the group's funding. Worse, it charged that foundation support had discouraged SSOC from developing a radical analysis since to do so likely would offend its financial patrons.[63]

The core of this section of the critique consisted of a detailed discussion of SSOC's regionalism, southern pride, and emphasis on southern distinctiveness. Primary among the complaints was that SSOC's regional orientation served to divide the movement. "A regionally separate organization of white students in the South," the critique declared, "is an anachronism left over from an earlier state in our own development as radicals. Perpetuation of SSOC would encourage a localism which hinders political development of the movement in the South as a functioning and organic part of the national movement." While it would have made sense for SSOC to take up white organizing in response to SNCC's decision to work exclusively among African Americans, the critique contended that SSOC shied away from this work because of "The white Southern activists' . . . feeling that the 'redneck' was the racist enemy." Thus rather than focusing on the white communities of the South and the concerns of its residents, SSOC instead turned its attention to issues that were not peculiar to the South, such as university reform and the Vietnam War. "To justify its existence as a regional organization while participating in these national activities," the resolution argued, "SSOC had to develop a rhetoric which emphasized the uniqueness of the South and Southern consciousness." According to SSOC's opponents, this inhibited the development of a radical movement in the South since it "mislead [sic] Southern white students to believe that there existed a logical analysis of their society and its liberation separate from that of the U.S. as a whole. (The implication would be that Northerners, not capitalists, are the enemy.)" As a result, SSOC became an increasingly insular organization, unable to recognize that its work was part of a larger struggle. "With no national perspective to give a coherent unity to programs . . . the concept of the South as a special problem reinforced an incipient tendency toward localism, already present in SSOC's liberal politics."[64]

SSOC's proponents found much to dislike about the critique. That northern SDSers were among those criticizing SSOC particularly was disturbing. Many SSOC backers believed that these non-southerners and non-SSOC members had no right to judge the group, and they thus dismissed the entire meeting as an undemocratic farce, a situation that "was so unrepresentative that it wasn't even worth getting uptight about," Mike Eisenscher dryly noted. Most SSOC supporters, though, took the debate quite seriously. Ed Hamlett assailed SSOC's critics for being hypocritical since they were attacking the group for " 'mistakes' it had made that national SDS had been 'guilty' of in the past and that local SDS groups had been 'guilty' of in the past few months—liberalism and accepting tainted money." Everett Long, a SSOC activist in Mississippi, believed that the detractors misleadingly had framed the critique around trivial or outdated statements that it claimed to be the group's official positions. As he later lamented, "it was unfair to accept something that somebody wrote a year or even six months previously as indicative of where the organization stood."[65] SSOC supporters also took exception to the resolution's disparagement of the group's work. For instance, while

SSOC believed its support of local activist groups was one of its most important functions, the resolution chided the organization for devoting so much time to " 'servicing' local groups, rather than developing political direction." Cutting remarks such as these deeply angered SSOC's supporters. At one point in the meeting, Ann Johnson, a SSOC activist from North Carolina, became so incensed with the cold, condescending, and unfeeling attitude of many of the pro-SDS people that she approached Mark Rudd, removed her shirt, and challenged him to do the same to prove he was human.[66]

Rudd kept his shirt on. And the pro-SDS faction brushed aside her criticism and continued to argue relentlessly for SSOC's dissolution. Hamlett commented that the pro-SDS activists "were generally more single-minded and took themselves more seriously than I have ever seen them." Outnumbered by those determined to destroy SSOC, the group's advocates faced an uphill battle to save the organization. The likelihood of SSOC's survival diminished with each passing hour of the debate, as the pro-SDS faction wore down many of the attendees, winning some undecided people to its position and driving others away from the meeting altogether. Hamlett, Sue Thrasher, and Gene Guerrero, all of whom were SSOC founders, attended the meeting with the vague hope that their presence would bolster the group's defense. But they quickly deemed the situation hopeless and chose to remain silent and in the background during the debate. As Thrasher sadly recalled, "we just felt totally impotent. . . . [E]arly on it became clear that it was an absolute madness, that the context of the thing was something else entirely and that there was no way to affect it."[67]

Although the pro-SDS faction held the upper hand throughout the discussion, SSOC's supporters believed they possibly could prevent the group's dissolution by restricting the vote on the issue to SSOC members. They argued that only the membership could authorize the group to disband. But others responded that SSOC's fate was of significance to people who were not part of the organization but who had an active interest in the southern movement. Ultimately, the two sides agreed that anyone who plausibly could claim to be working actively in the South had a right to vote. While the handful of southern PL and SDS members at the meeting claimed this right, an uncertain number of the approximately 25 northern activists in attendance also insisted on voting on SSOC's future.[68] Although SSOC's backers were infuriated that these northerners intended to cast votes, this quickly became a non-issue when it became clear that most of the SSOC members present planned to vote the group out of existence. Recognizing that SSOC would not survive the vote, the CP members in attendance led a desperate attempt to push through the SOC proposal, hoping that at least a reconstituted SSOC could be salvaged. Their effort failed. Some time before noon on Sunday, June 8, the activists who had gathered at Mt. Beulah voted to disband SSOC.

Specifically, they gave their approval to a seven-step implementation plan outlined in "Resolution: To Dissolve SSOC," the first step of which was "That this membership convention of the Southern Student Organizing Committee, Inc., officially declares the Southern Student Organizing Committee to be abolished." The other measures called for the liquidation of its assets, the shuttering of the Nashville office, the ending of its fundraising work, and the election of a five-member committee to oversee the dissolution process. Five years and two months after its creation amid great hopes for the future, a divided and dispirited SSOC ceased to exist.[69]

The End of SSOC

In many respects, the Mt. Beulah meeting was anti-climatic. Given the turmoil that had racked the group in the weeks leading up to the conference, it is not surprising that SSOC expired that weekend in Mississippi. Both SSOC's defenders and critics were in agreement on this point. The pro-SDS faction, of course, had condemned the organization and worked to destroy it. In its view, the student group needed to be replaced, David Simpson later explained, by "a more working-class organization that really had a chance to change the country." While SSOC's supporters were not happy to see the group demolished, neither were most of them interested in working to save it in its current form. Referring to the anti-SDS faction, Mike Eisenscher emphasized "that no one was for maintaining SSOC as it had been, all were for changes to get away from staff domination, lack of politics, etc." The SOC proposal was meant as an antidote for these problems. Yet many of SSOC's advocates were unenthusiastic about this proposal. Rather than working hard to help the group re-focus itself, Mike Welch remembers that he and his peers were content simply to let SSOC die, so long as the group did not endorse SDS as the appropriate organization to take its place in the South. "Most of us had kind of run our course as students, and we were also looking towards other things we wanted to do," he recalls. "And there was nobody who seemed prepared to try to fight for and maintain a SSOC, but we weren't prepared to support the idea that this stuff coming as the general SDS position was correct."[70]

Welch's remarks raise the issue of why so few people interested in saving SSOC, in any form, turned out for the conference. Somewhere between 60 and 80 SSOC members attended the meeting, and most of them sided with SDS and the SSOC faction that wanted to destroy the group. Conspicuous by their absence were some of the group's staffers and most of its rank-and-file members. Where were these people? Why did they not rush to Mt. Beulah to save the organization? SSOC's leadership was partially responsible for their absence. Though Mike Eisenscher exaggerated when he complained that "mobilization of the rank and file was nil," it was true that SSOC's leaders did a poor job of publicizing the meeting among the membership. The factional strife had made it difficult for the Nashville office to operate normally, and the printing problem that derailed plans for a special issue of the *Phoenix* detailing the conflict left the conference announcement as the only communication between the staff and the membership about the meeting. And despite Welch's and the other SSOC leaders' fear that the group's opponents were going to try to force its dissolution at Mt. Beulah, the announcement did not convey a sense of urgency about the meeting or SSOC's future.[71]

But even if the leadership had worked harder to recruit supporters to the meeting, it is unlikely that they would have persuaded many of them to journey to Mississippi to defend SSOC. The vast majority of the group's members disliked the kind of long, drawn-out, theoretical, and sectarian debates that they knew would characterize the conference. Such discussions simply were of no interest to most members, young people who had joined SSOC to work for progressive causes, not to hammer out abstract position papers or to discover the "correct line" on black nationalism. Ed Hamlett believed that it was probably just as well that more of the membership did not show up for the meeting. "I reckon that had he been there," Hamlett wrote of the average SSOC member, "he would have been turned off by what he saw and heard."[72]

The busy SSOC members at Furman University were a case in point. Preoccupied with trying to stir activism on campus and in nearby Greenville, they had no time or inclination to participate in policy debates, theoretical arguments, or political discussions within the larger organization. The little they knew about the growing factional rift in SSOC puzzled them. "These disputes were rather obscure from where we were at Furman," Steve Compton remembers. "We didn't really grasp all of the stuff going on at the national level." Or care about it. Speaking of their work at Furman, the group's founder, Jack Sullivan, recalls, "We were so preoccupied with these immediate issues—they were *real* issues—that we really didn't care . . . about those kinds of disputes." Thus, it was not surprising that no one from the chapter went to Mt. Beulah. Nor was it striking the chapter greeted news of SSOC's dissolution with a shrug. The Furman students did not believe that the sectarian warfare, most of which was incomprehensible to them, had anything to do with their chapter. And so even after learning that SSOC had voted itself out of existence, they retained the SSOC name for their group and continued to work as they always had. Under Compton's leadership, the Furman SSOC chapter persisted into the spring of 1970, the last of the organization's chapters to survive.[73]

For SSOC members and staffers whose commitment to SSOC ran deeper, the decision not to go to Mt. Beulah turned on their feeling that the meeting was likely to result in SSOC's demise. With the organization badly divided and with both SDS and PL intent on destroying it, they believed that SSOC's death was inevitable; the conference simply would make it official. "I didn't think anything good was going to come from [the conference]," David Doggett remembers. "It really wasn't even worth going to the meeting; it was a foregone thing.... SSOC was going to bite the dust." Bruce Smith, who had been a leading SSOC activist in Virginia, first in Lynchburg and then in Richmond, recollects that though his long involvement in SSOC made him "feel like I should have gone," he ultimately decided not to because "it sort of seemed like you were going to your own funeral." By the time of the Mt. Beulah gathering—ten weeks after SDS severed relations with SSOC, seven weeks after the anti-SDS faction's attempt to fire Wells—SSOC's supporters concluded that their struggle was in vain. Countless hours of debate and discussion made clear to them that SSOC, as it existed, could not survive. With people like Smith and Doggett certain that they could not affect the proceedings, with other SSOC members uninterested in the meeting altogether, and with those who were in attendance either determined to destroy SSOC or disinclined to push very hard for its survival, SSOC's fate was sealed before the first session at Mt. Beulah even had begun. Yet though few people fought hard to keep the group alive, many expressed anger at the outcome of the meeting. Some, like Ed Hamlett, directly blamed SDS. SSOC, he argued, "was sacrificed on the alter... of some faction or other of SDS.... I thought we were the sacrificial lamb." Others more indirectly insinuated that SDS was responsible for SSOC's collapse. Tom Gardner, who much later characterized his decision not to go to Mt. Beulah as "one of the biggest regrets of my entire life," believes that he could have gone "toe-to-toe with Klonsky or Rudd on arguing about Marxism-Leninism." Though the discussions at the meeting were less about economic theory and more about "what SSOC was and wasn't and what it did and didn't do," in Mike Welch's words, Gardner's statement reveals his sense that had he been there to stand-up to SDS, SSOC would have lived to see another day.[74]

SDS rightly deserved blame for helping SSOC into the grave. SSOC very well might have survived beyond June 1969 had it not been caught in the crossfire of the factional battle raging within SDS. Nonetheless, it is easy to overestimate the role SDS played in the group's dissolution. There was no inherent reason why SDS's descent into factionalism should have affected SSOC so adversely. Nor was it predetermined that once SDS severed its ties to the southern group that SSOC would splinter and move to dissolve itself. That it did so was a consequence of SSOC's own internal weaknesses. Ongoing disputes over the group's goals, strategy, and constituency hampered its ability to craft a coherent program of action, created insurmountable rifts among the leaders, and alienated many rank-and-file members. Further, the problems within SSOC itself made it vulnerable to attack; more than that, they invited outside interference by suggesting that SSOC would be an easy conquest. While SDS set in motion the disintegration process by cutting ties to the group and strongly urging it to dissolve, SSOC's internal problems, several years in the making, deprived it of the support and unity necessary to fend off the SDS attack, guaranteeing that the SDS assault would culminate in its collapse.

SSOC's intra-organizational difficulties are not hard to discover. In fact, the competing factions of SDS and the pro-SDS faction in SSOC identified many of the group's problems in their various critiques between March and June 1969. However, these problems were troublesome for different reasons from those suggested by the factions. Southern distinctiveness was not, as the group's critics charged, a deficient or bankrupt idea. A heightened awareness of regional differences, as Sue Thrasher and David Nolan argued in the wake of the break-up, had allowed SSOC activists to "come to grips with their own personal identity as southerners while still constituting a legitimate part of the national movement." But a belief that the South was distinctive caused turmoil in SSOC because some activists thought the group had exaggerated its significance and allowed it to shape its image. They worried that SSOC's regional orientation and its use of Confederate rhetoric and symbolism created the impression among progressive whites that the group was not interested in civil rights issues. As a white student organization, these activists believed the group had the obligation to make clear its support for black equality, and they feared that southern distinctiveness prevented it from doing so.[75]

SSOC's active pursuit of foundation funds also had long-term costs for the group. While the charge that liberal philanthropic organizations' financial support of SSOC gave them influence over SSOC's agenda was baseless, SSOC's reliance on outside funding meant that the staff did not, of necessity, have to cultivate especially close ties with the members. The membership dues, which were only a few dollars and not always collected, and the income from the group's literature program, accounted for just a small portion of the group's budget. For the fiscal year 1967, for example, dues and publications sales comprised less than four percent of the group's overall receipts. However, pro-SDS activists failed to recognize that though SSOC's dependence on external funding sources damaged staff-member relations, this relationship already had begun to founder over more significant issues. By 1969, SSOC's factionalization, more than anything else, had heightened local activists' sense of distance from the staff. To campus-based activists, the staff seemed to have turned inward. From their perspective, the leaders appeared to be absorbed in unintelligible and irrelevant theoretical disputes among themselves and with other organizations. It was sadly ironic that Lyn Wells was at the center of the group's factional split. Rather than working in the field to attract students to the group and encouraging them to found chapters, she became a leading partisan in the group's factional fight and focused much of her attention on the struggle for control of the organization. Additionally, some charged that SSOC's acceptance of foundation funds had made the staff dismissive of the membership because these funds, and not the members' support, were the basis of the staff's power and authority. Tom Gardner later mused that the flow of foundation money into the group enabled the staff to "set SSOC up as a little mini-foundation. So that you could have Virginia SSOC, Mississippi SSOC, West Virginia SSOC, Appalachia SSOC—all these little mini-SSOCs—doing their own thing, getting money from the national SSOC—foundation money being channeled through to the regional offices." For rank-and-file SSOC members, the central office was important not as a source of inspiration, ideas, or emotional support but as a source of money.[76]

SSOC's indifference toward ideological and political matters also created vexing internal problems for the organization. The SDS factions pointedly had criticized the group for its low level of ideological development and for its failure to develop—or to develop quickly enough—a radical analysis of the South and America. But these analytical failings did not, as the factions insisted, constitute damning evidence of SSOC's inability to work effectively in the South. Nor was SSOC completely bereft of a philosophical orientation. Throughout its life, a clear and consistent vision undergirded SSOC's activism: the creation of a more humane, peaceful, equitable, and integrated society. The problem for SSOC was that it was not easy to articulate this vision in the face of the SDS challenge. Unlike the factions of SDS, whose rigid ideologies were readily explained, SSOC's vision of a better society was more of a worldview or a way of life that drew on the activists' lived experiences in the South and their discovery of and identification with southern dissenters of earlier generations. While such a vision consistently animated SSOC's work, it was no match for the SDS factions' ideologies.

Over the years, some SSOC members had worried that the group's vision of a better South was insufficient for sustaining a movement for social change, and they questioned whether the absence of a well-formed, radical critique of society would create difficulties for the group. The absence of this type of critique, the activists believed, could be traced back to the circumstances of the group's founding, namely, that the moral power of the civil rights movement, not the discovery of radical political or social theories, had inspired the activism of SSOC's founders. "Southern students do not come to be radicals through an intellectual process," Thrasher and Nolan wrote in the *Great Speckled Bird* shortly after SSOC collapsed. "They become radicals because at some point in their personal history they have had to come face to face with the reality of the myth of their existence—in the early 60's it came in the civil rights movement." Howard Romaine lamented that the "naiveté of most of the students we work with leaves us in an intellectual vacuum so to speak. This means we have no shared radical political analysis of America." As SSOC moved beyond civil rights to take up issues of peace, university reform, and labor organizing, the absence of such an analysis increasingly troubled some in the group. "Do we have an overall political vision and strategy?" wondered Gene Guerrero in early 1966. "Should we have one?" More pointedly, Jim Williams, who had been involved with SSOC since its founding, warned in 1967 that for the group to continue without "an independent ideology and analysis . . . is simply to offer ourselves up on a silver platter to all comers. The present format says outright, we have nothing to offer—come take us away."[77]

Williams's assessment was unduly harsh given SSOC's success in reaching out to progressive and liberal-leaning whites across the region without a clear ideological focus. Moreover, most SSOC members did not believe that developing a sharper analytical and ideological focus was a priority. Ed Hamlett speculated this was because SSOC activists felt inferior to and intimidated by their peers in SDS. Contrasting the membership of the two groups, he suggested that "SSOC people generally saw themselves as activists, administrators, catalysts, servicers, 'Paul Reveeres [sic],' . . . but rarely as ideologues or intellectuals. Yankee SDSers, on the other hand, were thought of as theorists, analysts, intellectuals . . . though a bit lacking in soul, roots, and at times patronizing and paternalistic." SSOC activists' sense of difference grew out of an awareness that their lives and experiences in the South did not make it easy for them to turn to abstract issues. Howard Romaine observed that "We weren't part of a historical tradition of critical left thinking in the way of [Tom] Hayden and those SDS people." Many had grown up in culture-bound communities in which their progressive sentiments isolated them from their friends and their families. They did not have role models in dissent. And while as activists they sought out the relatively few older white dissidents in the region—the people who became SSOC elders—usually there was no one in their youths to nourish their unorthodox views.[78]

Additionally, SSOC members could justify their disdain for analytical discussions by invoking the needs of the movement in the South. They deemed it vastly more important to try to win white support for desegregation, build opposition to the war, and induce southern schools to abolish in loco parentis regulations than to write theoretical papers on the connection between students and the working class or hold meetings on the future of the southern movement. By insisting that they were too busy to craft an analytical framework for SSOC's actions, the activists were sending the northerners the message that, as Tom Gardner later put it, "you got the luxury sitting around arguing about ideology because you're not confronted with what we're confronted with in the South." Or, as Howard Romaine suggested in 1969, the challenge of organizing in the South "left little time for the luxury of massive fights over minute (relatively speaking) theoretical and tactical differences." SSOC defenders would have agreed whole-heartedly with SDS sympathizer Bob Goodman, who, in an article on the Mt. Beulah meeting in the *Great Speckled Bird*, said SSOC's supporters "would rather be active and numerous than theoretically correct, and incline toward the view that anything (any movement, any action) is better than nothing." What Goodman intended as a criticism SSOC's supporters considered one of the group's virtues. As Howard Romaine later noted, what counted was "not what you believe but what you do."[79]

Although most in SSOC agreed that deeds mattered more than words, action more than analysis, some had become convinced by 1969 that the group was too unfocused in its work. Their efforts to address this issue, however, served to deepen SSOC's problems. Attempts to develop theoretical underpinnings for the group's actions accelerated SSOC's factionalization and all but guaranteed its demise because there was no agreement over what these underpinnings should be. Lyn Wells and her supporters found analytical clarity in the pronouncements of the RYM faction of SDS, and they so completely identified with its goal of building a revolutionary movement of workers and students that they pushed for SSOC to dissolve. Others did not share Wells's enthusiasm for RYM and instead insisted that SSOC develop its own analytical approach. And they refused to develop this approach more quickly to satisfy the group's critics. As Gardner angrily counseled the group in April 1969, "We shouldn't try to whip together the scanty knowledge we have at this point into some instant analysis just because the grand tribunal of revolution, SDS, has told us to turn in our term paper by the end of the month or we just won't make the grade."[80]

By June, though, most of the group's backers simply believed that it was not worth the energy to try to re-develop SSOC. Regardless of whether SSOC lived or died in the spring of 1969, they believed the organization was more a part of their past than their future. They were ready to move on to new challenges and new issues, and they had no interest in fighting to save an organization in which they were feeling less and less invested with each passing day. Many believed that the organization had outlived its usefulness and doubted the need for SSOC to continue. By 1969, Sue Thrasher wrote in retrospect, "history had overtaken us." While the group had succeeded in its initial goal of "provid[ing] an avenue for white southern students to enter the Freedom Movement," by decade's end the movement "had progressed from being a southern movement to end segregation to a national movement that also opposed the war in Vietnam." "It may have been that SSOC had done what it was supposed to do," she later concluded. "Maybe it should have turned into the southern wing of SDS . . . since it was anti-war and since it was more broadly political. . . . I mean, there was no particular reason for a southern student organizing committee."[81]

And so after June 8, 1969, SSOC ceased to exist, a victim of both SDS's factional struggle and its own internal problems. After the break-up, the membership scattered in many different directions, with some joining other activist groups, some restricting their activism to their localities, and some withdrawing from activist causes altogether. SDS, with its abstractions and theories, did not fair any better than SSOC. Not only did it did not replace SSOC in the South, but a mere two weeks after Mt. Beulah it imploded, ripped apart by the competing factions. Indeed, that no organization ever succeeded SSOC highlighted the group's accomplishments. For more than five years, SSOC was the primary organizational expression of white student activism in the South. Although it never rivaled SDS or SNCC in prestige or size, it was the only organization to take seriously the challenge to organize for progressive causes among southern whites. Its efforts were not always successful. But it did leave its mark on the predominantly white campuses of the region: it popularized the antiwar movement among southern students, raised their awareness of the struggles of working-class whites, convinced many of the virtues of the civil rights movement, and, perhaps most significantly, was the first voice of protest at some schools, thereby legitimizing dissent on these campuses. The presence of a SSOC chapter or even just contact with SSOC travelers encouraged white students to consider alternatives to the status quo. On a personal level, this knowledge had the potential to change lives, for, as Anne Romaine explained, it encouraged people "to be who they wanted to be and relate in a more progressive way politically in the South." Whether the subject was civil rights, Vietnam, rural poverty, university reform, or union rights, Romaine spoke for many of the activists when she remarked that SSOC gave people "courage to think about these issues and to view them in different ways than their families had." By bringing young white southerners together in an organization of their own making, SSOC gave them the confidence to speak out for social change and to demand that society live up to their ideals. For those whose lives SSOC had touched, the sense that change was always possible, and that one did not have to accept things as they were, persisted long after the group disappeared.[82]

Appendix
"We'll Take Our Stand"
Nashville, April 4, 1964

It has been 35 years since a group of young intellectuals calling themselves the Southern "Fugitive Group" met here in Nashville and declared their hopes of stopping the clock and preventing social, spiritual and economic forces which are today still coming of age in the South. They wrote a statement called "I'll Take My Stand" in which they endorsed the feudal agrarian aristocratic order of the South and opposed what they saw coming in the new order—widespread industrialization and urbanization with democracy and equality for all people.

> We do hereby declare, as Southern students from most of the Southern states, representing different economic, ethnic and religious backgrounds, growing from birthdays in the Depression years and the War years, that we will here <u>take</u> <u>our</u> <u>stand</u> in the determination to build together a New South which brings democracy and justice for all its people.

Just a few years after the Fugitives took their stand, Franklin D. Roosevelt assumed the presidency of the U.S. and called our Southland "America's number one domestic problem." He talked about the needs of those Americans who were ill-housed, ill-fed, ill-clothed. Today, in 1964, when a majority of our nation is living in affluence which makes the specters of poverty and racism tenfold more inexcusable, a Southern is in the White House. Yet the struggle for equal opportunity for all men—white and non-white, young and old, man and woman, is by no means completed. Our Southland is still the leading sufferer and battleground of the war against racism, poverty, injustice and autocracy. It is our intention to win that struggle in our Southland in our lifetime—tomorrow is not soon enough.

> Our Southland is coming of age, they say. But we both hope and fear for her new industries and her new cities, for we also are aware of new slums, newly unemployed, new injustices, new political guile—and the Old as well. Is our dream of democracy to be dashed just as were Jefferson's dreams and the Populist struggles lost in the blend of feudal

power, racial fear and industrial oligarchic opportunism? Only as we dare to create new movements, new politics, and new institutions can our hopes prevail.

We hereby take <u>our</u> stand to start with our college communities and to confront them and their surrounding communities and to move from here out through all the states of the South—and to tell the Truth that must ultimately make us free. The Freedom movement for an <u>end</u> to segregation inspires us all to make our voices heard for a <u>beginning</u> of a true democracy in the South for all people. We pledge together to work in all communities across the South to create non-violent political and direct action movements dedicated to the sort of social change throughout the South and nation which is necessary to achieve our stated goals.

Our region must be an <u>exemplar</u> of the national goals we all believe in, rather than a deterrent to them:

— Not only an end to segregation and racism but the rise of full and equal opportunity for all;
— An end to personal poverty and deprivation;
— An end to the "public poverty" which leaves us without decent schools, parks, medical care, housing, and communities;
— A democratic society where politics poses meaningful dialogue and choices about issues that affect men's lives, not manipulation by vested elites;
— A place where industries and large cities can blend into farms and natural rural splendor to provide meaningful work and leisure opportunities for all—the sort of society we can all live in and believe in.

We as young Southerners hereby pledge to take <u>our</u> stand now together here to work for a new order, a new South, a place which embodies our ideals for all the world to emulate, not ridicule. We find our destiny as individuals in the South in our hopes and our work together as brothers.

Notes

Introduction

1. *Southern Patriot*, December 1964.
2. Interview, Sue Thrasher (interviewed by Ronald Grele), 15 December 1984, Columbia Oral History Project, Columbia University, New York, New York.
3. "The Role of Culture in the Movement," Southern Student Organizing Committee Thirtieth Reunion and Conference, 9 April 1994, Charlottesville, Virginia, in the author's possession.
4. Will D. Campbell to Deborah Cole, 30 September 1964, box 52, folder 20, Papers of Will D. Campbell, McCain Library and Archives, University Libraries, The University of Southern Mississippi, Hattiesburg, Mississippi.
5. Interview, Ronda Stilley Kotelchuck, 4 September 1994.
6. "1968–1969—Zenith & Dissolution," SSOC Reunion, 6 July 2002, Nashville, Tennessee, in the author's possession.
7. Jennifer Frost, *"An Interracial Movement of the Poor": Community Organizing and the New Left in the 1960s* (New York: New York University Press, 2001), 5.
8. *The Progressive*, 61:7 (1997), 40.
9. Kirkpatrick Sale, *SDS* (New York: Random House, 1973), 537; Clayborne Carson, *In Struggle: SNCC and the Black Awakening of the 1960s* (Cambridge, Mass.: Harvard University Press, 1981), 102; Alice Echols, *Daring to Be Bad: Radical Feminism in America, 1967–1975* (Minneapolis, Minn.: University of Minnesota Press, 1989), 316 n.52; William A. Link, *William Friday: Power, Purpose, and American Higher Education* (Chapel Hill: University of North Carolina Press, 1995), 142–143.
10. John Shelton Reed, "Flagging Energy," in John Shelton Reed, *Kicking Back: Further Dispatches from the South* (Columbia: University of Missouri Press, 1995), 48. The perceived novelty of SSOC has brought it to the attention of others, though they are more judicious in their remarks than Reed. See, for instance, Charles Reagan Wilson, "Creating New Symbols from Old Conflicts," *Reckon: The Magazine of Southern Culture* 1, no. 1 & 2 (1995): 17, 20.
11. Sara Evans, *Personal Politics: The Roots of Women's Liberation in the Civil Rights Movement and the New Left* (New York: Alfred A. Knopf, 1979); Bryant Simon, "Southern Student Organizing Committee: A New Rebel

Yell in Dixie" (Honors Essay: University of North Carolina at Chapel Hill, 1983).

12. Harlon Joye, "Dixie's New Left," *Trans-action,* 7, no. 11 (1970), 50–56, 62; Christian Greene, " 'We'll Take Our Stand': Race, Class, and Gender in the Southern Student Organizing Committee, 1964–1969," in *Hidden Histories of Women in the New South,* Virginia Bernhard, Betty Brandon, Elizabeth Fox-Genovese, Theda Perdue, and Elizabeth Hayes Turner, eds. (Columbia: University of Missouri Press, 1994), 173–203.

13. For examples of this scholarship, see Terry H. Anderson, *The Movement and the Sixties* (New York: Oxford University Press, 1995); James Miller, *"Democracy Is in the Streets": From Port Huron to the Siege of Chicago* (Cambridge, Mass.: Harvard University Press, 1994 [first edition: New York: Simon and Schuster, 1987]); Kenneth J. Heineman, *Campus Wars: The Peace Movement at American State Universities in the Vietnam Era* (New York: New York University Press, 1993); Nancy Zaroulis and Gerald Sullivan, *Who Spoke Up?: American Protest against the War in Vietnam, 1963–1975* (New York: Holt, Rinehart and Winston, 1984); Charles DeBenedetti and Charles Chatfield, assisting author, *An American Ordeal: The Antiwar Movement of the Vietnam Era* (Syracuse, N.Y.: Syracuse University Press, 1990); Tom Wells, *The War Within: America's Battle Over Vietnam* (Berkeley: University of California Press, 1994); Blanche Linden-Ward and Carol Hurd Green, *American Women in the 1960s: Changing the Future* (New York: Twayne Publishers, 1993).

14. Works that ignore white student civil rights activists in the South include Taylor Branch, *Pillar of Fire: America in the King Years, 1963–1965* (New York: Simon and Schuster, 1998); John Dittmer, *Local People: The Struggle for Civil Rights in Mississippi* (Urbana: University of Illinois Press, 1994); Fred Powledge, *Free At Last?: The Civil Rights Movement and the People Who Made It* (New York: Little, Brown and Company, 1991); Doug McAdam, *Freedom Summer* (New York: Oxford University Press, 1988); Howell Raines, *My Soul Is Rested: Movement Days in the Deep South Remembered* (New York: Penguin Books, 1983 [first edition: New York: G. P. Putnam and Sons, 1977]). In his excellent bibliographic essay at the end of his important book on the civil rights movement in Mississippi, Charles Payne is critical of top-down, white-focused civil rights history. Yet his essay ignores the contributions to the movement of grass-roots white activists. Charles Payne, *I've Got the Light of Freedom: The Organizing Tradition and the Mississippi Freedom Struggle* (Berkeley: University of California Press, 1995), 413–441.

15. The number of works that focus on young progressive whites in the South has begun to grow in recent years. Among these works are Jeffrey A. Turner, "From the Sit-Ins to Vietnam: The Evolution of Student Activism on Southern College Campuses, 1960–1970," *History of Higher Education Annual,* 21 (2001): 103–135; Constance Curry et al., *Deep in Our Hearts: Nine White Women in the Freedom Movement* (Athens: University of Georgia Press, 2000); Stephen Eugene Parr, "The Forgotten Radicals: The New Left in the Deep South, Florida State University, 1960–1972" (Ph.D. Dissertation: The Florida State

University, 2000); William J. Billingsley, *Communists on Campus: Race, Politics and the Public University in Sixties North Carolina* (Athens: University of Georgia Press, 1999); Douglas C. Rossinow, *The Politics of Authenticity: Liberalism, Christianity, and the New Left in America* (New York: Columbia University Press, 1998).

16. On the challenges and opportunities posed by interviewing social and political activists, see Kim Lacy Rogers, "Memory, Struggle, and Power: On Interviewing Political Activists," *Oral History Review*, 15 (Spring 1987): 165–184; Kim Lacy Rogers, "Oral History and the History of the Civil Rights Movement," *Journal of American History*, 75, no. 2 (1988): 567–576; Bret Eynon, "Community in Motion: The Free Speech Movement, Civil Rights, and the Roots of the New Left," *Oral History Review* 17, no. 1 (1989): 39–69; Bret Eynon, "Cast Upon The Shore: Oral History and New Scholarship on the Movements of the 1960s," *Journal of American History*, 83, no. 2 (1996): 560–570.

17. For particularly insightful discussions of the method and practice of oral history, see Peter Friedlander, *The Emergence of a UAW Local, 1936–1939: A Study in Class and Culture* (Pittsburgh: University of Pittsburgh Press, 1975), esp. xi–xxxiii; Paul Thompson, *The Voice of the Past* (New York: Oxford University Press, 1978); Charles W. Joyner, "Oral History as Communicative Event: A Folkloristic Perspective," *Oral History Review*, 7 (1979): 47–52; Luisa Passerini, "Work Ideology and Consensus under Italian Fascism," *History Workshop*, 8 (1979): 84–92; Samuel Schrager, "What Is Social in Oral History?" *International Journal of Oral History*, 4, no. 2 (1983): 76–98; Ronald J. Grele, "Movement without Aim: Methodological and Theoretical Problems in Oral History," in *Envelopes of Sound: Six Practitioners Discuss the Method, Theory and Practice of Oral History and Oral Testimony*, Ronald J. Grele, ed. (Chicago: Precedent Publishing, 1985), 127–154; John Murphy, "The Voice of Memory: History, Autobiography, and Oral History," *Historical Studies* 22 (1986): 157–175; Jean Peneff, "Myth in Life Stories," *The Myths We Live By*, Raphael Samuel and Paul Thompson, eds. (London: Routledge, 1990), 36–48; Michael Frisch, *A Shared Authority: Essays on the Craft and Meaning of Oral and Public History* (Albany: State University of New York Press, 1990); Alessandro Portelli, *The Death of Luigi Trastulli and Other Stories: Form and Meaning in Oral History* (Albany: State University of New York Press, 1991); Valerie Raleigh Yow, *Recording Oral History: A Practical Guide for Social Scientists* (Thousand Oaks, Calif.: Sage Publications, 1994); Alessandro Portelli, *The Battle of Valle Giulia: Oral History and the Art of Dialogue* (Madison: University of Wisconsin Press, 1997).

Chapter 1

1. John Robert Zellner, "Report on White Southern student project (School year of 1961–62)," folder 5, Sam Shirah Papers, Wisconsin Historical Society, Madison, Wisconsin.

2. Sam Shirah, "A Proposal for Expanded Work Among Southern White Students and an Appalachian Project," folder 5, Shirah Papers.
3. Howard Zinn, *SNCC: The New Abolitionists* (Boston, Mass.: Beacon Press, 1964), 174–180. On William Moore and his Freedom Walk, see Murray Kempton, "Pilgrimage to Jackson," *New Republic*, 148:19 (11 May 1963), 14–16; *Time*, 81:19 (10 May 1963), 18–19; Mary Stanton, *Freedom Walk: Mississippi or Bust* (Jackson: University Press of Mississippi, 2003), 3–90; Taylor Branch, *Parting the Waters: American in the King Years, 1954–1963* (New York: Simon and Schuster, 1988), 748–751.
4. *Newsweek*, 61:19 (13 May 1963), 28–29; *New York Times*, 2 May 1963; *Student Voice*, 5:18 (29 July 1964); *Woodstock (New York) Times*, 17 January 1980; Anne Braden, untitled essay, 1964, folder 4, box 62, Carl and Anne Braden Papers, 1928–1972, Wisconsin Historical Society, Madison, Wisconsin; Zinn, *SNCC*, 174–180; Pat Watters, *Down to Now: Reflections on the Southern Civil Rights Movement* (New York: Pantheon Books, 1971), 243–260; James Forman, *The Making of Black Revolutionaries*, second edition (Seattle, Wa.: Open Hand, 1985 [first edition: New York: Macmillan, 1972]), 308–310; Branch, *Parting the Waters*, 754; Charles Moore (text by Michael S. Durham), *Powerful Days: The Civil Rights Movement Photography of Charles Moore* (New York: Stewart, Tabori & Chang, 1991), 71–87; Stanton, *Freedom Walk*, 119–127; Diane McWhorter, *Carry Me Home: Birmingham, Alabama: The Climactic Battle of the Civil Rights Revolution* (New York: Simon and Schuster, 2001), 370, 373.
5. Shirah quoted in Anne Braden, untitled essay, 1964, folder 4, box 62, Braden Papers.
6. On SCEF and the Bradens, see Catherine Fosl, *Subversive Southerner: Anne Braden and the Struggle for Racial Justice in the Cold War South* (New York: Palgrave Macmillan, 2002); John Egerton, *Speak Now against the Day: The Generation before the Civil Rights Movement in the South* (Chapel Hill: University of North Carolina Press, 1994); David L. Chappell, *Inside Agitators: White Southerners in the Civil Rights Movement* (Baltimore, Md.: The Johns Hopkins University Press, 1994), 41–47; Linda Reed, *Simple Decency & Common Sense: The Southern Conference Movement, 1938–1963* (Bloomington: Indiana University Press, 1991), esp. 129–184; Irwin Klibaner, *Conscience of a Troubled South: The Southern Conference Educational Fund, 1946–1966* (Brooklyn, N.Y.: Carlson Publishing, 1989 [originally published 1971]); Aldon Morris, *The Origins of the Civil Rights Movement: Black Communities Organizing for Change* (New York: The Free Press, 1984), 166–173; Anthony P. Dunbar, *Against the Grain: Southern Radicals and Prophets, 1929–1959* (Charlottesville: University Press of Virginia, 1981), 187–219; Clayborne Carson, *In Struggle: SNCC and the Black Awakening of the 1960s* (Cambridge, Mass.: Harvard University Press, 1981), 51–52; Anne Braden, *The Wall Between* (New York: Monthly Review Press, 1958).
7. Chappell, *Inside Agitators*, 46–47; interview, Anne Braden, 25 April 1993.

8. *Southern Patriot,* March 1961 and May 1962; Interview, Braden. On the movement at the University of Texas, see Douglas C. Rossinow, *The Politics of Authenticity: Liberalism, Christianity, and the New Left in America* (New York: Columbia University Press, 1998); Martin Kuhlman, "Direct Action at the University of Texas During the Civil Rights Movement, 1960–1965," *Southwestern Historical Quarterly,* 98:4 (1995), 550–566; and Doug Rossinow, " 'The Break-through to New Life': Christianity and the Emergence of the New Left in Austin, Texas, 1956–1964," *American Quarterly* 46, no. 3 (September 1994): 309–340.
9. Anne Braden to Ella Baker, 22 November 1960; and Anne Braden to Marion S. Barry, Jr., 22 November 1960, folder 3, box 62, Braden Papers; Braden quoted in Klibaner, *Conscience of a Troubled South,* 186; Carson, *In Struggle,* 52–53.
10. Southern Conference Educational Fund, Inc., "Proposal for a Student Project," 18 July 1961, box 62, folder 3, Braden Papers (quote); Klibaner, *Conscience of a Troubled South,* 184–187.
11. Anne Braden to Ella Baker, 22 November 1960; Anne Braden to Sandra Cason, 15 November 1960; Anne Braden to Rebecca M. Owen, 28 March 1961; Anne Braden to James Dombrowski, 8 May 1961, folder 3, box 62, Braden Papers (quote); Fosl, *Subversive Southerner,* 254–255; Klibaner, *Conscience of a Troubled South,* 185–187; Carson, *In Struggle,* 25; Zinn, *SNCC,* 35. Sandra Cason, better known as Casey Hayden, tells the story of her involvement in the civil rights movement in Constance Curry et al., *Deep in Our Hearts: Nine White Women in the Freedom Movement* (Athens: University of Georgia Press, 2000).
12. Interview, Ed Hamlett, 6 May 1993.
13. "Agents of Change: Southern White Students in the Civil Rights Movement," A Public Seminar, 7 April 1994, Charlottesville, Virginia (quote); Fosl, *Subversive Southerner,* 275–277; Zinn, *SNCC,* 168–170 in the author's possession.
14. Ibid.
15. John Robert Zellner, "Report on White Southern student project (School year of 1961–62)," folder 5, Shirah Papers (quote); "Agents of Change," A Public Seminar, 7 April 1994; Carson, *In Struggle,* 45–55; Zinn, *SNCC,* 170–171. The demonstration in McComb had occurred to protest the shooting death of Herbert Lee, who was involved in SNCC's voter registration campaign in Amite County, by Mississippi state representative E. H. Hurst, as well as to protest the decision of the principal of the black high school not to readmit two students, Brenda Travis and Ike Lewis, who had been in arrested in an earlier demonstration in McComb. On McComb, see John Dittmer, *Local People: The Struggle for Civil Rights in Mississippi* (Urbana: University of Illinois Press, 1994), 99–115; Charles M. Payne, *I've Got the Light of Freedom: The Organizing Tradition and the Mississippi Freedom Struggle* (Berkeley: University of California Press, 1995), 111–131; Henry Hampton and Steve Fayer, with Sarah Flynn, *Voices of Freedom: An Oral History of the Civil Rights Movement from the 1950s through the 1980s* (New York: Bantam Books, 1990), 139–147; Carson, *In Struggle,* 45–55.

16. Hugh Davis Graham and Nancy Diamond, *The Rise of American Research Universities: Elites and Challengers in the Postwar Era* (Baltimore, Md.: The Johns Hopkins University Press, 1997), 57. On federal funding of higher education during the post–World War II years, see Graham and Diamond, *The Rise of American Research Universities,* esp. 26–83; James Ridgeway, *The Closed Corporation: American Universities in Crisis* (New York: Random House, 1968), 4–10, 125–150, 223–239; Richard M. Freeland, *Academia's Golden Age: Universities in Massachusetts, 1945–1970* (New York: Oxford University Press, 1992), 91; Alice M. Rivlin, *The Role of the Federal Government in Financing Higher Education* (Washington, D.C.: Brookings Institution, 1961); Chester E. Finn, *Scholars, Dollars, and Bureaucrats* (Washington, D.C.: Brookings Institution, 1978); and Clarence L. Mohr, "World War II and the Transformation of Southern Higher Education," in *Remaking Dixie: The Impact of World War II on the American South,* Neil R. McMillen, ed. (Jackson: University Press of Mississippi, 1997), 47–49.
17. Kenneth J. Heineman, *Campus Wars: The Peace Movement at American State Universities in the Vietnam Era* (New York: New York University Press, 1993), 19; Terry H. Anderson, *The Movement and the Sixties* (New York: Oxford University Press, 1995), 89–90, 95–96; Freeland, *Academia's Golden Age,* 87–88, 93–97; Jan G. Owen, "Shannon's University: A History of the University of Virginia, 1959–1974" (unpublished Ph.D. Dissertation: Columbia University, 1993), 337; Seymour E. Harris, *A Statistical Portrait of Higher Education* (New York: McGraw-Hill, 1972), 276–277, 284–287. By 1970, nearly 2,000,000 students attended college in the South, a 700 percent increase over the number of matriculated students in the region on the eve of America's entry into World War II. Mohr, "World War II and the Transformation of Southern Higher Education," 47.
18. Helen Lefkowitz Horowitz, *Campus Life: Undergraduate Cultures from the End of the Eighteenth Century to the Present* (New York: Alfred A. Knopf, 1987), 168–172; David D. Henry, *Challenges Past, Challenges Present: An Analysis of American Higher Education since 1930* (San Francisco: Jossey-Bass, 1975), 92–98; Todd Gitlin, *The Sixties: Years of Hope, Days of Rage* (New York: Bantam Books, 1987), 11–30; James Miller, *"Democracy Is in the Streets": From Port Huron to the Siege of Chicago,* 2nd edition (Cambridge, Mass.: Harvard University Press, 1994 [first edition: New York: Simon and Schuster, 1987]), 21–61; W. J. Rorabaugh, "Challenging Authority, Seeking Community, and Empowerment in the New Left, Black Power, and Feminism," *Journal of Policy History* 8, no. 1 (1996): 110–113. Their activism took many forms. In 1957, students at the University of California at Berkeley created SLATE, a student political party that called for the elimination of compulsory R.O.T.C. and that in 1959 nearly elected its candidate president of the student body. At the University of Michigan, several students who later were among the founders of the Students for a Democratic Society, including Al Haber, Sharon Jeffrey, Bob Ross, and Tom Hayden,

were among the students who formed the Political Issues Club, established VOICE, a student political party, and helped to elect Haber to the Student Government Council in 1958. At Harvard in 1959, students organized a chapter of Student SANE, the anti-nuclear group, and that same year about 100 Chicago-area students founded the Student Peace Union, a group that would become a fast-growing national organization in the early 1960s. Heineman, *Campus Wars,* 115–116; Gitlin, *The Sixties,* 86–87; Maurice Isserman, *If I Had a Hammer . . . : The Death of the Old Left and the Birth of the New Left* (New York: Basic Books, 1987), 194–195, 204; W. J. Rorabaugh, *Berkeley at War: The 1960s* (New York: Oxford University Press, 1989), 14–16; Miller, *"Democracy Is in the Streets,"* 24–61.

19. Interview, Dorothy Dawson Burlage, 4 September 1994; interview, Robb Burlage, 4 September 1994; *The Austin American,* 5, 8, 9, 10, and 11 May 1957; *Time,* 69:20 (20 May 1957), 50; Lewis L. Gould and Melissa R. Sneed, "Without Pride or Apology: The University of Texas at Austin, Racial Integration, and the Barbara Smith Case, *Southwestern Historical Quarterly* 103, no. 1 (1999): 67–87; Curry et al., *Deep in Our Hearts,* 97–98.

20. Interview, Dorothy Dawson Burlage; Curry et al., *Deep in Our Hearts,* 96, 338–340; Sara Evans, *Personal Politics: The Roots of Women's Liberation in the Civil Rights Movement & the New Left* (New York: Alfred A. Knopf, 1979), 33; Rossinow, " 'The Break-through to New Life,' " 313–326; Doug Rossinow, "Secular and Christian Liberalisms and Student Activism in Early Cold War Texas," delivered at the American Historical Association One Hundred Eighth Annual Meeting, Chicago, Ill., 5–8 January 1995.

21. Horowitz, *Campus Life,* 172; Willie Morris, *North Toward Home* (Boston, Mass.: Houghton Mifflin, 1967), 188–191, 256–257 (quote); *The Austin American,* 24 March 1959; *Daily Texan,* 6 and 17 February, 24 March, and 22 May 1959; interview, Robb Burlage, 22 April 1993 and 4 September 1994. Burlage lost his editorship as a consequence of his being placed on disciplinary probation for accumulating an excessive number of parking violations on campus. Burlage and others, however, believed that his crusading editorials had inspired the administration to seek a "back door" way to silence him.

22. During the 1961–62 school year, Zellner visited 15 predominantly white colleges and universities, including Birmingham Southern College, the University of North Carolina, and Emory, Vanderbilt, and Tulane universities. John Robert Zellner, "Report on White Southern student project (School year of 1961–62)," 19 May 1962, folder 5, Shirah Papers; Bob Zellner to SNCC, "Progress Report," 5 November 1961, folder 3, Shirah Papers; "Agents of Change," A Public Seminar, 7 April 1994; interview, Braden.

23. John Robert Zellner, "Report on White Southern student project (School year of 1961–62)," 19 May 1962, folder 5, Shirah Papers; "Agents of Change," A Public Seminar, 7 April 1994.

24. *Southern Patriot,* January 1964. Zellner was not the only white activist to face the dilemma of trying to organize whites. In 1962 and 1963, as part of a special YWCA human-relations project, Sandra Cason (Casey Hayden) and Mary King traveled the South to talk to students about racial issues. On campuses throughout the region, they encountered hostile, indifferent, and scared students and faculty who had little interest in debating the merits of desegregation. Mary King, *Freedom Song: A Personal Story of the 1960s Civil Rights Movement* (New York: Quill, 1987), 36–37, 59–66.
25. John Robert Zellner, "Report on White Southern student project (School year of 1961–62)," 19 May 1962, folder 5, Shirah Papers.
26. Interview, Charles McDew, 24 April 1993; interview, Sue Thrasher, 1 September 1994.
27. Interview, John Lewis, 30 March 1995; *Southern Patriot,* June 1963; Anne Braden, untitled essay, 1964, folder 4, box 62, Braden Papers (quote). During his stay in Danville, Shirah was arrested and roughed up by local police officials. An unknown third person complained to the FBI about the mistreatment of Shirah, and the bureau responded by interviewing Shirah and briefly investigating the case. This episode is detailed in Shirah's FBI file, which Jules Tygiel generously has shared with the author. Federal Bureau of Investigation Files and Records, Samuel Curtis Shirah, Jr., file, no. 44-22674, in the author's possession.
28. *Southern Patriot,* February 1964; Klibaner, *Conscience of a Troubled South,* 210; Anne Braden, untitled essay, 1964, folder 4, box 62, Braden Papers.
29. Carson, *In Struggle,* 90–95; King, *Freedom Song,* 79–119; Fred Powledge, *Free At Last?: The Civil Rights Movement and the People Who Made It* (New York: Little, Brown and Company, 1991), 496–542; Harvard Sitkoff, *The Struggle for Black Equality, 1954–1980* (New York: Hill and Wang, 1981), 127–145, 155–166; Howell Raines, *My Soul Is Rested: Movement Days in the Deep South Remembered* (New York: Penguin Books, 1983 [first edition: G. P. Putnam and Sons, 1977]), 139–185; Hampton and Fayer, *Voices of Freedom,* 123–138, 159–176; Branch, *Parting the Waters,* 708–711, 725–731, 734–802, 846–892; David J. Garrow, *Bearing the Cross: Martin Luther King, Jr., and the Southern Christian Leadership Conference* (New York: Vintage Books, 1988 [first edition: New York: William Morrow and Company, 1986]), 231–286.
30. *Southern Patriot,* May 1964; *New Rebel,* 1:1 (27 May 1964), 3–4; Bryant Simon, "Southern Student Organizing Committee: A New Rebel Yell in Dixie" (Honors Essay: University of North Carolina at Chapel Hill, 1983), 5; Klibaner, *Conscience of a Troubled South,* 209–210; Fosl, *Subversive Southerner,* 279–280; "Minutes of the Meeting of the SNCC Executive Committee," 27–31 December 1963, A:II:4, reel 3, *Student Nonviolent Coordinating Committee Papers, 1959–1972* (Sanford, N.C.: Microfilm Corporation of America, 1982) (quote); Sam Shirah, "A Proposal For Expanded Work Among Southern White Students and an Appalachian Project," folder 5, Shirah Papers.

31. Sam Shirah, "A Proposal For Expanded Work Among Southern White Students and an Appalachian Project," folder 5, Sam Shirah, "Field Reports," Fall 1963, folder 6, "SNCC Worker Ordered Off Alabama Campus," News Release, 16 September 1963, folder 2, Todd Gitlin to Sam Shirah, 4 October 1963, folder 3, "Charges Against SNCC Worker Dropped," News Release, 5 October 1963, folder 2, all in Shirah Papers; Sam Shirah, "Campus Visit Report (University of South Carolina)," A:IV:332, reel 9, SNCC Papers; *Gamecock*, 18 October 1963; *Student Voice*, 4:4 (11 November 1963).
32. Sam Shirah, "Campus Visit Report (Florida State University)" and Sam Shirah, "Campus Visit Report (Florida A&M University)," A:IV:332, reel 9, SNCC Papers; *Tallahassee Democrat*, 12 November 1963; *St. Petersburg Times*, 13 November 1963.
33. "Minutes of the Meeting of the SNCC Executive Committee," 27–31 December 1963, A:II:4, reel 3, SNCC Papers; interview, Hamlett.
34. "Minutes of the Meeting of the SNCC Executive Committee," 27–31 December 1963, A:II:4, reel 3, SNCC Papers; Ed Hamlett to Sam Shirah, 28 February 1964, folder 3, Shirah Papers; interview, Hamlett; interview, Ed Hamlett and Howard Romaine, 5 April 1990 and ca. 23–24 January 1992, in the author's possession; Sam Shirah, "Campus Visit Report (Huntingdon College)," and Sam Shirah, "Campus Visit Report (University of Southern Mississippi)," A:IV:332, reel 9, SNCC Papers; "Re: Report to Sam Shirah for the week of March 1–7, 1964 from Ed Hamlett," 6 March 1964 and "Re: Field Report—Ed Hamlett, Campus traveler for the White Southern Student project of SNCC," 31 March 1964, Edwin Hamlett Papers (uncatalogued), Special Collections and University Archives, The Jean and Alexander Heard Library, Vanderbilt University, Nashville, Tennessee (also in box 2, David and Ronda Kotelchuck Papers (Southern Student Organizing Committee Papers, 1959–1977) and Steve Wise Papers (Southern Student Organizing Committee Additional Papers, 1963–1979 *and* Southern Student Organizing Committee Papers, 1959–1977), Special Collections, University of Virginia Library, Charlottesville, Virginia); Sam Shirah, "Report to SCEF and SNCC, In Re: White Southern Student Project (School Year 1963–1964)," 15 May 1964, folder 6, Shirah Papers.
35. Sam Shirah, "Report to SCEF and SNCC, In Re: White Southern Student Project (School Year 1963–1964)," 15 May 1964, folder 6 and Sam Shirah, "A Proposal For Expanded Work Among Southern White Students and an Appalachian Project," folder 5, Shirah Papers. Ed Hamlett also began to advocate that SNCC focus more on organizing white students. Ed Hamlett, "Prospectus for Campus Oriented Program," n.d. (ca. 1964), Appendix A:611, reel 70, SNCC Papers. Initially, Shirah believed the "natural place" for organizing poor whites was in Appalachia, partly because he already had done some work there and partly because it was likely SDS would help fund the project given its interest in the area (SDS previously had focused attention on miners in

Hazard, Kentucky, as part of its Economic and Research Action Project). Sam Shirah, "A Proposal For Expanded Work Among Southern White Students and an Appalachian Project," folder 5, Shirah Papers; interview, Nelson Blackstock, 8 August 1995. On ERAP, see Kirkpatrick Sale, *SDS* (New York: Random House, 1973), 95–150; and Miller, *"Democracy Is in the Streets,"* esp. 184–217.

36. On the Nashville movement generally, see *Southern Patriot,* April 1962; Zinn, *SNCC,* 19–23; Powledge, *Free At Last?,* 203–210; Raines, *My Soul Is Rested,* 98–100; Morris, *The Origins of the Civil Rights Movement,* 174–178, 205–213; Hampton and Fayer, *Voices of Freedom,* 53–61, 65–67; Branch, *Parting the Waters,* 260–264, 278–280, 295, 345, 379–380; and Forman, *The Making of Black Revolutionaries,* 145–157 (quote from 146). As a result of his involvement in the sit-ins, James Lawson was expelled from the Vanderbilt Divinity School, precipitating a serious crisis at the university and bringing the school unwanted national publicity. On the crisis see Paul Conkin, assisted by Henry Lee Swint and Patricia S. Miletich, *Gone with the Ivy: A Biography of Vanderbilt University* (Knoxville: University of Tennessee Press, 1985), 547–571.
37. *Southern Patriot,* April 1962 (quote) and January 1963; Powledge, *Free At Last?,* 449–451; interview, David Kotelchuck, 31 August 1994. On the Fayette County project, see Forman, *The Making of Black Revolutionaries,* 116–145; Richard A. Couto, *Lifting the Veil: A Political History of Struggles for Emancipation* (Knoxville: University of Tennessee Press, 1993); Robert Hamburger, *Our Portion of Hell. Fayette County, Tennessee: An Oral History of the Struggle for Civil Rights* (New York: Links Books, 1973); Fayette County Project Volunteers, *Step by Step: Evolution and Operation of the Cornell Students' Civil-Rights Project in Tennessee, Summer, 1964* (New York: W. W. Norton & Company, 1965).
38. Interview, David Kotelchuck; Danny Lyon, *Memories of the Southern Civil Rights Movement* (Chapel Hill: University of North Carolina Press, 1992), 49–51, 188; Conkin, *Gone with the Ivy,* 574–575 (quote); *Nashville Tennessean,* 9 December 1962 and *Vanderbilt Hustler,* 4 and 11 January 1963, folder 1, Kotelchuck Papers; Branscomb quoted in Conkin, *Gone with the Ivy,* 575. On the Nashville demonstrations prior to the Kotelchuck incident, see *Student Voice,* 3:4 (December 1962).
39. For instance, in March 1963 approximately 75 black and white students took part in a freedom march through downtown to protest the continued segregation of numerous commercial establishments. *Nashville Banner,* 23 March 1963, folder 1, Kotelchuck Papers.
40. Scarritt's official name is Scarritt College for Christian Workers. Conkin, *Gone with the Ivy,* 294–295; J. Anthony Lukas, *Don't Shoot We Are Your Children!* (New York: Random House, 1971 [first edition: New York: Random House, 1968]), 140 (quote); Alice G. Knotts, "Methodist Women Integrate Schools and Housing," in *Women in the Civil Rights Movement: Trailblazers and Torchbearers, 1941–1965* (Brooklyn, N.Y.: Carlson Publishing, 1990), 253.

41. *Southern Patriot,* May 1963; interview, Sue Thrasher (interviewed by Ronald Grele), 15 December 1984, Columbia Oral History Project, Columbia University, New York, New York; interview, Thrasher, 1 September 1994 (quote); Lukas, *Don't Shoot We Are Your Children!,* 143–145; Curry et al., *Deep in Our Hearts,* 226; Evans, *Personal Politics,* 28–35.
42. *Southern Patriot,* May 1963; interview, Thrasher, 15 December 1984; interview, Thrasher, 1 September 1994 (quote); Lukas, *Don't Shoot We Are Your Children!,* 143–145; Curry et al., *Deep in Our Hearts,* 220–222.
43. Ron K. Parker, "The Southern Student Organizing Committee," 1, Hamlett Papers (also in box 2, Kotelchuck Papers and Wise Papers); interview, Thrasher, 1 September 1994.
44. Parker quoted in "Agents of Change," A Public Seminar, 7 April 1994; interview, Chuck Myers, 11 December 2002; Parker, "The Southern Student Organizing Committee," 1–2, Hamlett Papers; *Southern Patriot,* February 1964; Ron K. Parker, informational letter on Campus Grill to interested students, 14 October 1963, Gregory T. Armstrong, "Dear Colleague," 7 November 1963, "Please Do Not Patronize The CAMPUS GRILL" flyer, November 1963, "Please Continue To Withhold Patronage From The CAMPUS GRILL" flyer, November 1963, and *Vanderbilt Hustler,* 1, 8, and 15 November 1963, all in folder 2, Kotelchuck Papers.
45. Parker, "The Southern Student Organizing Committee," 2, Hamlett Papers; interview, Myers, 11 December 2002; interview, Thrasher, 1 September 1994; interview, Kotelchuck, 31 August 1994.
46. *Southern Patriot,* May 1964; interview, Thrasher interview, 15 December 1984; interview, Thrasher, 1 September 1994; Hamlett, "Questionnaire, "Southern Student Organizing Committee Thirtieth Reunion and Conference, April 1994, Charlottesville, Virginia, in the author's possession; "Burlage Response to Gitlin Weekend Memo of 10 Feb," 12 February 1964, folder 5, box 2 (quote) and Robb and Dorothy Burlage to Todd Gitlin, Lee Webb, Rennie Davis, et al., 31 January 1964, folder 2, box 5 (quote), Robb Burlage Papers, Wisconsin Historical Society, Madison, Wisconsin. In response to Burlage's entreaties, Gitlin, SDS's president, wrote Shirah to suggest that if a white student organization is created in the South, it should work closely with SDS since SDS could provide this new entity with literature, help it to make contacts, and perhaps even be able to help fund the group. Todd Gitlin to Sam Shirah, 4 and 25 February 1964, folder 3, Shirah Papers.
47. Parker, "The Southern Student Organizing Committee," 3, Hamlett Papers; "Re: Field Report—Ed Hamlett, Campus traveler for the White Southern Student project of SNCC," 31 March 1964, Hamlett Papers; interview, Thrasher, 1 September 1994; interview, Thrasher, 15 December 1984; Lukas, *Don't Shoot We Are Your Children!,* 153–154; David Kotelchuck, "Meeting Notes," February 1964, folder 4,

Kotelchuck Papers (quote); "Notes on the First SSOC Conference," Hamlett Papers.

48. Interview, Thrasher, 1 September 1994; Parker, "The Southern Student Organizing Committee," 3, Hamlett Papers; Klibaner, *Conscience of a Troubled South*, 212–213; interview, Anne Braden, 25 April 1993 (quote); Jim Williams, "Tentative Agenda for Atlanta Meeting, February 22 [1964], "Proposed Agenda Check List," ca. February 1964, and "Minutes for Evening Session," 21 February 1964 (quote), folder 2, Shirah Papers; Robb and Dorothy Burlage to Todd Gitlin, Lee Webb, Rennie Davis, et. al., 31 January 1964 (quote), folder 5, box 2, Robb Burlage Papers.

Chapter 2

1. *Southern Patriot*, April 1964.
2. Interview, Sue Thrasher, 1 September 1994.
3. Interview, Anne Braden, 25 April 1993; interview, Thrasher, 1 September 1994; Ron K. Parker, "The Southern Student Organizing Committee," 3, Edwin Hamlett Papers (uncatalogued), Special Collections and University Archives, The Jean and Alexander Heard Library, Vanderbilt University, Nashville, Tennessee (also in box 2, David and Ronda Kotelchuck Papers (Southern Student Organizing Committee Papers, 1959–1977) and Steve Wise Papers (Southern Student Organizing Committee Additional Papers, 1963–1979 *and* Southern Student Organizing Committee Papers, 1959–1977), Special Collections, University of Virginia Library, Charlottesville, Virginia); "Executive [Committee] Minutes," 19 April 1964, A:III:1, reel 3, *The Student Nonviolent Coordinating Committee Papers, 1959–1972* (Sanford, N.C.: Microfilm Corporation of America, 1982); Constance Curry to Interested Students, 11 March 1964, Constance Curry Papers (uncatalogued), Special Collections, University of Virginia Library, Charlottesville, Virginia (quote).
4. "Mailing List of Students Attending the Nashville Conference, SSOC" and "Participants at Nashville Conference," Curry Papers; interview, Nelson Blackstock, 8 August 1995; interview, Bruce Smith, 2 April 1995; interview, Braden; interview, Thrasher, 1 September 1994. Also in attendance were students from Western Kentucky University in Bowling Green, the University of Texas at Austin, Emory University, Georgia Tech University, Clemson University, the University of Louisville, and Martin College in Pulaski, Tennessee.
5. *Southern Patriot*, April 1964; "Notes on the First SSOC Conference," Hamlett Papers.
6. Like SSOC, the Students for a Democratic Society was hampered throughout its existence by a failure to provide definite answers to important questions that had arisen at its founding. In particular, James Miller has shown that the failure of the group's founders to adequately define the concept of "participatory democracy" allowed later members to interpret its meaning as they saw fit, thereby depriving the group any

long-term organizational coherence. James Miller, "*Democracy Is in the Streets*": *From Port Huron to the Siege of Chicago*, second edition (Cambridge, Mass.: Harvard University Press, 1994 [first edition: New York: Simon and Schuster, 1987]), 152.

7. Terry H. Anderson, *The Movement and the Sixties* (New York: Oxford University Press, 1995), xv. I use the term "generation" cautiously, to suggest that certain identifiable characteristics and factors distinguished students active in SSOC during one time period from those in another. This is not to say that the generations were neatly defined, separated by crisp divisions and sharp breaks. Rather, there was some continuity between the generations. For instance, the goals and strategies of the generations did, to some extent, overlap. And while turnover within the group was marked, some individuals' participation in SSOC spanned several years and both generations. Still, the salient point here is that SSOC was not a well-defined group comprised of an undifferentiated mass of student activists but a flexible, ever-changing organization that reflected the views, attitudes, goals, and backgrounds of the students who directed it at any given point in time.

8. Different generations also emerged in other activists groups of the day. See Clayborne Carson, *In Struggle: SNCC and the Black Awakening of the 1960s* (Cambridge, Mass.: Harvard University Press, 1981), 1–4, 119–152, 191–243, 298–303; Miller, "*Democracy Is in the Streets*," 179, 196, 218–259; Todd Gitlin, *The Sixties: Years of Hope, Days of Rage* (New York: Bantam Books, 1987), 26–28, 65–67, 73–75, 83, 186, 225–226, 229–230, 306–308; Annie Gottlieb, *Do You Believe in Magic?: The Second Coming of the Sixties Generation* (New York: Time Books, 1987), 9–12.

9. Sue Thrasher, "Dear Friend," 16 March 1964, Bruce Smith Papers (Southern Student Organizing Committee Papers, 1959–1977), Special Collections, University of Virginia Library, Charlottesville, Virginia; "Agenda [April 3–5, 1964]," Curry Papers; Southern Patriot, April 1964 (quote).

10. "Agenda [April 3–5, 1964]," Curry Papers; Sue Thrasher to Conference Attendees, 16 March 1964, Smith Papers; interview, Thrasher 1 September 1994.

11. "Agenda [April 3–5, 1964]," Curry Papers. According to Carl Braden, "The name Southern Student *Organizing Committee* was suggested deliberately. It shows that the group's purpose is to organize now and decide later where its permanent berth will be. . . . This leaves great flexibility in dealing with those you seek to organize." To Braden, the important point was that white students had begun to organize. Not until the group was firmly established, he believed, should it worry about its formal organizational structure or its relationship with other activist organizations. Anne [and Carl] Braden to Continuations Committee of SSOC, 17 April 1964, Curry Papers.

12. Interview, Thrasher, 1 September 1994; Jerry Gainey to Hayes Mizell, 16 April 1964, Curry Papers.

13. Interview, Blackstock; interview, Smith; interview, Cathy Cade (interviewed by Ronald Grele), 5 January 1985, Columbia Oral History Project, Columbia University, New York, New York; Cathy Cade, *A Lesbian Photo Album: The Lives of Seven Lesbian Feminists* (Oakland, Calif.: Waterwoman Books, 1987), 84–85; Cathy Cade, correspondence with author, 18 April 1994.
14. Unidentified SSOC activist, 1964, quoted in "Material from SOC [sic] Files," in the possession of Harlon Joye, who the author wishes to thank for generously providing access to portions of this material.
15. Hamlett and Grogan quoted in "Agents of Change: Southern White Students in the Civil Rights Movement," A Public Seminar, 7 April 1994, Charlottesville, Virginia, in the author's possession; interview, Dan Harmeling, 14 March 1995; interview, Thrasher, 1 September 1994.
16. Parker, "The Southern Student Organizing Committee," 4, 11, Hamlett Papers; Sue Thrasher to SSOC, 25 May 1964, in the author's possession.
17. Gitlin, *The Sixties,* 100–101 (quote), 112, 213; Miller, *"Democracy Is in the Streets,"* 77, 122, 178, 187; Sara Evans, *Personal Politics: The Roots of Women's Liberation in the Civil Rights Movement & the New Left* (New York: Alfred A. Knopf, 1979), 33–34, 48, 51–55, 103, 109; interview, Robb Burlage, 22 April 1993 and 4 September 1994; interview, Dorothy Dawson Burlage, 4 September 1994.
18. Interview, Robb Burlage, 22 April 1993 and 4 September 1994. Burlage recalled that he was attracted to Harvard partly because of the presence of political scientist and Austin native V. O. Key, whose recently published *Southern Politics in State and Nation* (New York: Knopf, 1949) strongly influenced Burlage's understanding of the region's political history.
19. Robb Burlage, "For Dixie With Love and Squalor," Prospectus and Introduction for an SDS Pamphlet, 1962, in the possession of Dorothy Burlage.
20. Robb Burlage, "The South As An Underdeveloped Country," distributed by Students for a Democratic Society, 1962, 4B:46, reel 36, *Students for a Democratic Society Papers, 1958–1970* (Glen Rock, N.J.: Microfilm Corporation of America, 1977). Roosevelt had asserted that the South was "the nation's number one economic problem" during a speech he gave in Georgia in 1938 to announce the publication of the *Report on the Economic Conditions of the South,* a grim portrait of the region during the Depression compiled by his administration. Patricia Sullivan, *Days of Hope: Race and Democracy in the New Deal Era* (Chapel Hill: University of North Carolina Press, 1996), 65.
21. Interview, Robb Burlage, 22 April 1993 and 4 September 1994; interview, Dorothy Dawson Burlage; "We'll Take Our Stand," Hamlett Papers.
22. The opening section of the Port Huron Statement speaks of "people of this generation . . . looking uncomfortably to the world we inherit" and hoping to "chang[e] the conditions of humanity in the late twentieth century." Port Huron Statement, quoted in Miller, *"Democracy Is in the Streets,"* 329, 331.

23. "We'll Take Our Stand," Hamlett Papers. In her brief but perceptive study of SSOC, Christina Greene argues that the "use of the term 'brothers' [in the manifesto] indicated a male-centered perspective that manifested itself not simply in language but throughout the organization." While this gendered reading of "We'll Take Our Stand" allows for a deeper understanding of the document, it has led Greene to overstate the "male-centeredness" of SSOC. Not only were women involved in the founding of SSOC, but numerous women held important organizational roles in the group throughout its existence. Additionally, by 1969, SSOC, led by its women members, had become outspoken in its support for the burgeoning women's liberation movement. Christina Greene, "'We'll Take Our Stand': Race, Class, and Gender in the Southern Student Organizing Committee, 1964–1969," in *Hidden Histories of Women in the New South*, Virginia Bernhard, Betty Brandon, Elizabeth Fox-Genovese, Theda Perdue, and Elizabeth Hayes Turner, eds. (Columbia: University of Missouri Press, 1994), 178–179. For more on gender relations and issues in SSOC, see chapter 7.
24. Twelve Southerners, *I'll Take My Stand: The South and the Agrarian Tradition* (New York: Harper and Brothers, 1930). Ransom, Donaldson, Tate, and Warren also had been key members of the Fugitive group, the collection of writers who in the early 1920s published the *Fugitive,* a magazine of criticism and poetry. On the Fugitives and the Nashville Agrarians, see Thomas Daniel Young, *Waking Their Neighbors: The Nashville Agrarians Rediscovered* (Athens: University of Georgia Press, 1982); William C. Harvard and Walter Sullivan, eds., *A Band of Prophets: The Vanderbilt Agrarians after Fifty Years* (Baton Rouge: Louisiana State University Press, 1982); Louis D. Rubin, Jr., *The Wary Fugitives: Four Poets and the South* (Baton Rouge: Louisiana State University Press, 1977); Garvin Davenport, *The Myth of Southern History: Historical Consciousness in Twentieth-Century Southern Literature* (Nashville, Tenn.: Vanderbilt University Press 1967), 44–82; and Alexander Karenakis, *Tillers of a Myth: Southern Agrarians as Social and Literary Critics* (Madison: University of Wisconsin Press, 1966).
25. Interview, Robb Burlage, 22 April 1993; "We'll Take Our Stand," Hamlett Papers; H. L. Mencken, "Uprising in the Confederacy," *American Mercury* 22 (March 1931): 381. Mencken also attacked the Agrarians in "The South Astir," *Virginia Quarterly Review* 11 (January 1935): 47–60. "We'll Take Our Stand" did not, as Christina Greene suggests, oppose industrialization and urbanization for the South. Rather, the manifesto, as well as most SSOC members, held a more ambivalent view toward these processes. SSOC sought the eradication of racism, poverty, discrimination, alienation—conditions not necessarily a consequence of urbanization and industrialization. Most SSOC members sought to make industrial society more equitable, to make it a better place to live. As Burlage wrote in "We'll Take Our Stand," SSOC envisioned the South as a place where the rural and the urban, the agricultural and the industrial, co-existed and "provided meaningful work and

leisure opportunities for all." Greene, "'We'll Take Our Stand'," 178–179.
26. Interview, Robb Burlage, 22 April 1993; interview, John Lewis, 30 March 1995.
27. Interview, Gene Guerrero, 26 March 1993; "Agents of Change," A Public Seminar, 7 April 1994, Charlottesville, Virginia; interview, Howard Romaine, 17–18 March 1993; interview, Roger Hickey, 26 March 1993; interview, Tom Gardner, 2 September 1994; interview, Bill Leary, 2 April 1995; interview, Steve Wise, 18 March 1993; C. Vann Woodward, *The Strange Career of Jim Crow* (New York: Oxford University Press, 1955); C. Vann Woodward, "The Search for Southern Identity," in C. Vann Woodward, *The Burden of Southern History* (Baton Rouge: Louisiana State University Press, 1960), 3–25 (originally published in *The Virginia Quarterly Review* XXXII [1956]: 258–267).
28. The works by Woodward cited in the SSOC "Bibliography on Southern History" were *Origins of the New South, The Strange Career of Jim Crow, Reunion and Reaction, A Southern Philosophy* (edited by Woodward), *The Burden of Southern History,* and "Age of Reinterpretation." Other works on the list included W. J. Cash's *Mind of the South,* V. O. Key's *Southern Politics in State and Nation,* and Kenneth Stampp's *The Peculiar Institution.* "Bibliography on Southern History," undated, Hamlett Papers.
29. Thomas N. Gardner, speech at the University of Virginia, 20 November 1990, in the author's possession; Anne Braden to the Southern Student Organizing Committee Thirtieth Reunion and Conference, 5 April 1994, in the author's possession. The League of Young Southerners was an organization of young southern whites who openly opposed segregation and racism during the late 1930s and early 1940s. The League had grown out of the Council of Young Southerners, which in the mid-1930s was a branch of the Southern Conference for Human Welfare. Robin D. G. Kelley, *Hammer and Hoe: Alabama Communists during the Great Depression* (Chapel Hill: University of North Carolina Press, 1990), 197–203.
30. Interview, Tom Gardner, 5 May 1993.
31. Sue Thrasher to SSOC, 25 May 1964, in the author's possession. The Continuations Committee had sixteen members—one person from each of the 15 schools represented at the Nashville conference as well as member-at-large Sue Thrasher. Parker "The Southern Student Organizing Committee," 11, Hamlett Papers.
32. Ron K. Parker and David Kotelchuck, "Southern Student Organizing Committee," Curry Papers (quotes). By the end of the Continuations Committee meeting the students had suggested numerous specific actions local groups could take, including creating alternative student newspapers, leading off-campus seminars, coordinating action with community-based organizations such as women strike for peace clubs and human relations groups, and becoming involved in a wide array of community issues, ranging from employment and housing discrimination to

police brutality and juvenile delinquency. Parker "The Southern Student Organizing Committee," 8–9, Hamlett Papers.
33. Parker and Kotelchuck, "Southern Student Organizing Committee," Curry Papers; Sue Thrasher to SSOC, 25 May 1964, in the author's possession.
34. Cathy Cade to unidentified SNCC worker, December 1964, in "Material from SOC [sic] Files," in the possession of Harlon Joye.
35. Interview, Bob Moses, 18 April 1995; interview, Ed Hamlett, 6 May 1993; Parker "The Southern Student Organizing Committee," 17, Hamlett Papers; Anne Braden, "The Southern Freedom Movement in Perspective," *Monthly Review,* 17:3 (July–August 1965), 66; "Agents of Change," A Public Seminar, 7 April 1994, Charlottesville, Virginia. On the evolution of SNCC, see Carson, *In Struggle,* 1–82, 96–152. Other white students argued that SSOC should retain its independence because they doubted that SNCC had a strong enough financial base to support the new organization. And without sufficient funds, they feared SSOC would quickly wither away. Parker "The Southern Student Organizing Committee," 17.
36. *Southern Patriot,* April 1964; Parker "The Southern Student Organizing Committee," 17, Hamlett Papers; Roy Money to [Sue Thrasher], Spring 1964, and Cathy Cade to unidentified SNCC worker, December 1964, in "Material from SOC [sic] Files," in the possession of Harlon Joye.
37. Anne [and Carl] Braden to Continuations Committee of SSOC, 17 April 1964, Curry Papers.
38. Ibid.; interview, Braden; Anne Braden to the Southern Student Organizing Committee Thirtieth Reunion and Conference, 5 April 1994, in the author's possession (quote).
39. "Notes on SNCC ex. [ecutive committee] meeting, April 19, 1964," folder 5, box 1, Southern Student Organizing Committee Papers, Archives and Library, The Martin Luther King, Jr. Center for Nonviolent Social Change, Atlanta, Georgia; "Executive [Committee] Minutes," 19 April 1964, A:III:1, reel 3, SNCC Papers.
40. Sue Thrasher to Connie Curry, 23 April 1964, Curry Papers; interview, Thrasher, 1 September 1994; Constance Curry et al., *Deep in Our Hearts: Nine White Women in the Freedom Movement* (Athens: University of Georgia Press, 2000), 232.
41. Stanley Wise quoted in Greene, "'We'll Take Our Stand,'" 182; interview, Lewis, 30 March 1995; interview, Sue Thrasher (interviewed by Ronald Grele), 15 December 1984, Columbia Oral History Project, Columbia University, New York, New York. It is unclear what position Bob Moses took on SSOC-SNCC relations. Moses was revered by the young whites in SSOC, and they typically gave great weight to his thoughts and ideas. More than three decades after SSOC's founding, Moses recalls that he thought "that SNCC really represented one of the few institutions, young as it was, that young black people could say they owned, that belonged to them. So I thought that the idea of trying to merge this into a wider student movement that would somehow include

some few whites was just the wrong way to go." Anne Braden and Sue Thrasher also recollect that Moses opposed a uniting of the two groups, with Thrasher recalling that his support was instrumental in getting SSOC off the ground. The minutes of the Executive Committee meeting, however, indicate that Moses did not support an autonomous SSOC and quote him as saying at one point that SSOC should "be part of SNCC and have representation in SNCC.... SNCC has to cut across racial lines." Regardless of his stance on relations between the groups, Moses was unwavering in his support of the students' effort to organize among whites. Interview, Moses, 18 April 1995; interview, Braden, 25 April 1993; interview, Thrasher, 15 December 1984; "Executive [Committee] Minutes," 19 April 1964, A:III:1, reel 3, SNCC Papers; Carson, *In Struggle*, 103.

42. Braden also recalls Forman commenting that the students in SSOC wanted to be independent because " 'they want their own power base.' " Anne Braden to the Southern Student Organizing Committee Thirtieth Reunion and Conference, 5 April 1994, in the author's possession. Although she considered Forman's view to be cynical and unfair, she did believe that the students "want[ed] their own thing. It was nice to have your own office in Nashville, and I guess it did mean a lot to people as they came along." Interview, Braden.

43. Sue Thrasher to SSOC, 25 May 1964, in the author's possession (quote); "Notes on SNCC ex. meeting, April 19, 1964," folder 5, box 1, SSOC Papers, King Center; Ed Hamlett to Howard Romaine, April 1964, and Cathy Cade to unidentified SNCC worker, December 1964, in "Material from SOC [*sic*] Files," in the possession of Harlon Joye; *Southern Patriot*, April 1964; interview, Thrasher, 15 December 1984; Carson, *In Struggle*, 102–103; Sue Thrasher to Connie Curry, 23 April 1964, Curry Papers.

44. "Second SSOC Conference" and Sue Thrasher to SSOC, April 1964, Curry Papers; Parker "The Southern Student Organizing Committee," 3–4, 10, Hamlett Papers.

45. Parker "The Southern Student Organizing Committee," 10, Hamlett Papers; "Second SSOC Conference," Curry Papers; Parker and Kotelchuck, "Southern Student Organizing Committee," Curry Papers; interview, Guerrero; interview, Thrasher, 1 September 1994. In addition to electing officers, the students appointed a seven-member Executive Committee—Hamlett, Shirah, and one person to represent each of five different groups of southern states: Roy Money of Vanderbilt University (Georgia, Florida, and South Carolina), Ben Spalding of the University of North Carolina (Virginia, North Carolina, and Maryland), Jim Williams of the University of Louisville (Kentucky, Tennessee, and Arkansas), Charles Smith of the University of Texas (Texas and Oklahoma), and Cathy Cade of Tulane University (Alabama, Mississippi, and Louisiana). Parker "The Southern Student Organizing Committee," 11, Kotelchuck Papers, box 2.

46. "Second SSOC Conference," Curry Papers; Sue Thrasher to SSOC, 25 May 1964, in the author's possession; Jim Williams, "Needed: A

South-Wide Student Paper," folder [6], Sam Shirah Papers, Wisconsin Historical Society, Madison, Wisconsin; *New Rebel,* 1:1 (27 May 1964), in the SSOC Papers, King Center.

47. "Second SSOC Conference," Curry Papers; Parker "The Southern Student Organizing Committee," 10, Hamlett Papers; "Agents of Change," A Public Seminar, 7 April 1994, Charlottesville, Virginia. SSOC was not the first organization to use clasping black and white hands to symbolize its commitment to racial equality. Shortly after its founding, for instance, SNCC had used a similar image as the letterhead of the organization's correspondence. In 1961, Dorothy Burlage, then a Harvard student and the head of the Boston chapter of the Northern Student Movement, a predominantly white student group that worked to build northern support for SNCC, designed and sold buttons featuring this logo to help raise funds for SNCC. And in an earlier era, packinghouse organizers in Chicago had used the symbol to convey their opposition to racial discrimination. Dorothy Burlage to author, 1 December 1997, in the author's possession; Roger Horowitz, *"Negro and White, Unite and Fight!": A Social History of Industrial Unionism in Meatpacking, 1930–1990* (Urbana: University of Illinois Press, 1997), 4.

48. Parker "The Southern Student Organizing Committee," 10, Hamlett Papers; "Second SSOC Conference," Curry Papers; Todd Gitlin to Robb and Dorothy Burlage, 4–5 February 1964, folder 5, box 2, Robb Burlage Papers, Wisconsin Historical Society, Madison, Wisconsin; Todd Gitlin to SSOC, 7 May 1964, in the author's possession. SDS's southern presence in late 1963 and early 1964 was limited to the chapter at the University of Texas. The group's largest chapters were at schools in the Midwest and the Northeast. Kirkpatrick Sale, *SDS* (New York: Random House, 1973), 116–119.

49. Todd Gitlin to SSOC, 7 May 1964, in the author's possession.

50. "Second SSOC Conference," Curry Papers; "A Resolution Concerning SDS's Role in the South and the Relationship Between SDS and the Southern Student's [*sic*] Organizing Committee," submitted by the Executive Committee of the Southern Student's [*sic*] Organizing Committee, May 1964, 2A:115, reel 9, SDS Papers.

51. Interview, Thrasher, 15 December 1984; Hamlett quoted in J. Anthony Lukas, *Don't Shoot—We Are Your Children!,* third edition (New York: Random House, 1971 [originally published 1968]), 153.

52. *National Guardian,* 23 October 1965.

53. Charles Smith to SSOC, May 1964, Curry Papers; "A Resolution Concerning SDS's Role in the South and the Relationship Between SDS and the Southern Student's Organizing Committee," submitted by the Executive Committee of the Southern Student's Organizing Committee, May 1964, 2A:115 reel 9, SDS Papers; William J. Billingsley, "The New Left in the New South: SDS and the Organization of Southern Political Communities," presented at the Eighty-Seventh Annual Meeting of the Organization of American Historians, 16 April 1994, 20, 26–27, in the author's possession.

54. In her autobiography, Virginia Durr, who by the 1960s was in her third decade of civil rights activism, wrote movingly about how stepping "outside the magic circle" to embrace black equality forever altered her relations with other white southerners. Hollinger F. Barnard, ed., *Outside the Magic Circle: The Autobiography of Virginia Foster Durr* (Tuscaloosa: University of Alabama Press, 1985).

55. Richard Flacks, himself an influential early SDS member, has promoted this view of SDS's membership. See, for instance, Richard Flacks, "Who Protests: The Social Bases of the Student Movement," in *Protest!: Student Activism in America*, Julian Foster and Durward Long, eds. (New York: William Morrow and Co., 1970), 134–157. Doug McAdam attributes similar characteristics to the northern students who came to Mississippi for Freedom Summer in 1964. Doug McAdam, *Freedom Summer* (New York: Oxford University Press, 1988), esp. 41–65. Kenneth Heineman, on the other hand, suggests that such a view of student activists describes only a small number of students, specifically those in leadership positions and those who attended elite institutions. At northern public universities, for instance, he found that rank-and-file activists were not secular, not wealthy, and not from liberal families. By choosing, however, to focus exclusively on northern universities—ignoring what he calls the "relatively inactive Southern universities"—Heineman misses an opportunity to explore characteristics and patterns among student activists that cut across regions. Kenneth J. Heineman, *Campus Wars: The Peace Movement at American State Universities in the Vietnam Era* (New York: New York University Press, 1993), 4–5, 79–83 (quote p. 5).

56. In this respect, SSOC's first generation was similar to what Todd Gitlin has called the "pre-Vietnam New Left": "Itself ignited by the civil rights movement, it was the small motor that turned the larger motor of the mass student movement of the late Sixties." Gitlin, *The Sixties*, 26–28, 65–67, 73–75, 83 (quote p. 26).

57. This is not to suggest that there was a direct causal link between religious faith and civil rights activism. As Doug Rossinow has written about student activists at the University of Texas at Austin, "Christianity did not *cause* [his emphasis] the involvement of these young people in civil rights protest or in the new left. A compound of political and cultural forces moved them in those directions, a compound that was forged in the heat of activism. However, the important role of Christian religion and Christian institutions in the emergence of white youth activism in Austin during this time is unmistakable." Doug Rossinow, " 'The Break-through to New Life': Christianity and the Emergence of the New Left in Austin, Texas, 1956–1964," *American Quarterly*, 46, no. 3 (September 1994): 329.

58. Ed Hamlett, biographical letter or speech written in Biloxi, Mississippi, 10 August 1964, Hamlett Papers; Ed Hamlett to Henry Bucker, 29 September 1966, Hamlett Papers; "Agents of Change: A Public Seminar," 7 April 1994, Charlottesville, Virginia.

59. Interview, Anne Cooke Romaine, 19 March 1993.
60. Interview, Thrasher, 15 December 1984; Lukas, *Don't Shoot—We Are Your Children!*, 130–131, 136; Curry et al., *Deep in Our Hearts*, 209–217.
61. Interview, Guerrero.
62. Interview, Steve Wise.
63. Interview, Smith.
64. Ed Ball, "The White Issue," *Village Voice*, 18 May 1993. Virginia Durr recalled that one of the biggest challenges she and other white liberals faced was recognizing that they had been raised in a racist environment. Once they recognized and accepted this fact about their upbringing, it became possible to break away from their racist roots. Barnard, *Outside the Magic Circle*, 18–19, 31–32, 44, 56–59, 104, 308–311. Lewis M. Killian expresses a similar sentiment in *Black and White: Reflections of a White Southern Sociologist* (Dix Hills, N.Y.: General Hall, 1994), esp. 7–15. Fred Hobson offers an insightful overview of white southerners who have written about their conversion experience in *But Now I See: The White Southern Racial Conversion Narrative* (Baton Rouge: Louisiana State University Press, 1999).
65. Ed Hamlett, biographical letter or speech written in Biloxi, Mississippi, 10 August 1964, Hamlett Papers; interview, Howard Romaine and Ed Hamlett, ca. 23–24 January 1992, in the author's possession.
66. Interview, Guerrero.
67. Interview, Cathy Cade, 5 January 1985, (quote); Cathy Cade, *A Lesbian Photo Album*, 72–85; Evans, *Personal Politics*, 35; Cathy Cade correspondence with author, 18 April 1994.
68. Interview, Anne Cooke Romaine; "Agents of Change: A Public Seminar," 7 April 1994, Charlottesville, Virginia.
69. Interview, Anne Cooke Romaine; interview, Howard Romaine; interview, Howard Romaine and Ed Hamlett, April 1990. Southwestern at Memphis has since been renamed Rhodes College.
70. Interview, Blackstock. The editor of the campus newspaper, the *Red and Black*, noted that the peace marchers were all outspoken integrationists, and as they marched into Athens robed Klansmen harassed and heckled them from a truck. The *Red and Black*, 29 October 1963.
71. Author's notes on Sue Thrasher, comments on "Movement Women: Perspectives on Race, Civil Rights and Feminism," *Activism & Transformation: A Symposium on the Civil Rights Act of 1964*, Connecticut College, New London, Connecticut, 4 November 1994; Curry et al., *Deep in Our Hearts*, 227.
72. Parker, "The Southern Student Organizing Committee," 25–26, Hamlett Papers.
73. "Agents of Change: A Public Seminar," 7 April 1994, Charlottesville, Virginia; interview, Nan Grogan Orrock, 14 November 1994; Patrick Henry Bass, *Like a Mighty Stream: The March on Washington, August 28, 1963* (Philadelphia, Penn.: Running Press, 2002), 142–145.

74. Interview, Smith. On white fears about the March on Washington, see Taylor Branch, *Parting the Waters: America in the King Years, 1954–1963* (New York: Simon and Schuster, 1988), 871–872, 876.
75. Interview, Thrasher, 15 December 1984.
76. Curry et al., *Deep in Our Hearts: Nine White Women in the Freedom Movement,* 76; interview, Dorothy Dawson Burlage; *Southern Patriot,* September 1964.
77. Interview, Guerrero; "Agents of Change: A Public Seminar," 7 April 1994, Charlottesville, Virginia.
78. Interview, Howard Romaine; interview, Bob Hall, 16 April 1993.
79. Exchange of letters between Hamlett and his parents between fall 1963 and summer 1965, Hamlett Papers.
80. Interview, Thrasher, 15 December 1984; Lukas, 148–149 (quotes).
81. Sue Thrasher to SSOC, 25 May 1964, in the author's possession.

Chapter 3

1. Interview, Sue Thrasher, 1 September 1994.
2. *New Rebel,* 1:1 (27 May 1964), in the Southern Student Organizing Committee Papers, Archives and Library, The Martin Luther King, Jr. Center for Nonviolent Social Change, Atlanta, Georgia.
3. Ron K. Parker, "The Southern Student Organizing Committee," 24, Edwin Hamlett Papers (uncatalogued), Special Collections and University Archives, The Jean and Alexander Heard Library, Vanderbilt University, Nashville, Tennessee (also in box 2, David and Ronda Kotelchuck Papers (Southern Student Organizing Committee Papers, 1959–1977) and Steve Wise Papers (Southern Student Organizing Committee Additional Papers, 1963–1979 *and* Southern Student Organizing Committee Papers, 1959–1977), Special Collections, University of Virginia Library, Charlottesville, Virginia); Sue Thrasher to Franklin D. Roosevelt III, 5 December 1964, folder 3, Southern Student Organizing Committee Papers, 1964–1968, Special Collections, University of Virginia Library, Charlottesville, Virginia; Sue Thrasher to SSOC, 25 May 1964, in the author's possession; interview, Sue Thrasher (interviewed by Ronald Grele), 15 December 1984, Columbia Oral History Project, Columbia University, New York, New York; interview, Thrasher, 1 September 1994; Clayborne Carson, *In Struggle: SNCC and the Black Awakening of the 1960s* (Cambridge, Mass.: Harvard University Press, 1981), 102–103; Bryant Simon, "Southern Student Organizing Committee: A New Rebel Yell in Dixie" (Honors Essay: University of North Carolina at Chapel Hill, 1983), 23; *New Rebel,* 1:1 (27 May 1964), and "Notes on SNCC meeting, May 7, 1964," folder 5, box 1, in the SSOC Papers, King Center.
4. While Freedom Summer is a prominent touchstone in the historiography of the civil rights movement, few scholars have given any attention to the White Folks Project or to the southern whites who participated in it. Works that ignore the White Folks Project do a particular disservice for

they create the perception that the only whites involved in Freedom Summer were middle- and upper-class northern and western students who went South for the summer. Among these studies are Howard Zinn, *SNCC: The New Abolitionists* (Boston, Mass.: Beacon Press, 1964); Doug McAdam, *Freedom Summer* (New York: Oxford University Press, 1988); Nicolaus Mills, *Like a Holy Crusade: Mississippi 1964—The Turning of the Civil Rights Movement in America* (Chicago, Ill.: Ivan R. Dee, 1992); Charles M. Payne, *I've Got the Light of Freedom: The Organizing Tradition and the Mississippi Freedom Struggle* (Berkeley: University of California Press, 1995); and the film "Freedom on My Mind," Connie Field and Marilyn Mulford producers (Berkeley, Calif.: Clarity Educational Productions, 1994). Studies that briefly treat the White Folks Project include Elizabeth Sutherland, ed., *Letters from Mississippi* (New York: McGraw-Hill, 1965), 154, 159–160; Mary Aickin Rothschild, *A Case of Black and White: Northern Volunteers and the Southern Freedom Summers, 1964–1965* (Westport, Conn.: Greenwood Press, 1982), 34–35, 67–68; and John Dittmer, *Local People: The Struggle for Civil Rights in Mississippi* (Urbana: University of Illinois Press, 1994), 244, 252. In their limited discussions of the project, these works often gloss over its complexities and offer factually incorrect information. For instance, Rothschild writes that all of the southern volunteers "came from large 'New South' metropolitan areas or attended Ivy League schools" (34–35) when, in fact, some of the southern participants arrived in Mississippi from such decidedly non-metropolitan areas as Athens, Georgia, and Sewannee, Tennessee. More thorough considerations of the White Folks Project are offered by Len Holt, *The Summer that Didn't End* (New York: William Morrow and Co., 1965), 132–141; Irwin Klibaner, *Conscience of a Troubled South: The Southern Conference Educational Fund, 1946–1966* (Brooklyn, N.Y.: Carlson Publishing, 1989), 217–219; Carson, *In Struggle*, 118–119; and Robert Pardun, *Prairie Radical: A Journey through the Sixties* (Los Gatos, Ca.: Shire Press, 2001).

5. *Student Voice*, 4:4 (11 November 1963), 5:8 (3 March 1964), and Special Issue (spring 1964); Carson, *In Struggle*, 97–103; McAdam, *Freedom Summer*, 35–41; Dittmer, *Local People*, 200–210; Payne, *I've Got the Light of Freedom*, 290–298; David Harris, *Dreams Die Hard* (New York: St. Martin's/Marek, 1982), 30–54; Richard Cummings, *The Pied Piper: Allard K. Lowenstein and the Liberal Dream* (New York: Grove Press, 1985), 223–257; William H. Chafe, *Never Stop Running: Allard Lowenstein and the Struggle to Save American Liberalism* (New York: Basic Books, 1993), 180–196; William H. Chafe, "The Personal and the Political: Two Case Studies," in *U.S. History as Women's History: New Feminist Essays*, Linda K. Kerber, Alice Kessler-Harris, and Kathryn Kish Sklar, eds. (Chapel Hill: University of North Carolina Press, 1995), 206–207.

6. Sue Thrasher to SSOC, 25 May 1964, in the author's possession; Ed Hamlett to Howard Romaine, April 1964, and Sue Thrasher to unknown correspondent, 20 May 1964, in "Material from SOC [*sic*] Files," in the

possession of Harlon Joye; Christina Greene, "'We'll Take Our Stand': Race, Class, and Gender in the Southern Student Organizing Committee, 1964–1969," in *Hidden Histories of Women in the New South,* Virginia Bernhard, Betty Brandon, Elizabeth Fox-Genovese, Theda Perdue, and Elizabeth Hayes Turner, eds. (Columbia: University of Missouri Press, 1994), 184.

7. Pardun, *Prairie Radical,* 58; "Mississippi Summer Project, Summer 1964 Volunteer Applications," Hamlett Papers.

8. Pardun, *Prairie Radical,* 56; Douglas Tiberus [*sic*], "Involvement in Mississippi," *Oikoumene Communique* (publication of University of Arkansas-Fayetteville Disciples, Episcopalians, Methodists, and Presbyterians), 1:5 (15 October 1964), Douglas Tiberiis White Folks Project Collection (unprocessed), McCain Library and Archives, University Libraries, The University of Southern Mississippi, Hattiesburg, Mississippi.

9. Ed Hamlett, "White Folk's Project," Hamlett Papers (also located in Ed King Collection, L. Zenobia Coleman Library, Tougaloo College, Tougaloo, Mississippi); "Investigative Report, Oxford, Ohio and Jackson, Mississippi," 26 June 1964, folder 10, box 135, Series II, Sub-Series 9, Sovereignty Commission, Paul B. Johnson Family Papers, McCain Library and Archives, University Libraries, The University of Southern Mississippi, Hattiesburg, Mississippi; interview, Thrasher, 15 December 1984; Author's notes on Anne Braden, "Remembering Freedom Summer," presented at the Southern Historical Association 60th annual meeting, 9 November 1994, Louisville, Kentucky. On the training sessions for Freedom Summer, see Carson, *In Struggle,* 111–114; Dittmer, *Local People,* 242–246; McAdam, *Freedom Summer,* 66–75; James F. Findlay, Jr., *Church People in the Struggle: The National Council of Churches and the Black Freedom Movement, 1950–1970* (New York: Oxford University Press, 1993), 84–87.

10. "We'll Take Our Stand," Hamlett Papers; Sam Shirah, "A Proposal for Expanded Work Among Southern White Students and an Appalachian Project," n.d. (ca. mid-1963), folder 5, Sam Shirah Papers, Wisconsin Historical Society, Madison, Wisconsin; *Southern Patriot,* April 1964; Ron K. Parker and David Kotelchuck, "Southern Student Organizing Committee," Constance Curry Papers (uncatalouged), Special Collections, University of Virginia Library, Charlottesville, Virginia; *Nashville Banner,* 29 June 1964; interview, Thrasher, 15 December 1984; Pardun, *Prairie Radical,* 59–60, 62; Holt, *The Summer that Didn't End,* 137–138.

11. Ed Hamlett to Mike Welch, 12 March 1968, Hamlett Papers. Among the seven activists who worked in the Jackson area were Howard Romaine, Ed Hamlett, Charles Smith, Bob Bailey, Dick Jewett, and Sue Thrasher.

12. According to the 1960 census, whites comprised 87 percent (38,383) and non-whites 13 percent (5,670) of Biloxi's population. In Gulfport, whites made up 79 percent (23,834) and non-whites 21 percent (6,370) of the population. For the state as a whole, whites were 58 percent

(1,257,407) and non-whites 42 percent (920,734) of the population. U.S. Bureau of the Census, *Census of Population: 1960,* volume 1 (Characteristics of Population), part 26: Mississippi (Washington, D.C.: U.S. Government Printing Office, 1963), 112, 147.
13. Robert Moses, "Letter from a Mississippi Jail Cell," quoted in Zinn, *SNCC,* 76. It is unclear who the original Gulf Coast volunteers were since more than 18 activists worked in Biloxi and Gulfport at some point during the summer. For instance, Hamlett, Smith, Bailey, and Thrasher all spent time working on the Gulf Coast in addition to their work in and around Jackson. Other Gulf Coast volunteers included Sam Shirah, Gene Guerrero, Bruce Maxwell, Soren Sorenson, Nelson Blackstock, Jeff Powers, Michael Waddell, Diane Burrows, Douglas Tiberiis, Liz Krohne, Judy Schiffer, Robert Pardun, Douglas Tiberus, and Lon Clay Hill. Grenville Whitman, John Parkmon, Tom Hill, Maryka Matthews, Marjorie Henderson, Anne Strickland, and John Strickland (no relation) were also volunteers who more than likely spent most of their time in Biloxi and Gulfport.
14. Interview, Howard Romaine, 17–18 March 1993; Hamlett, "White Folk's Project," Hamlett Papers; "White Community Project Workshop," Camp Landon, Gulfport, Mississippi, July 27–28, 1964, folder 1, box 5, Southern Student Organizing Committee and Thomas N. Gardner Papers, 1948–1994 [1958–1975], Special Collections, University of Virginia Library, Charlottesville, Virginia; Holt, *The Summer that Didn't End,* 137; "Mississippi Summer Project: The White Community Too," folder 3, box 61, Carl Braden and Anne Braden Papers, 1928–1972, Wisconsin Historical Society, Madison, Wisconsin.
15. "White Community Project Workshop," July 27–28, 1964, folder 1, box 5; SSOC and Gardner Papers; interview, Ed Hamlett, 6 May 1993; *New Rebel,* 1:2 (October 1964).
16. *New Rebel,* 1:2 (October 1964). Interview, Thrasher, 1 September 1994; interview, Thrasher, 15 December 1984 (quote).
17. Ed Hamlett [?], "Notes on White Folks Project," Douglas Tiberiis White Folks Project Collection; *Student Voice,* 5:16 (15 July 1964); "Wats Reports," A:VII:7, reel 15 and "Mississippi Summer Project: Running Summary of Incidents," A:XV:157, reel 38, *The Student Nonviolent Coordinating Committee Papers, 1959–1972* (Sanford, N.C.: Microfilm Corporation of America, 1982); Pardun, *Prairie Radical,* 61–62; Constance Curry et al., *Deep in Our Hearts: Nine White Women in the Freedom Movement* (Athens: University of Georgia Press, 2000), 235; Ed Hamlett to author, 3 April 2003.
18. Interview, Thrasher, 1 September 1994; interview, Gene Guerrero, 26 March 1993; *New Rebel,* 1:2 (October 1964); "Mississippi Project," A:XV:157, reel 38, SNCC Papers; "Wats line Digest," July 19, 1964, folder 39, box 2, SSOC Papers, King Center; *Student Voice,* 5:18 (29 July 1964); Charles Smith quoted in Holt, *The Summer that Didn't End,* 138.
19. Sorenson quoted in Sutherland, *Letters from Mississippi,* 159; interview, Hamlett; Ed Hamlett, "White Folk's Project," Hamlett Papers.

20. Ibid.; interview, Nelson Blackstock, 8 August 1995; Bruce Maxwell, "We Must Be Allies . . . : Race Has Led Us Both to Poverty," A:XV:225, reel 40, SNCC Papers.
21. Maxwell, "We Must Be Allies . . . ," A:XV:225, reel 40, SNCC Papers; "White Community Project Workshop," July 27–28, 1964, folder 1, box 5, SSOC and Gardner Papers; Hamlett, "White Folk's Project," Hamlett Papers; Holt, *The Summer that Didn't End*, 137–138; interview, Thrasher, 15 December 1984.
22. Notes on 5 July 1964 meeting of volunteers, Hamlett White Folks Project Collection.
23. Klibaner, *Conscience of a Troubled South*, 218–219; Maxwell, "We Must Be Allies . . . ," A:XV:225, reel 40, SNCC Papers; *Southern Patriot*, October 1964; "Mississippi Summer Project," A:XV:157, reel 38, SNCC Papers; *New Rebel*, 1:2 (October 1964); Hamlett, "White Folk's Project," Hamlett Papers; Carson, *In Struggle*, 118–119.
24. Interview, Guerrero; interview, Blackstock; interview, Hamlett; interview, Thrasher, 15 December 1984 (quotes) and 1 September 1994; Curry et al., *Deep in Our Hearts*, 236.
25. Hamlett, "White Folk's Project" and Ed Hamlett to Bob Moses, ca. July 1964, Hamlett Papers; "White Community Project Workshop," July 27–28, 1964, folder 1, box 5, SSOC and Gardner Papers; "Bob Williams, 25 (?)," 22 July 1964, Hamlett White Folks Project Collection; *Free Biloxi Herald*, 14 August 1964, generously loaned to the author by Harlon Joye; "Wats Report," A:VII:7, reel 15, SNCC Papers; Holt, *The Summer that Didn't End*, 133–134; Pardun, *Prairie Radical*, 63, 66; Mary Stanton, *Freedom Walk: Mississippi or Bust* (Jackson: University Press of Mississippi, 2003), 175.
26. Anne Braden, "The Southern Freedom Movement in Perspective," *Monthly Review*, 17:3 (July-August 1965), 68; interview, Thrasher, 1 September 1994; Hamlett quoted in Klibaner, *Conscience of a Troubled South*, 219; Maxwell, "We Must Be Allies . . . ," A:XV:225, reel 40, SNCC Papers. COFO, in fact, did agree to continue and to expand the White Folks Project in the fall of 1964. *Southern Patriot*, November 1964.
27. Archie Allen to Philip Hirschkop, ca. September 1964, folder 3, Southern Student Organizing Papers, Special Collections, University of Virginia Library, Charlottesville, Virginia; interview, Thrasher, 15 December 1984; interview, Thrasher, 1 September 1994.
28. Sue Thrasher to SSOC, 25 May 1964, in the author's possession; Todd Gitlin to SSOC, 7 May 1964, in the author's possession; C. Clark Kissinger to Sue Thrasher, 9 September 1964, 2A:115, reel 9, SDS Papers; Parker, "The Southern Student Organizing Committee," 25–26, Hamlett Papers.
29. Quote from William J. Billingsley, "The New Left in the New South: SDS and the Organization of Southern Political Communities," delivered at the Organization of American Historians' Eighty-Seventh Annual meeting, 16 April 1994, Atlanta, Georgia, 22.

30. Archie Allen to Sue Thrasher, 13 October 1964 and 17 October 1964, folder 3, SSOC Papers, University of Virginia; Archie Allen, untitled essay, ca. November 1964 (quote), folder 2, SSOC Papers, University of Virginia (a later version would appear as "Report of Field Secretary," *Newsletter: Southern Student Organizing Committee,* 1:3 [December 1964]); Parker, "The Southern Student Organizing Committee," 12–13, Hamlett Papers; Kathy Barrett to Clark Kissinger, 25 December 1964 and 11 January 1965 and Sue Thrasher to Clark Kissinger, 30 December 1964, 2A:115, reel 9, SDS Papers. Archie Allen's exact salary as campus traveler is not known, but according to SSOC treasurer Ron Parker, Gene Guerrero and Kathy Barrett each earned $25 per week to campus travel for SSOC. Parker, "The Southern Student Organizing Committee," 13.

31. M. Hayes Mizell to Steven McNichols, 5 October 1964, Curry Papers; "Southern Student Organizing Committee: Income, September 1, 1964–September 1, 1965," file "SSOC 7th 1965," box 2S427, Field Foundation Archives, 1940–1990, The Center for American History, The University of Texas at Austin, Austin, Texas; Parker, "The Southern Student Organizing Committee," 24, Hamlett Papers; Archie Allen, "Fundraising Report for SSOC," ca. December 1964, folder 7 and Sue Thrasher to Franklin Delano Roosevelt III, 5 December 1964, Deborah Allen to Archie Allen, 8 December 1964, and Archie Allen to Leon Davis, 6 January 1965, folder 3, all in SSOC Papers, University of Virginia; interview, Thrasher, 15 December 1994; interview, Thrasher, 15 December 1984 (quote). Among the union officials Allen met with were Elliot Godoff of Local 1199, the hospital worker's union; Sam Meyers, president of United Autoworkers Local 259; Frank Brown, Vice President of District 65, Retail and Wholesale Department Store Union; Morris Pizer, International President, United Furniture Workers; and Tom DeLorenzo, Assistant Regional Director, Region 9A, United Autoworkers.

32. Archie Allen, "Fundraising Report for SSOC," ca. December 1964, folder 7, and Archie Allen to Phillip Hirschkop, ca. September 1964, folder 3, SSOC Papers, University of Virginia; Simon, "Southern Student Organizing Committee," 24.

33. On SCEF-SRC relations, see John Egerton, *Speak Now against the Day: The Generation before the Civil Rights Movement in the South* (Chapel Hill: University of North Carolina Press, 1994), 312–316, 355–356, 432–442, 564–566. General treatments of SCEF include Linda Reed, *Simple Decency & Common Sense: The Southern Conference Movement, 1938–1963* (Bloomington: Indiana University Press, 1991), esp. 129–184; Klibaner, *Conscience of a Troubled South;* Anthony P. Dunbar, *Against the Grain: Southern Radicals and Prophets, 1929–1959* (Charlottesville: University Press of Virginia, 1981), 187–219; and David L. Chappell, *Inside Agitators: White Southerners in the Civil Rights Movement* (Baltimore, Md.: The Johns Hopkins University Press, 1994), 41–47. General works on the SRC are fewer in number and tend to focus on the years prior to 1954. See, for instance, Martin Sosna, *In Search of*

the Silent South: Southern Liberals and the Race Issue (New York: Columbia University Press, 1977), 152–164; John Kneebone, *Southern Liberal Journalists and the Issue of Race, 1920–1944* (Chapel Hill: University of North Carolina Press, 1985), 202–222; Julia Anne McDonough, "Men and Women of Goodwill: A History of the Commission on Interracial Cooperation and the Southern Regional Council, 1919–1954" (Ph.D. Dissertation: University of Virginia, 1993); and Egerton, *Speak Now against the Day,* esp. 301–312, 432–437, 446–448, 481–483, and 564–568.

34. "Southern Student Organizing Committee: Income, September 1, 1964–September 1, 1965," file "SSOC 7th 1965, box 2S427, Field Foundation Archives; Sue Thrasher to Phillip J. Hirschkop, 30 October 1964, folder 3, SSOC Papers, University of Virginia; Sue Thrasher to M. Hayes Mizell, 14 October 1964, Curry Papers; interview, Thrasher, 1 September 1994 (quote).

35. Ron Parker to Hayes Mizell, 20 September 1964, Curry Papers; Sue Thrasher to C. Clark Kissinger, 10 October 1964, 2A:115, reel 9, SDS Papers (quote); Deborah M. Cole to Rev. Will D. Campbell, 9 October 1964, folder 20, box 52, Ed Hamlett White Folks Project Collection (unprocessed), McCain Library and Archives, University Libraries, The University of Southern Mississippi, Hattiesburg, Mississippi; Simon, "Southern Student Organizing Committee," 24–25; interview, Thrasher, 1 September 1994; interview, Thrasher, 15 December 1984. SSOC was endorsed by others besides the SRC. One of the first individuals to offer support for the group was Will Campbell, the iconoclastic Baptist preacher who at the time worked for the Committee of Southern Churchmen. Archie Allen to Phillip Hirschkop, 1 October 1964, folder 3, SSOC Papers, University of Virginia.

36. Sue Thrasher to Gene Guerrero, 13 December 1964 and "SSOC History (recollected from lost notes)," in "Material from SOC [*sic*] Files," in the possession of Harlon Joye; Gene Guerrero to author, 17 February 2003.

37. Parker, "The Southern Student Organizing Committee," 21, Hamlett Papers; interview, Thrasher, 1 September 1994; interview, Thrasher, 15 December 1984.

38. Interview, Thrasher, 15 December 1984; Parker, "The Southern Student Organizing Committee," 23, Hamlett Papers.

39. Ibid.

40. Interview, David Kotelchuck, 31 August 1994; interview, Thrasher, 1 September 1994; interview, Thrasher, 15 December 1984; Parker, "The Southern Student Organizing Committee," 22–23, Hamlett Papers; Ron Parker to M. Hayes Mizell, 20 September 1964, Curry Papers; Archie Allen to Sue Thrasher, Gene Guerrero, Ed Hamlett, etc., 11 August 1964, folder 3, SSOC Papers, University of Virginia.

41. Parker, "The Southern Student Organizing Committee," 12–13, 21–22, Hamlett Papers; *National Guardian,* 19 December 1964; Simon, "Southern Student Organizing Committee," 25–27.

42. Sue Thrasher to Anne and Carl Braden, 12 January 1965, folder 3, box 61, Braden Papers; interview, Thrasher, 15 December 1984; interview, Thrasher, 1 September 1994.
43. Interview, Thrasher, 1 September 1994; interview, Thrasher, 15 December 1984.
44. Interview, Lawrence Goodwyn, 6 March 1995; interview, Blackstock, 8 August 1995; "Conference Agenda," Curry Papers; *Southern Patriot*, December 1964; *1965 Conference Report*, Hamlett Papers; *Newsletter: Southern Student Organizing Committee*, 1:3 (December 1964); Parker, "Southern Student Organizing Committee," 13–14, Hamlett Papers; "Report on Southwide Fall Conference," in *1965–66 Proposal*, folder 1, SSOC Papers, University of Virginia; *Student Voice*, 5:23 (25 November 1964); interview, Thrasher, 15 December 1984.
45. *Southern Patriot*, December 1964; Archie Allen, untitled essay, ca. November 1964, folder 2, Southern Student Organizing Papers, University of Virginia (quote); Sue Thrasher to Nelson Blackstock, ca. September 1964, Nelson Blackstock Papers (uncatalogued), Alderman Library, University of Virginia; *Southern Student Organizing Committee Newsletter*, 1:3 (December 1964).
46. Parker, "Southern Student Organizing Committee," 14, Hamlett Papers; "Report on Southwide Fall Conference," in *1965–66 Proposal*, folder 1, SSOC Papers, University of Virginia; *Newsletter: Southern Student Organizing Committee*, 1:3 (December 1964); *Southern Patriot*, December 1964. Connie Curry to unidentified SNCC worker, December 1964, in "Material from SOC [*sic*] Files," in the possession of Harlon Joye.
47. *Newsletter: Southern Student Organizing Committee*, 1:3 (December 1964); Hamlett quoted in "Agents of Change: Southern White Students in the Civil Rights Movement," A Public Seminar, 7 April 1994, Charlottesville, Virginia, in the author's possession; unnamed SSOC activist quoted in November 1964 letter to unnamed philanthropic foundation representative, in "Material from SOC [*sic*] Files," in the possession of Harlon Joye; interview, Anne Braden, 25 April 1993. SSOC's move toward biracialism is discussed in Parker, "The Southern Student Organizing Committee," 14, Hamlett Papers; "Report on Southwide Fall Conference," in *1965–66 Proposal*, folder 1, SSOC Papers, University of Virginia; *Southern Patriot*, April 1965; Anne Braden, "The Southern Freedom Movement in Perspective," *Monthly Review*, 17:3 (July–August 1965), 66–67. On SNCC's Waveland Retreat, see Carson, *In Struggle*, 140–148.
48. Parker, "The Southern Student Organizing Committee," 14–15, Hamlett Papers; *Newsletter: Southern Student Organizing Committee*, 1:3 (December 1964); *Southern Patriot*, December 1964; Hamlett quoted in "Agents of Change," A Public Seminar, 7 April 1994.
49. Interview, Dan Harmeling, 14 March 1995; interview, Blackstock; Archie Allen to Sue Thrasher, 17 October 1964, folder 3, SSOC Papers, University of Virginia (also located in "Material from SOC [*sic*] Files," in the possession of Harlon Joye).

50. *Southern Patriot*, December 1964; "Agents of Change," A Public Seminar, 7 April 1994; interview, Braden; interview, Guerrero. Although SSOC immediately removed *New Rebel* from the masthead of the newsletter, a new title did not take its place until the *New South Student* was unveiled in October 1965. Until then, the publication simply was called *Newsletter: Southern Student Organizing Committee*.
51. Parker, "The Southern Student Organizing Committee," 15, Hamlett Papers; Sue Thrasher to Anne Braden, "After-Thoughts," November 1964, folder 3, box 61, Braden Papers.
52. Hayes Mizell to Steven McNichols, 5 October 1964, Curry Papers.
53. Ibid.; Parker, "The Southern Student Organizing Committee," 16–17, Hamlett Papers; *Newsletter: Southern Student Organizing Committee*, 1:3 (December 1964) and 2:1 (January 1965); "Christmas Project," in *1965 Conference Report*, Hamlett Papers; interview, Anne Cooke Romaine, 19 March 1993; interview, Tom Gardner, 2 September 1994; interview, David Nolan, 13–14 March 1995.
54. J. Anthony Lukas, *Don't Shoot—We Are Your Children!*, third edition (New York: Random House, 1971 [first edition: New York: Random House, 1968]), 156; Laurel (Mississippi) *Leader-Call*, 23, 24, 26, and 28 December 1964; *Newsletter: Southern Student Organizing Committee*, 1:3 (December 1964) and 2:1 (January 1965); "Agents of Change," A Public Seminar, 7 April 1994; interview, Nan Grogan Orrock, 14 November 1994; Ed Hamlett to John Doar, ca. December 1964, "Material from SOC [*sic*] Files," in the possession of Harlon Joye.
55. C. Clark Kissinger to Sue Thrasher, 17 December 1964, 2A:115, reel 9, SDS Papers.
56. Hayes Mizell to Steven McNichols, 5 October 1964, Curry Papers.
57. C. Clark Kissinger to Sue Thrasher, 17 December 1964, SDS Papers; Sue Thrasher to Archie Allen, October 1964, in "Material from SOC [*sic*] Files," in the possession of Harlon Joye.
58. Sue Thrasher to Anne Braden, 12 January 1965, folder 3, box 61, Braden Papers.

Chapter 4

1. Marshall B. Jones, "Berkeley of the South: A History of the Student Movement at the University of Florida, 1963–1968," 7, 14, George A. Smathers Libraries, University of Florida, Gainesville, Florida.
2. *1965 Conference Report*, Edwin Hamlett Papers (uncatalogued), Special Collections and University Archives, The Jean and Alexander Heard Library, Vanderbilt University, Nashville, Tennessee.
3. *Newsletter: Southern Student Organizing Committee*, 1:3 (December 1964). Maryville students also organized and hosted a human relations conference in late October 1964 for students from several Tennessee schools. *Highland Echo* (Maryville College), 24 October 1964.
4. Archie Allen to Sue Thrasher, 13 October 1964 and 17 October 1964, folder 3, Southern Student Organizing Committee Papers, 1964–1968,

Special Collections, University of Virginia Library, Charlottesville, Virginia.
5. *Southern Patriot,* May 1964.
6. Ibid.; *New Rebel,* 1:1 (27 May 1964), in the Southern Student Organizing Committee Papers, Archives and Library, The Martin Luther King, Jr. Center for Nonviolent Social Change, Atlanta, Georgia; *Newsletter: Southern Student Organizing Committee,* 1:3 (December 1964); and *Student Voice,* 5:10 (5 May 1964) and 5:11 (19 May 1964).
7. Interview, David Kotelchuck, 31 August 1994; *Nashville Tennessean,* 4 May 1964; *Southern Patriot,* May 1964 (quote); *New Rebel,* 1:1 (27 May 1964), in the SSOC Papers, King Center; *Student Voice,* 5:10 (5 May 1964) and 5:11 (19 May 1964). Protest card handed out by sip-in participants located in David and Ronda Kotelchuck Papers (Southern Student Organizing Committee Papers, 1959–1977), Special Collections, University of Virginia Library, Charlottesville, Virginia.
8. *Southern Patriot,* May 1964; *Newsletter: Southern Student Organizing Committee,* 1:3 (December 1964); interview, David Kotelchuck.
9. *Newsletter: Southern Student Organizing Committee,* 2:1 (January 1965).
10. *Newsletter: Southern Student Organizing Committee,* 2:1 (January 1965) and 2:4 (May 1965); *New South Student,* 3:2 (February 1966).
11. *Red and Black,* 1 August 1963, 29 October 1963, 12 November 1963, and 26 May 1964. On the desegregation of the University of Georgia, see Robert A. Pratt, *We Shall Not Be Moved: The Desegregation of the University of Georgia* (Athens: University of Georgia Press, 2002); Thomas G. Dyer, *The University of Georgia: A Bicentennial History, 1785–1985* (Athens: University of Georgia Press, 1985), 303–334; Calvin Trillin, *An Education in Georgia: The Integration of Charlayne Hunter and Hamilton Holmes* (New York: Viking Press, 1963); Charlayne Hunter-Gault, *In My Place* (New York: Farrar Straus Giroux, 1992), esp. 169–246; Robert Cohen, " 'Two, Four, Six, Eight, We Don't Want to Integrate': White Student Attitudes toward the University of Georgia's Desegregation," *The Georgia Historical Quarterly,* LXXX, no. 3 (Fall 1996): 616–645.
12. "We Believe Segregation Is Wrong: Join Us To Fight It" (quote), "Mass Protest Sunday," and William Tate to Nelson Perry Blackstock, 19 February 1964 (quote), Nelson Blackstock Papers (uncatalogued), Special Collections, University of Virginia Library, Charlottesville, Virginia; *Red and Black,* 25 February 1964; interview, Nelson Blackstock, 8 August 1995.
13. *New Rebel,* 1:1 (27 May 1964), in the SSOC Papers, King Center; *Florida Alligator,* 24 and 28 February 1964; Jones, "Berkeley of the South," 21–22; Stuart Howard Landers, "The Gainesville Women for Equal Rights, 1963–1978" (Master's Thesis: University of Florida, 1995), 32–33; interview, Mike Lozoff, 26 September 1996; interview, Marilyn Sokalof, 30 September 1996; interview, Dan Harmeling, 14 March 1995.

14. *Florida Alligator,* 6, 13, 15, and 30 January 1964 and 19 January 1965; Judy Benninger to Carl Braden, 17 December 1963, folder 1, box 42, Carl Braden and Anne Braden Papers, 1928–1972, Wisconsin Historical Society, Madison, Wisconsin; Jones, "Berkeley of the South," 22–26; interview, Sokalof.
15. *Florida Alligator,* 20, 21, 24, and 28 February 1964, 2 and 4 March 1964, and 19 January 1965; *New Rebel,* 1:1 (27 May 1964), in the SSOC Papers, King Center; Jones, "Berkeley of the South," 21–33.
16. *Newsletter: Southern Student Organizing Committee,* 2:2 (February 1965); interview, Sokalof. On the Free Speech Movement, see W. J. Rorabaugh, *Berkeley at War: The 1960s* (New York: Oxford University Press, 1989).
17. *Newsletter: Southern Student Organizing Committee,* 2:1 (January 1965) and 2:2 (February 1965); "The History of the Freedom Party," folder 4, box 61, Braden Papers; *Florida Alligator,* 22, 26 (quote), 29 January, and 3, 5, 8 (quote) February 1965; Jones, "Berkeley of the South," 42–44.
18. *Newsletter: Southern Student Organizing Committee,* 2:2 (February 1965); "The History of the Freedom Party," folder 4, box 61, Braden Papers; *Florida Alligator,* 12 and 15 February and 10 March (quote) 1965; Jones, *Berkeley of the South,* 42–44. For an extended treatment on Jim Harmeling's career at Florida after the Freedom Party's demise, see Jones, "Berkeley of the South," 176–190.
19. *Newsletter: Southern Student Organizing Committee,* 2:1 (January 1965); "Letter to Supporters," ca. January 1965, box 17, The Papers of the Boyte Family, Special Collections Library, Duke University, Durham, North Carolina.
20. *1965–66 Proposal,* folder 1, and "Southern Student Organizing Committee Spring Conference, March 19–21, 1965, Atlanta, Georgia," folder 2, SSOC Papers, University of Virginia; "Southern Student Organizing Committee Spring Conference, March 19–21, 1965, Atlanta, Georgia: Tentative Agenda," Constance Curry Papers (uncatalogued), Special Collections, University of Virginia Library, Charlottesville, Virginia; "Report on Spring Conference," Bruce Smith Papers (Southern Student Organizing Committee Papers, 1959–1977), Special Collections, University of Virginia Library, Charlottesville, Virginia; *Newsletter: Southern Student Organizing Committee,* 2:3 (April 1965). SSOC had invited Martin Luther King, Jr., to address the conference, but he begged off, responding by letter that he needed to preach at his church that weekend. In a handwritten note at the bottom of the letter, though, King suggested that the group ask his executive assistant, Andrew Young, to speak in his place since "he is an excellent speaker and quite familiar with SSOC." Martin Luther King, Jr., to Sue Thrasher, 23 February 1965, in the author's possession.
21. *Newsletter: Southern Student Organizing Committee,* 2:3 (April 1965) and 2:4 (May 1965); *Southern Patriot,* January and March 1966.
22. *1965–66 Proposal,* folder 1 SSOC Papers, University of Virginia; "Report on Spring Conference," Smith Papers; *Newsletter: Southern Student*

Organizing Committee, 2:3 (April 1965); "Southern Student Organizing Committee Spring Conference," March 19–21, 1965, Atlanta, Georgia, Curry Papers; *Southern Patriot,* April 1965.
23. Interview, Roger Hickey, 26 March 1993. Virginia students Howard Romaine, Steve Wise, and Tom Gardner served as SSOC chairman successively from 1965 to 1968.
24. *Human Relations: A News Letter Written and Distributed by the Jefferson Chapter of the Council on Human Relations,* no. 1 (20 July 1961), Special Collections, Special Collections, University of Virginia Library, Charlottesville, Virginia; Bryan Kay, "The History of Desegregation at the University of Virginia, 1950–1969" (Undergraduate Honors Essay: University of Virginia, 1979), 71–76.
25. Kay, "The History of Desegregation at the University of Virginia, 1950–1969," 71–76; Virginius Dabney, *Mr. Jefferson's University: A History* (Charlottesville: University Press of Virginia, 1981), 480; Tom Gardner, "Culture, Identity, and Resistance: Reflections on Counterhegemonic Steps by White Southern U.S. Activists in the 1960s" (unpublished seminar paper: University of Massachusetts, 1994), in the author's possession; interview, Tom Gardner, 2 September 1994 (quote); interview, Bill Leary, 2 April 1995 (quote); interview, Anne Cooke Romaine, 19 March 1993; interview, David Nolan, 13–14 March 1995. That white thugs had assaulted Gaston and the president of the local NAACP as they picketed the segregated Buddy's Restaurant near campus in 1963 also increased his stature among the students. Not only had he backed his words with actions by taking part in the demonstration, but as a young professor he had put his career at risk by bringing, from the administration's perspective, unfavorable attention to the university. Kay, "The History of Desegregation at the University of Virginia, 1950–1969," 91–93; Paul Gaston, "Sitting in in the Sixties," Lecture at the University of Virginia, October 1985, in the author's possession.
26. *Daily Progress,* 26 March 1963; *Cavalier Daily,* 27 March 1963, 16 February 1965, and 4 May 1965; *Human Relations,* no. 2 (31 July 1961); Kay, "The History of Desegregation at the University of Virginia, 1950–1969," 86–87; interview, Leary; interview, Nolan.
27. Interview, Howard Romaine, 17–18 March 1993; interview, Nolan; interview, Anne Cooke Romaine; Archie Allen, untitled essay, ca. November 1964, folder 2, SSOC Papers, University of Virginia (a later version would appear as "Report of Field Secretary," *Newsletter: Southern Student Organizing Committee,* 1:3 [December 1964]); David Nolan, "A Personal History of the Virginia Students' Civil Rights Committee," folder 7, box 6, David Nolan Papers, Wisconsin Historical Society, Madison, Wisconsin; "Southwide Fall Conference, Southern Student Organizing Committee, Old Gammon Theological Seminary, Atlanta, Georgia, November 13–15, 1964," folder 2, SSOC Papers, University of Virginia; *Newsletter: Southern Student Organizing Committee,* 2: 1 (January 1965); Roger Hickey, speech at the University of Virginia, 20 November 1990, in the author's possession.

28. *Cavalier Daily,* 10 and 17 December 1964, 7 January 1965, and 17, 18, and 19 March 1965; *Newsletter: Southern Student Organizing Committee,* 2:1 (January 1965); interview, Nolan; Dabney, *Mr. Jefferson's University,* 480.
29. *Newsletter: Southern Student Organizing Committee,* 2:1 (January 1965); *New South Student,* 3:6 (September 1966);*Cavalier Daily,* 14 October 1965.
30. The six targeted Southside counties were Lunenburg, Powhatan, Nottoway, Amelia, Brunswick, and Dinwiddie. *Southern Patriot,* April and September (quote) 1965; *New South Student,* 2:6 (November 1965); "Southside Project," 1966, Hamlett Papers; "A Summer Project for Virginia in 1965," folder 11, box 6, and David Nolan, "A Personal History of the Virginia Students' Civil Rights Committee," folder 7, box 6, Nolan Papers; "The Story of VSCRC," ca. September 1965, Steve Wise Papers (Southern Student Organizing Committee Additional Papers, 1963–1979 *and* Southern Student Organizing Committee Papers, 1959–1977), Special Collections, University of Virginia Library, Charlottesville, Virginia (quote); "The Virginia Project," Southern Student Organizing Committee Thirtieth Reunion and Conference, 9 April 1994, Charlottesville, Virginia, in the author's possession; interview, Stanley Wise, 16 March 1995; interview, Howard Romaine; interview, Anne Cooke Romaine; interview, Nolan; interview, Gardner, 2 September 1994; interview, Nan Grogan Orrock, 14 November 1994.
31. VSCRC fundraising letter, 26 August 1965 and "Report on the Amelia-Nottoway Medical Program, Summer 1965," Kotelchuck Papers; "Minutes of VSCRC Staff Meeting," Blackstone (Nottoway Co.), Virginia, 20 July 1965, and Lawrenceville (Brunswick Co.), Virginia, 16 December 1965, folder 6, box 6, Nolan Papers; *The New Virginia: Virginia Students' Civil Rights Committee Newsletter,* 1:4 (December 1965), Smith Papers.
32. "Progress Report and Future Program Developments of the Virginia Students' Civil Rights Committee," ca. February 1966, Smith Papers; "Minutes of VSCRC Staff Meetings," Amelia Courthouse (Amelia Co.), Virginia, 15 July 1965 and Blackstone (Nottoway Co.), Virginia, 20 July 1965, folder 6, box 6, Nolan Papers; *The New Virginia: The Virginia Students' Civil Rights Committee Newsletter,* 1:1 (September 1965), A:xv:253, reel 4, *Student Nonviolent Coordinating Committee Papers, 1959–1972* (Sanford, N.C.: Microfilm Corporation of America, 1982); VSCRC fundraising letter, 26 August 1965 and "Amelia Freedom Rally," 14 August 1965, Kotelchuck Papers; "The Virginia Project," 9 April 1994; interview, Leary; interview, Bruce Smith, 2 April 1995; interview, Hickey; interview, Anne Cooke Romaine; interview, Orrock; interview, Nolan.
33. "The Story of VSCRC," ca. September 1965, Wise Papers; "Minutes of VSCRC Staff Meetings," location unknown, 30 November 1965 and Lawrenceville (Brunswick Co.), Virginia, 16 December 1965, folder 6,

box 6, Nolan Papers; *The New Virginia: Virginia Students' Civil Rights Committee Newsletter,* 1:2 (October 1965) and 1:4 (December 1965), and Dick Langley to Bruce Smith, 30 March 1966, Smith Papers; "Voter Registration Information, Amelia County, Virginia," July 1965, Kotelchuck Papers; "The Virginia Project," 9 April 1994; interview, Smith; interview, Bill and Betsy Jean Towe, 16 November 1994; interview, Orrock; interview, Nolan.

34. Interview, Orrock, 14 November 1994; interview, Nolan; interview, Stanley Wise; interview, Bill and Betsy Jean Towe; "Minutes of VSCRC Staff Meetings," location unknown, 30 November 1965, Lawrenceville (Brunswick Co.), Virginia, 16 December 1965, Victoria (Lunenburg Co.), Virginia, 30 December 1965, Victoria, Virginia, 6 July 1966, folder 6, box 6, Nolan Papers; "Booknotes," 1966, folder 6, box 7, Nolan Papers; *Richmond Times-Dispatch,* 11 October 1965. Approximately 25 students took part in the second summer project, as VSCRC expanded into several new counties, including Charlotte, Southampton, and Mecklenburg. "Virginia Students' Civil Rights Committee County Assignments," 1966, folder 11, box 6, Nolan Papers.

35. Jones, "Berkeley of the South," 88–89.

36. *New South Student,* 2:5 (October 1965) and 2:6 (November 1965); *Emory Wheel,* 1, 15, 29 October and 4 November 1965; interview, Jody Palmour, 2 and 18 April 1995; Nancy Diamond, "Catching Up: The Advance of Emory University since World War II," *History of Higher Education Annual,* 19 (1999): 159.

37. *New South Student,* 3:1 (January 1966); Jones, "Berkeley of the South," 53–55, 60–61, 89–91; interview, David Kotelchuck; "Walk for Peace" flyer, March 1966, Kotelchuck Papers.

38. Interview, Leary; interview, Tom Gardner, 2 September 1994 (quotes); interview, Nolan; Tom Gardner, "The Southern Student Organizing Committee 1964–1970," (October 1970), 26, in the author's possession; Paul Gaston to the author, 17 February 1999, in the author's possession.

39. *Cavalier Daily,* 17 February 1966; *Richmond Times-Dispatch,* 13 February 1966; *New South Student,* 3:4 (April 1966); interview, Hickey, 26 March 1993; interview, Leary; Bob Dewart, "Master Plan B (quote)," 21 July 1966, folder 9, box 6, and David Nolan, "Booknotes," folder 7, box 6, Nolan Papers.

40. *New South Student,* 2:6 (November 1965).

Chapter 5

1. Will D. Campbell to George Loft, 9 October 1968, box 2S427, file "Southern Student Organizing Committee, Fall 1968," Field Foundation Archives, 1940–1990, The Center for American History, The University of Texas at Austin, Austin, Texas.

2. Interview, Tom Gardner, 2 September 1994.

3. Sue Thrasher to Carl and Anne Braden, 25 May 1965, folder 3, box 61, Carl Braden and Anne Braden Papers, 1928–1972, Wisconsin Historical Society, Madison, Wisconsin.
4. Bryant Simon, "Southern Student Organizing Committee: A New Rebel Yell in Dixie" (Honors Essay: University of North Carolina, 1983), 47–49; "Southern Student Organizing Committee Statement of Cash Receipts and Disbursements For the year ended August 31, 1966," file "Southern Student Organizing Committee November 1966," box 2S427, Field Foundation Archives, 1940–1990.
5. Marilyn B. Young, *The Vietnam Wars, 1945–1990* (New York: HarperCollins, 1991), 116–171.
6. Interview, Howard Romaine, 18–19 March 1993; unnamed student to SSOC, April 1966 and notes on "Spring 1965—Vietnam Workshop," in "Materials from SOC [sic] Files," in the possession of Harlon Joye.
7. Interview, Ed Hamlett, 6 May 1993.
8. J. Anthony Lukas, *Don't Shoot—We Are Your Children!*, third edition (New York: Random House, 1971 [first edition: New York: Random House, 1968]), 161; *New South Student*, 3:2 (February 1966).
9. Interview, Bruce Smith, 2 April 1995; Bruce Smith to David Nolan, 8 February 1967, Bruce Smith Papers (Southern Student Organizing Committee Papers, 1959–1977), Special Collections, University of Virginia Library, Charlottesville, Virginia; notes on "Highlander Workshop—Summer 1965—Vietnam Workshop," in "Material from SOC [sic] Files," in the possession of Harlon Joye.
10. Archie Allen, untitled essay, ca. November 1964, folder 2, Southern Student Organizing Committee Papers, 1964–1968, University of Virginia (a later version appeared as "Report of Field Secretary," *Newsletter: Southern Student Organizing Committee*, 1:3 [December 1964]).
11. Robert William Dewart, "Questionnaire," Southern Student Organizing Committee Thirtieth Reunion and Conference, April 1994, Charlottesville, Virginia, in the author's possession.
12. Interview, Lyn Wells, 16 November 1994; interview, Howard Romaine.
13. *New South Student*, 2:5 (October 1965), 2:6 (November 1965), and 2:7 (December 1965). With the October 1965 issue, the first of the new school year, SSOC had redesigned its newsletter. In addition to renaming it and thus completing the process set in motion earlier in the year when it discarded the name the *New Rebel* to show its sensitivity to black students, the newsletter also presented either a photograph, drawing, or graphic design on the cover instead of a simple hand-written title or the first page of text.
14. Todd Gitlin, *The Sixties: Years of Hope, Days of Rage* (New York: Bantam Books, 1987), 181; *Report on Spring Conference*, in *1965–66 Proposal*, folder 1, SSOC Papers, University of Virginia; *New South Student*, 2:5 (October 1965), 2:6 (November 1965), and 2:7 (December 1965). Because the group chose not to participate in the 1965 antiwar march, a number of SSOC activists, including Sue Thrasher and Gene Guerrero,

took part in the march as individuals rather than as representatives of SSOC. Interview, Sue Thrasher (interviewed by Ronald Grele), 15 December 1984, Columbia Oral History Project, Columbia University, New York, New York.
15. *New South Student,* 2:6 (November 1965).
16. Protests occurred elsewhere as well; antiwar rallies took place in Jackson, Mississippi, and on the grounds of the University of Florida and Florida State University, and the University of Arkansas Little Rock hosted that state's first teach-in. *New Left Notes,* 1:3 (4 February 1966) and 1:5 (18 February 1966); *New South Student,* 3:2 (February 1966); *Daily Progress,* 14 February 1966; *Richmond Times-Dispatch,* 13 February 1966; *Atlanta Constitution,* 12 and 14 February 1966; *Atlanta Journal and Constitution,* 13 February 1966; Bryant Simon, "Southern Student Organizing Committee: A New Rebel Yell in Dixie," 47.
17. *Southern Patriot,* April 1966.
18. *New South Student,* 3:4 (April 1966) (quote); *Southern Patriot,* April and June 1966; *New Left Notes,* 1:15 (29 April 1966) and 1:20 (3 June 1966); "If You Would," *Worklist Mailing,* 1:3 (April 1966), Southern Student Organizing Committee and Thomas N. Gardner Papers, 1948–1994 [1958–1975], Special Collections, University of Virginia Library, Charlottesville, Virginia; interview, Mike Welch, 1 September 1994. On the Selective Service College Qualification Test, see Michael S. Foley, *Confronting the War Machine: Draft Resistance during the Vietnam War* (Chapel Hill: University of North Carolina Press, 2003), 38–40; Nancy Zaroulis and Gerald Sullivan, *Who Spoke Up?: American Protest against the War in Vietnam, 1963–1975* (New York: Doubleday & Co., 1984), 85–86; Tom Wells, *The War Within: America's Battle over Vietnam* (Berkeley, Calif.: University of California Press, 1994), 82–83; Charles DeBenedetti and Charles Chatfield, assisting author, *An American Ordeal: The Antiwar Movement of the Vietnam Era* (Syracuse, N.Y.: Syracuse University Press, 1987), 165–167. Nationally, by the end of 1966, 17 percent of the men who took the exam failed. In the South, 48 percent of the test takers failed the exam. *New South Student,* 4: 1 (February 1967). The argument that working-class youths were over-represented in the Vietnam-era military is made by Christian G. Appy, *Working-Class War: American Combat Soldiers and Vietnam* (Chapel Hill: University of North Carolina Press, 1993). James Fallows offered a personal and, at the time, highly controversial perspective on the subject in "What Did You Do In The Class War, Daddy?" *Washington Monthly* 7, no. 8 (October 1975), 5–19.
19. "Southern Student Organizing Committee: 1966–67 Proposal," file "Southern Student Organizing Committee, November 1966," box 2S427, Field Foundation Archives; "North Nashville Project: Proposal for Action," and "A Letter to Friends About SSOC," Smith Papers; Ray Payne, "A Working Paper on Community Organizing in Vine City, Atlanta," n.d., David and Ronda Kotelchuck Papers (Southern Student Organizing Committee Papers, 1959–1977), Special Collections,

University of Virginia Library, Charlottesville, Virginia; *New South Student,* 3:2 (February 1966) (quote); *New Left Notes,* 1:5 (18 February 1966); Simon, "Southern Student Organizing Committee," 47–49; interview, Ronda Stilley Kotelchuck, 4 September 1994. On the Vine City project, see Clayborne Carson, *In Struggle: SNCC and the Black Awakening of the 1960s* (Cambridge, Mass.: Harvard University Press, 1981), 191–200.

20. Interview, Anne Cooke Romaine, 19 March 1993; interview, Gene Guerrero, 26 March 1993; J. Anthony Lukas, *Don't Shoot—We Are Your Children!,* 159.
21. Interview, Howard Romaine; interview, Guerrero; interview, Anne Cooke Romaine; interview, Wells; Lukas, *Don't Shoot—We Are Your Children!,* 158–159. On the new folk music of the day, see Robert Cantwell, *When We Were Good* (Cambridge, Mass.: Harvard University Press, 1996), esp. 337–352.
22. Thomas L. Connelly, *Will Campbell and the Soul of the South* (New York: Continuum Publishing Co., 1982), 11; Ed Hamlett to John A. Strickland, ca. October 1964, Edwin Hamlett Papers (uncatalogued), Special Collections and University Archives, The Jean and Alexander Heard Library, Vanderbilt University, Nashville, Tennessee; Thrasher quoted in Lukas, *Don't Shoot—We Are Your Children!,* 159. The lyrics to the song about "Big Jim" Folsom relate his treatment of the young girl to the way the rich have always treated the poor:

> Well she was poor but she was honest,
> The Victim of a rich man's whim;
> Took a ride with that fine Christian gentleman
> > big Jim Folsom
> And she had a child by him.

> Well now he sits in the legislature
> Makin' laws for all mankind;
> While she roams the streets of Cullman, Alabama
> Sellin' grapes from her grape vine.

> Now you think this is my story
> But the worst is yet to come.
> While he sits up in the capital kissin' women
> He won't even name his son.

> Well now, the moral of the story
> Is don't ever take a ride
> With that fine Christian gentleman
> > big Jim Folsom
> And you'll be a virgin bride.

> And it's the rich what get the glory
> And the poor what get the blame;
> And it's the same the whole world over, over, over;
> Now ain't that a shame.

23. "The GTU Report," Membership Mailing Number One, 24 October 1966, folder 1, box 1, SSOC and Gardner Papers; David Nolan to author, 20 August 2003.
24. "Southern Student Organizing Committee: Income, September 1, 1964–September 1, 1965," file "SSOC 7th 1965," "Southern Student Organizing Committee: Statement of Cash Receipts and Disbursements for the year ended August 31, 1966," file "SSOC November 1966," and "Grant Proposal," winter 1965, file "Southern Student Organizing Committee, 1965," box 2S427, Field Foundation Archives (quote); "Meeting of Board of Directors for allocation of funding for 1966," file "The Field Foundation, 1965," box 2T121, Field Foundation Archives; Norman Thomas to Archie Eugene Allen, 3 February 1965, in the possession of Archie Allen; *New Left Notes*, 1:7 (4 March 1966); Ed Clark, "The South Must Be Won," *PL: Progressive Labor* 7, no. 3 (November 1969): 20–21; Ron Parker to "Friends of SSOC," 12 January 1965, Constance Curry Papers (uncatalogued), Special Collections, University of Virginia Library, Charlottesville, Virginia.
25. "Agents of Change: Southern White Students in the Civil Rights Movement," A Public Seminar, 7 April 1994, Charlottesville, Virginia, in the author's possession; interview, Anne Cooke Romaine; interview, Howard Romaine; interview, Bob Moses, 18 April 1995.
26. Interview, Anne Cooke Romaine; interview, Howard Romaine.
27. Interview, Anne Cooke Romaine; *New South Student*, 3:1 (January 1966); "Proposal for Southern Folk Cultural Revival Project," A:xvii:160, reel 44, *The Student Nonviolent Coordinating Committee Papers, 1959–1972* (Sanford, N.C.: Microfilm Corporation of America, 1982); "Southern Folk Festival," Spring 1965, folder 3, box 1, Papers of the United States National Student Association, 1955–1969, Archives and Library, The Martin Luther King, Jr. Center for Nonviolent Social Change, Atlanta, Georgia.
28. *Southern Patriot*, April 1966 (quote); *New South Student*, 3:3 (March 1966); Charles Joyner to Pat and Hayes Mizell, 20 May 1966, M. Hayes Mizell Papers, South Caroliniana Library, University of South Carolina, Columbia, South Carolina; interview, Anne Cooke Romaine; "Proposal for Southern Folk Cultural Revival Project," A:xvii:660, reel 44, SNCC Papers; Bernice Johnson Reagon, *If You Don't Go, Don't Hinder Me: The African American Sacred Song Tradition* (Lincoln: University of Nebraska Press, 2002), 76–77.
29. After 1967, the Southern Folk Foundation, a tax-exempt organization established by Romaine and Reagon, changed the festival's name to the Southern Folk Cultural Revival Project and ran it as an independent organization. While SSOC remained a steadfast supporter of the project, it no longer was instrumental to its success. Interview, Anne Cooke Romaine; interview, Earl Wilson, 11 November 1994; "Getting the Word Out: Membership Mailing Number 7," 18 April 1967, folder 28, box 38, Papers of the United States National Student Association, 1955–1969, The Martin Luther King, Jr. Center for Nonviolent Social

Change; Anne Romaine, "Fledging Southern Folk Festival Flapping Its Wings for the Leap!!", ca. June 1967, folder 2, box 4, SSOC and Gardner Papers; "Southern Student Organizing Committee: 1966–1967 Proposal," file "Southern Student Organizing Committee, November 1966," box 2S427, Field Foundation Archives; "Proposal for Southern Folk Cultural Revival Project," A:xvii:660, reel 44, SNCC Papers; *New South Student*, 3:3 (March 1966).

30. Insightful treatments of SNCC's turn toward racial separatism and discussions of the origins, meaning, and rise of Black Power include Stokely Carmichael and Charles V. Hamilton, *Black Power: The Politics of Liberation in America* (New York: Vintage Books, 1967); Allen J. Matusow, "From Civil Rights to Black Power: The Case of SNCC, 1960–1966," in Barton J. Bernstein and Allen J. Matusow, eds., *Twentieth-Century America: Recent Interpretations* (New York: Harcourt Brace Jovanovich, 1972), 494–520; James Forman, *The Making of Black Revolutionaries,* second edition (Seattle, Wa.: Open Hand, 1985 [first edition: New York: Macmillan, 1972]), 411–481; Carson, 191–243; William L. Van Deburg, *New Day in Babylon: The Black Power Movement and American Culture, 1965–1975* (Chicago: The University of Chicago Press), 1992.

31. Interview, Alan Levin, 26 December 1994; Ed Hamlett to Jane Stembridge, March 1966, in "Materials from SOC [*sic*] Files," in the possession of Harlon Joye; *New South Student*, 3:3 (March 1966).

32. Interview, Tom Gardner, 5 May 1993. Perhaps the most awkward and uncomfortable moment of the discussion occurred when some of the SNCC activists, in the course of pressing their point that blacks needed to unite and separate from whites in the movement, expressed their disapproval of the marriage of Janet Dewart, SSOC's lone black staffer, to Bob Dewart, a white University of Virginia student. Two of the black activists even threatened the Dewarts with violence. Stokely Carmichael disarmed the Dewarts' accusers, and the next day Bob Dewart passionately defended their marriage to the entire assemblage. Janet Dewart Bell to SSOClist, 12 May 2003; Tom Gardner to SSOC list, 11 May 2003; Tom Gardner to author, 26 April 1996, in the author's possession; interview, David Nolan, 13–14 March 1995.

33. Tom Gardner, "The Southern Student Organizing Committee, 1964–1970" (October 1970), 21–22, in the author's possession; Tom Gardner to author, 26 April 1996, in the author's possession.

34. Interview, Gardner, 2 September 1994 (first quote); Thomas N. Gardner, speech at the University of Virginia, 20 November 1990, in the author's possession (second quote).

35. Interview, Steve Wise, 18 March 1993.

36. Hamlett quoted in *New South Student*, 3:5 (May 1966).

37. "Southern Student Organizing Committee: 1966–67 Proposal," file "Southern Student Organizing Committee, November 1966, box 2S427, Field Foundation Archives (quote); Steve Wise, "Incomplete and Selective Notes on the Buckeye Cove Conference," 28 June 1966, folder 2, SSOC Papers, University of Virginia; interview, Tom Gardner, 2 September 1994.

38. Interview, Nolan.

39. Steve Wise, "Incomplete and Selective Notes on the Buckeye Cove Conference," 28 June 1966, folder 2, SSOC Papers, University of Virginia; Harlon E. Joye, "Dixie's New Left," *Trans-action*, 7, no. 11 (September 1970), 54.
40. Burlage quoted in Anne Braden to SSOC, June 1966, A:xvii:164, reel 44, SNCC Papers. Others to whom SSOC turned to for guidance included Rev. Edwin King, the chaplain at Mississippi's Tougaloo Southern Christian College and a veteran of the freedom movement in the Magnolia state; Connie Curry, the first full-time director of the NSA Southern Project; and Lillian Smith, the famed author of the southern classics *Strange Fruit* and *Killers of the Dream* and an ardent proponent of desegregation and black equality.
41. Will D. Campbell to George Croft, 9 October 1968, Field Foundation Archives. Works that treat Campbell's life and career include Connelly, *Will Campbell and the Soul of the South*; Charles W. Eagles, "The Closing of Mississippi Society: Will Campbell, *The $64,000 Question*, and Religious Emphasis Week at the University of Mississippi," *Journal of Southern History* 67, no. 2 (2001): 331–372; James F. Findlay, Jr., *Church People in the Struggle: The National Council of Churches and the Black Freedom Movement, 1950–1970* (New York: Oxford University Press, 1993), 22–27; John Egerton, *A Mind to Stay Here: Profiles from the South* (New York: Macmillan, 1970), 18–31. The books Campbell has authored remain the best guide to his beliefs and his career. See especially Will D. Campbell, *Brother to a Dragonfly* (New York: Seabury Press, 1977) and Will D. Campbell, *Forty Acres and a Goat: A Memoir* (Atlanta, Ga.: Peachtree Publishers, 1986).
42. Interview, Will D. Campbell, 18 March 1993.
43. Anne Braden to SSOC, June 1966, A:xvii:164, reel 44, SNCC Papers; *Southern Patriot*, June 1967; *Southern Patriot*, May 1966 (quote).
44. *Southern Patriot*, May 1966 and June 1967 (quote); interview, Anne Braden, 25 April 1993.
45. Anne Braden to SSOC, June 1966, A:xvii:164, reel 44, SNCC Papers.
46. Ibid.; *New South Student*, 3:7 (December 1966).
47. Interview, Thrasher, 15 December 1984; "Southern Student Organizing Committee: 1966–1967 Proposal," file "Southern Student Organizing Committee, November 1966," box 2S427, Field Foundation Archives.
48. "Southern Student Organizing Committee: 1966–1967 Proposal," file "Southern Student Organizing Committee, November 1966," box 2S427, Field Foundation Archives (quote); *New South Student*, 3:2 (February 1966); *Southern Patriot*, August 1966; Ronda Stilley to Bruce Smith, ca. April 1966, Smith Papers.
49. *New South Student*, 3:7 (December 1966).
50. Interview, Braden; Gardner, "The Southern Student Organizing Committee, 1964–1970," 16–17; Anne Braden to SSOC, June 1966, A:xvii:164, reel 44, SNCC Papers.
51. Sue Thrasher, "Some Assorted Thoughts of Sue Thrasher," ca. March 1966, Nelson Blackstock Papers (uncatalogued), Special Collections, University of Virginia Library, Charlottesville, Virginia (quote).

52. "The Meeting at Buckeye Cove, June 6–10, 1966," folder 2, "1966–67 Proposal," folder 1, Steve Wise, "Incomplete and Selective Notes on the Buckeye Cove Conference," 28 June 1966, folder 2, (quote), all in SSOC Papers, University of Virginia; "Constitution of the Southern Student Organizing Committee (adopted June 1966)," Steve Wise Papers (Southern Student Organizing Committee Additional Papers, 1963–1979 *and* Southern Student Organizing Committee Papers, 1959–1977), Special Collections, University of Virginia Library, Charlottesville, Virginia; Sue Thrasher to Leslie Dunbar, 29 October 1966, file "Southern Student Organizing Committee, 1965?", box 2S427, Field Foundation Archives; "Not Too Much: sporadic worklist mailing of the Southern Student Organizing Committee," 1:1 (7 March 1966), Smith Papers; *New South Student*, 3:2 (February 1966); interview, Nolan.
53. *New South Student*, 3:3 (March 1966); Minutes of SSOC Advisory Board Meeting, 4 March 1966, Nashville, Tennessee, Mizell Papers; Steve Wise, "Dear Friends," 14 September 1966, folder 1, box 6, Nolan Papers; "1966–67 Proposal," folder 1, SSOC Papers, University of Virginia; Nan Grogan to Gordon Patton, 2 October 1966, Wise Papers; Howard Romaine to Steve Wise, Roger Hickey, Tom Gardner and other UVa activists, ca. May 1966, folder 1, box 1, SSOC and Gardner Papers.
54. "Constitution of the Southern Student Organizing Committee (adopted June 1966)," Wise Papers; Minutes of SSOC Advisory Board Meeting, 4 March 1966, Nashville, Tennessee, Mizell Papers.

Chapter 6

1. Interview, Lyn Wells, 16 November 1994.
2. *Great Speckled Bird*, 24 May–6 June 1968.
3. Exchange of letters between Ed Hamlett and anonymous Birmingham student, July-September 1966, in "Material from SOC [*sic*] Files," in the possession of Harlon Joye.
4. Ibid.
5. The University of Virginia SSOC chapter formed when students in the leading activist organization on campus, Students for Social Action, decided to drop this local name and to rechristen themselves as a SSOC chapter. David Lubs to Tom Gardner, 13 September 1966 and Bob Dewart to Tom Gardner and Jody Palmour, 20 September 1966, folder 3, box 5, Southern Student Organizing Committee and Thomas N. Gardner Papers, 1948–1994 [1958–1975], Special Collections, University of Virginia Library, Charlottesville, Virginia. SSOC chapters subsequently emerged on a number of campuses, including the University of Florida, Memphis State University, Davidson College, Florida State University, Southwestern at Memphis, and the University of Miami. "Activities of the Southern Student Organizing Committee, September 1966–Sept. 1967," folder 1, box 4, SSOC and Gardner Papers.

6. Interview, Anne Cooke Romaine, 19 March 1993; interview, Howard Romaine, 17–18 March 1993; interview, Sue Thrasher, 1 September 1994; interview, Sue Thrasher (interviewed by Ronald Grele), 15 December 1984, Columbia Oral History Project, Columbia University, New York; J. Anthony Lukas, *Don't Shoot—We Are Your Children!*, third edition (New York: Random House, 1971 [first edition: New York: Random House, 1968]), 161–162; Ed Hamlett to Connie Curry, 12 December 1966, Constance Curry Papers (uncatalogued), Special Collections, University of Virginia Library, Charlottesville, Virginia; "Whither the Smart-Ass," Membership Mailing Number Four, 14 January 1967, Bruce Smith Papers (Southern Student Organizing Committee Papers, 1959–1977), Special Collections, University of Virginia Library, Charlottesville, Virginia.
7. Interview, Thrasher, 15 December 1984; interview, Ronda Stilley Kotelchuck, 4 September 1994.
8. Interview, Wells; Ronda Stilley to Nan and Gene Guerrero, 30 March 1967 (quote), Steve Wise Papers (Southern Student Organizing Committee Additional Papers, 1963–1979 *and* Southern Student Organizing Committee Papers, 1959–1977), Special Collections, University of Virginia Library, Charlottesville, Virginia.
9. Shirley Newton Bliley, "Questionnaire," Southern Student Organizing Committee Thirtieth Reunion and Conference, April 1994, Charlottesville, Virginia, in the author's possession; Shirley Newton Bliley, "The Sixties—A Personal Perspective," in the author's possession.
10. Interview, Mike Welch, 1 September 1994.
11. Interview, Jody Palmour, 2 and 18 April 1995.
12. *Southern Patriot*, December 1968, March 1969; interview, Everett Long, 12 November 1994. For background on Millsaps' history, see G. L. Harrell, *History of Millsaps College* (Jackson, Miss.: Millsaps College, 1943). For a brief discussion of the early civil rights movement and Millsaps, see John Dittmer, *Local People: The Struggle for Civil Rights in Mississippi* (Urbana: University of Illinois Press, 1994), 61–62; and Charles Marsh, *God's Long Summer: Stories of Faith and Civil Rights* (Princeton, N.J.: Princeton University Press, 1997), 120–121. On the events surrounding the establishment of Strike City, see Dittmer, *Local People*, 364–368.
13. Earle Johnston, Jr., "Memorandum to File," 20 June 1967, Mississippi State Sovereignty Commission Records, SCR ID# 3-11-0-26-1-1-1, Mississippi Department of Archives & History, Jackson, Mississippi; interview, David Doggett, 30 August 1994; Jackson *Clarion-Ledger*, 12 and 13 May 1967; *New South Student*, 4:4 (May 1967); Letter from Mark Matheny, President, Millsaps College Student Association, ca. May 1967, box 38, folder 4, Administrative Records, President's office (B. B. Graves, 1965–1970), Millsaps College Archives, Millsaps-Wilson Library, Millsaps College, Jackson, Mississippi; David Doggett, "*The Kudzu* Story: Underground in Mississippi," *Southern Exposure*, 2, no. 4 (1975), 88 (quote); David Doggett, "*The Kudzu:* Birth and Death in Underground

Mississippi," in *Voices from the Underground: Insider Histories of the Vietnam Era Underground Press,* Ken Wachsberger, ed. (Tempe, Ariz.: Mica Press, 1993), 213–232.
14. Interview, David Nolan, 13–14 March 1995.
15. Tom Gardner, "Culture, Identity, and Resistance: Reflections on Counterhegemonic Steps by White Southern U.S. Activists in the 1960s" (unpublished seminar paper: University of Massachusetts, 1994), in the author's possession (quote); Interview, Tom Gardner, 5 May 1993.
16. Interview, Gardner, 5 May 1993.
17. Interview, Earl Wilson, 11 November 1994.
18. Interview, Nolan; Lyn Wells, "American Women: Their Use and Abuse," box 17, The Papers of the Boyte Family, Special Collections Library, Duke University, Durham, North Carolina.
19. Interview, Wells.
20. Ibid.
21. Gardner, "Culture, Identity, and Resistance," 9; interview, Tom Gardner, 5 May 1993; interview, Mike Welch.
22. Interview, Doggett; interview, Palmour; interview, Gardner, 5 May 1993; *Southern Patriot,* March 1969; Shirley Newton Bliley, "Questionnaire," Southern Student Organizing Committee Thirtieth Reunion and Conference, April 1994, Charlottesville, Virginia, in the author's possession.
23. Exchange of letters between Nolan and his father, 30 April, 10 May, and 13 May 1965 and letter from grandmother, 7 September 1965, all in folder 8, box 6, David Nolan Papers, Wisconsin Historical Society, Madison, Wisconsin.
24. Doggett, *"The Kudzu,"* 213–215; interview, Doggett.
25. *Florida Alligator,* 29 March 1967; interview, Bruce Smith, 2 April 1995; "Alternative Tour of the University of Virginia," Southern Student Organizing Committee Thirtieth Reunion and Conference, 10 April 1994, Charlottesville, Virginia, in the author's possession.
26. Interview, Howard Romaine; interview, Gene Guerrero, 26 March 1993; *Great Speckled Bird,* 1:1 (March 15–28, 1968); *Emory Wheel,* 12 October 1967; Sally Gabb, "A Fowl in the Vortices of Consciousness: The Birth of the *Great Speckled Bird,*" in *Voices from the Underground,* Wachsberger, ed., 41–50; Patrick K. Frye, *"The Great Speckled Bird:* An Investigation of the Birth, Life, and Death of an Underground Newspaper" (unpublished M.A. thesis: University of Georgia, 1981); Laurence Leamer, *The Paper Revolutionaries: The Rise of the Underground Press* (New York: Simon and Schuster, 1972), 93–104.
27. Doggett previously had been involved in the publication of several short-lived alternative papers in Jackson. In early 1966, he and fellow Millsaps student Lee Makemson published the *Free! South Student,* a paper, they wrote in its introduction, "for and by students who possess a true moral conscience" and devoted to "the realistic discussion of any relevant moral problems (civil rights, administrative control of the private lives of students, the foreign policies of our country, sex and the student, etc.)." The next year he published one issue of the *Mockingbird* and in early

1968 he and several others at Millsaps published a few issues of the *Unicorn. Free! South Student,* 1 (15 February 1966), Ed King Collection, Archives, L. Zenobia Coleman Library, Tougaloo College, Tougaloo, Mississippi; David Doggett to SSOClist, 16 April 2003, in the author's possession.

28. *Kudzu,* 1:1 (18 September 1968); Leamer, *The Paper Revolutionaries,* 142–146; Doggett, *"The Kudzu,"* 213–232; Doggett, *"The Kudzu Story,"* 86–95; *Southern Patriot,* December 1968; Donald Cunnigen, "Standing at the Gates: The Civil Rights Movement and Liberal White Mississippi Students," *Journal of Mississippi History,* 62, no. 1 (2000): 1–19; Stephen Flinn Young, *"The Kudzu:* Sixties Generational Revolt—Even in Mississippi," *The Southern Quarterly,* 34, no. 3 (1996): 122–136; interview, Doggett; interview, Long; Cassell Carpenter to author, 27 July 2002; *Purple & White* (Millsaps College), 18 October 1968, Millsaps College Archives, Millsaps-Wilson Library, Millsaps College, Jackson, Mississippi; *Kudzu,* 1:3 (23 October 1968) and 1:9 (ca. 26 February 1969); *Jackson Daily News,* 18 October 1968; Jackson *Clarion-Ledger,* 10 October and 20 December 1968; *The Reflector* (Mississippi State University), 17 December 1968. The *Kudzu* and its staff also were the subject of intense interest to Mississippi State Sovereignty Commission officials. Investigators regularly obtained copies of the *Kudzu* in order to track activist whites in Jackson, and they maintained files on Doggett, Long, and the others.
29. Interview, Doggett; interview, Long; Doggett, *"The Kudzu,"* esp. 218–221, 225–232; Leamer, *Paper Revolutionaries,* 141–146.
30. Interview, Alan Levin, 26 December 1994; interview, Nancy Lewis, 30 September 1996; *New South Student,* 5:2 (March 1968); Alan Levin to SDS, 21 March 1967, 3:27, reel 22, *Students for a Democratic Society Papers, 1958–1970* (Glen Rock, N.J.: Microfilm Corporation of America, 1977).
31. On military developments and antiwar actions during 1966, see Marilyn B. Young, *The Vietnam Wars, 1945–1990* (New York: Harper Collins, 1991), 180–191; Nancy Zaroulis and Gerald Sullivan, *Who Spoke Up?: American Protest against the War in Vietnam, 1963–1975* (New York: Holt, Rinehart and Winston, 1984), 69–99; Tom Wells, *The War Within: America's Battle over Vietnam* (Berkeley: University of California Press, 1994), 67–101; Charles DeBenedetti and Charles Chatfield, assisting author, *An American Ordeal: The Antiwar Movement of the Vietnam Era* (Syracuse, N.Y.: Syracuse University Press, 1987), 141–167.
32. *New South Student,* 3:5 (May 1966).
33. "Executive Committee Minutes," 25–26 February 1967, Wise Papers; Brian Heggen, "Application for Employment," 24 February 1967, Smith Papers; *Nashville Tennessean,* 15, 16, and 17 March 1967; "Press Release: Stop the Genocide in Vietnam" and Ronda Stilley to Nan and Gene Guerrero, 30 March 1967, both in Wise Papers; Shirley Newton Bliley, "The Sixties—A Personal Perspective," in the author's possession; interview, Wilson; interview, Nolan.

34. Steve Wise to Lyn Wells, 30 March 1967, Ronda Stilley to Nan and Gene Guerrero, 30 March 1967, and Lyn Wells to Steve Wise, 25 March 1967, Wise Papers; interview, Wells; [Janet Dewart?] to David Nolan, 13 March 1967, folder 1, box 6, Nolan Papers.
35. *Florida Alligator,* 24 and 25 January 1967; "Executive Committee Minutes," 25–26 February 1967, Wise Papers; *Daily Progress,* 30 June 1967; *NET Journal: Homefront 1967,* directed by Robert D. Souier, 45 min., Logos Limited, 1987, videocassette (originally produced by WETA); *Cavalier Daily,* 15 and 17 February 1967; *New South Student,* 4:3 (April 1967).
36. "Changes," SSOC Membership Mailing Number Five, 14 February 1967, and "Executive Committee Minutes," 25–26 February 1967, Wise Papers; *New South Student,* 4:2 (March 1967); Jackson *Clarion-Ledger,* 24 and 25 February 1967; Charles Cobb to Tom Gardner, 25 January 1967, in author's possession.
37. Tom Gardner to Dave Dellinger, 28 July 1967, folder 2, box 1, SSOC and Gardner Papers; *New South Student,* 4:6 (November 1967); George Brosi, "The Bratislava Meeting or Me and the V.C.," 1 October 1967, George Brosi Papers, Wisconsin Historical Society, Madison, Wisconsin; interview, Tom Gardner, 2 September 1994; interview, George Brosi, 12 November 1994; interview, Wells; interview, Robert McMahon, 30 October 1994; *Worklist Mailing: Confront the Warmakers Oct. 21,* 1:2 (11 October 1967), in the possession of Dorothy Burlage; *Worklist Mailing: Individuals Against the Crime of Silence,* 1:3 (18 November 1967), Wise Papers; *Daily Progress,* 20 October 1967; Braynt Simon, "Southern Student Organizing Committee: A New Rebel Yell in Dixie" (Honors Essay: University of North Carolina at Chapel Hill, 1983), 75; Lukas, *Don't Shoot—We Are Your Children!,* 162. On the Bratislava meeting, see Christopher Jencks, "Limits of the New Left," *The New Republic,* 157, no. 17 (21 October 1967): 19–21; Raymond Mungo, *Famous Long Ago: My Life and Hard Times with Liberation News Service* (Boston, Mass.: Beacon Press, 1970), 9–11; Tom Hayden, *Reunion: A Memoir* (New York: Random House, 1988), 208; David T. Dellinger, *From Yale to Jail: The Life Story of a Moral Dissenter* (New York: Pantheon Books, 1993), 255–256; Mary Hershberger, *Traveling to Vietnam: American Peace Activists and the War* (Syracuse, N.Y.: Syracuse University Press, 1998), 138–142.
38. Interview, Guerrero.
39. Interview, Gardner, 2 September 1994; Tom Gardner, "Manpower Unchanneled," in *We Won't Go: Personal Accounts of War Objectors,* Alice Lynd, ed. (Boston, Mass.: Beacon Press, 1968), 3–14 (quotes from 6, 11).
40. Interview, Levin (quotes); *Southern Patriot,* March 1968.
41. Robert William Dewart, "Questionnaire"; *New Left Notes,* 1:42 (11 November 1966); *New South Student,* 4:1 (February 1967); Bob Dewart to Tom Gardner, 27 February 1968, folder 3, box 3, Gardner Papers; Southern Student Organizing Committee, *Prospectus, 1967–68,* 17, Wise Papers. On draft resisters and deserters from the armed forces

who migrated to Canada, see Renée G. Kasinsky, *Refugees from Militarism: Draft-Age Americans in Canada* (New Brunswick, N.J.: Transaction Books, 1976); and John Hagan, *Northern Passage: American Vietnam War Resisters in Canada* (Cambridge, Mass.: Harvard University Press, 2001).
42. *Worklist Mailing: SSOC Stops the Draft*, ca. January 1968, in the possession of Dorothy Burlage; *Phoenix*, January 1968, Wise Papers; *New South Student*, 4:7 (December 1967); *Daily Progress*, 18 November 1967; *Southern Patriot*, March 1968.
43. *Worklist Mailing*, September 1967; *Worklist Mailing: Confront the Warmakers Oct. 21*, 1:2 (11 October 1967), in the possession of Dorothy Burlage; *Southern Patriot*, March 1968; interview, Levin.
44. *North Carolina Anvil*, 27 January 1968, box 4, The Papers of the Boyte Family, Special Collections Library, Duke University, Durham, North Carolina; N.C. Student Committee Against the War, *Worklist Mailing #2*, 30 October 1967, and *Worklist Mailing #4*, 20 January 1968, in author's possession; *Southern Patriot*, March 1968; *Worklist Mailing: SSOC Stops the Draft*, ca. January 1968, in the possession of Dorothy Burlage; "North Carolina Report," ca. February 1968, Wise Papers; interview, Joseph Tieger, 26 December 1994.
45. Interview, Gardner, 5 May 1993; interview, Guerrero; interview, Nan Grogan Orrock, 14 November 1994; "Press Release: 75 Picketers Support Atlanta Draft Refuser Gene Guerrero," 4 December 1967, in the author's possession; Gene Guerrero, "Why I Must Refuse Induction," 4 December 1967, folder 8, box 1, SSOC and Gardner Papers; *Southern Patriot*, March 1968; *Phoenix*, 1:4 (December 1968).
46. Tom Gardner, Nancy Hodes, and David Nolan, "Florida Peace Tour, February 23–April 7, 1967," ca. May 1967, in the author's possession; interview, Nolan; interview, Gardner, 2 September 1994.
47. Interview, Nolan; interview, Gardner, 5 May 1993; interview, Gardner, 2 September 1994. Even while on the NSA Southern Project staff, Gardner's loyalties remained with SSOC. In fact, he had joined the Southern Project staff because in the fall of 1966 SSOC lacked the funds to hire him. Thus, he saw himself operating as a SSOC staffer from within the NSA. As such, he encouraged Southern Project people to support SSOC, shared the details of the Southern Project budget with his friends in SSOC, and not infrequently used credit cards intended for Southern Project business to fund various purchases for SSOC. Tom Gardner to SSOC, ca. October 1966, folder 1, box 1, SSOC and Gardner Papers; interview, Gardner, 2 September 1994.
48. Gardner, Hodes, and Nolan, "Florida Peace Tour" (quote); *New South Student*, 4:1 (February 1967); "Executive Committee Minutes," 25–26 February 1967, Smith Papers; Alan Levin to SDS, 21 March 1967, 3:27, reel 22, *Students for a Democratic Society Papers, 1958–1970* (Glen Rock, N.J.: Microfilm Corporation of America, 1977); *Florida Alligator*, 20 March 1967; Nancy Hodes, "Vietnam: The Choices We Made," Harvard/Radcliffe Class of 1968 Twenty-fifth Reunion Symposium,

7 June 1993, in the author's possession; interview, Gardner, 5 May 1993; interview, Nolan; interview, Nancy Hodes, 3 September 1994; interview, Bruce Smith, 2 April 1995. The campuses visited by the Peace Tour were Florida State University, the University of Florida, the University of South Florida in Tampa, the New College in Sarasota, the University of Miami, Jacksonville University, Florida Atlantic University in Boca Raton, Miami-Dade Junior College (North and South Campuses) and two high schools in the Miami area. Bob Dewart also was supposed to participate in the Florida Peace Tour, but at the last minute withdrew from it. Gardner, who initially was to focus on the connection between universities and U.S. foreign policy, altered his presentation to focus on American foreign policy more generally, the area Dewart was to speak on.

49. *Southern Patriot,* April 1967; Gardner, Hodes, and Nolan, "Florida Peace Tour"; Alan Levin to SDS, 21 March 1967, 3:27, reel 22, SDS Papers; Marshall B. Jones, "Berkeley of the South: A History of the Student Movement at the University of Florida, 1963–1968," 95, George A. Smathers Libraries, University of Florida, Gainesville, Florida; interview, Nolan; interview, Hodes. The key person in ensuring that the University of South Florida would be open to the activists was local television reporter Don Harris. Harris had taken an interest in the Peace Tour and had visited with the activists at one point during their sojourn. When campus administrators threatened to close the school to the Peace Tour, Harris made it clear that he was prepared to broadcast the story on local television. While it is unclear if he aired such a story, the Peace Tour did make the local news in Tampa. As a national news correspondent a decade later, Harris was killed prior to the mass suicide by the followers of Jim Jones in Jonestown, Guyana. Interview, Nolan; Kenneth Wooden, *The Children of Jonestown* (New York: McGraw-Hill, 1981).

50. Gardner, Hodes, and Nolan, "Florida Peace Tour"; *Florida Alligator,* 1 March 1967; interview, Hodes; interview, Nolan; unidentified college dean quoted in *Southern Patriot,* April 1967.

51. Ibid.; Miami *Herald,* 28 and 30 March 1967; "Peace Talkers Arrested at Dade Junior College," 30 March 1967, folder 8, box 6, SSOC and Gardner Papers.

52. "SSOC Peace Tour Comes to North Carolina," ca. November 1967, Wise Papers (quote); interview, Smith; Tom Gardner, "Proposal for Funding of Southern Peace Tours by the Southern Student Organizing Committee," Fall 1967, folder 4, box 4, SSOC and Gardner Papers; interview, Gardner, 2 September 1994. On the South Carolina Tour only, Steve Andors, a graduate student at Columbia University who previously had spent two years as a Peace Corps volunteer in Thailand, spoke on China in place of Nancy Hodes. *New South Student,* 5:6 (December 1968).

53. Randy Shannon to Steve Wise, 23 December 1967, Wise Papers; *New South Student,* 4:6 (November 1967); Tom Gardner, "The Southern Student Organizing Committee, 1964–1970" (October 1970), 29, in the author's possession; interview, Gardner, 2 September 1994; interview, Nolan.

54. Interview, Smith; interview, Gardner, 2 September 1994; interview, Wells; interview, Hodes; *Southern Patriot*, January 1968; Gardner, "The Southern Student Organizing Committee, 1964–1970," October 1970, in the author's possession.
55. Smith quoted in "Press Release," 11 November 1967, Lyn Wells Papers, Wisconsin Historical Society, Madison, Wisconsin; *Southern Patriot*, January 1968; interview, Gardner, 2 September 1994; interview, Smith.
56. Interview, Nolan.
57. Unidentified SSOC member quoted in "Report to Stockholders," Membership Mailing Number Three, 7 December 1966, folder 1, box, Gardner Papers.
58. Interview, Bob Moses, 18 April 1995.
59. *Southern Patriot*, June 1967. Two useful general works that consider organized labor's reluctance to work with African Americans or to support the civil rights movement are Alan Draper, *Conflict of Interests: Organized Labor and the Civil Rights Movement in the South, 1954–1968* (Ithaca, N.Y.: ILR Press, 1994); and Peter Levy, *The New Left and Labor in the 1960s* (Urbana: University of Illinois Press, 1994).
60. "1966-7 Proposal," Edwin Hamlett Papers (uncatalogued), Special Collections and University Archives, The Jean and Alexander Heard Library, Vanderbilt University, Nashville, Tennessee; Brandon quoted in Tom Gardner, "Solidarity Forever!", 4 May 1966, folder 9, box 4, SSOC and Gardner Papers. On the Duke and Emory union drives, see Peter Brandon and Nancy Park, "A Brief History of Duke Employees Local 77, AFSCME, AFL-CIO" (Nashville, Tenn.: Southern Student Organizing Committee, April 1966), Wise Papers; *New South Student*, 2:5 (October 1965) and 3:5 (May 1966); and Erik Ludwig, "Closing in on the 'Plantation': Coalition Building and the Role of Black Women's Grievances in Duke University Labor Disputes, 1965–1968," *Feminist Studies*, 25, no. 1 (1999): 79–94.
61. Gene Guerrero and Brian Peterson, "Report and Notes on the Students and Labor Conference," Wise Papers; *New South Student*, 3:2 (February 1966); Duke *Chronicle*, 22 April 1966; Gardner, "Solidarity Forever!", 4 May 1966, folder 9, box 4, SSOC and Gardner Papers; Gardner, "The Southern Student Organizing Committee, 1964–1970," 17–18; interview, Guerrero.
62. On migrant worker organizing, see the *New South Student*, 3:6 (September 1966), 3:6 (October 1966) [note: both the September and October 1966 *New South Student* were designated as the sixth issue of the third volume], 4:1 (February 1967), and 4:3 (April 1967); "Southern Student Organizing Committee: 1966–1967 Proposal," file "Southern Student Organizing Committee, November 1966," box 2S427, Field Foundation Archives, 1940–1990; interview, Bo Lozoff, 10 March 1995; interview, Smith; and interview, Mike Lozoff, 26 September 1996.
63. The seven Cone Mills plants were located in Reidsville, Salisbury, Haw River, and Greensboro. *Cone Workers Union Voice*, 9 May 1967, Wise Papers.

64. *New South Student,* 4:3 (April 1967); *Southern Patriot,* June 1967; *Cone Workers Union Voice,* February 1967, Peter Brandon, "A History of the Cone Textile Strike," Untitled Essay on Whiteville Campaign, 1967," Paul Swaity, "Address to Southern Conference of Textile Workers, Students, and Church Leaders," Greensboro, North Carolina, 23 April 1967, and Steve Wise, "A Report on a Strike of Textile Workers at the National Spinning Company, Columbus County, North Carolina," July 1967, all in Wise Papers. The history of unionism at Cone Mills is discussed in Timothy J. Minchin, *What Do We Need a Union For?: The TWUA in the South, 1945–1955* (Chapel Hill: University of North Carolina Press, 1997), *passim.*
65. *Southern Patriot,* June 1967; Brandon, "A History of the Cone Textile Strike," Wise, "A Report on a Strike of Textile Workers at the National Spinning Company," and Untitled Essay on Whiteville Campaign, 1967 (quote), all in Wise Papers; interview, Guerrero.
66. *New South Student,* 4:3 (April 1967); Brandon, "A History of the Cone Textile Strike" and Untitled Manuscript on Cone Mills Campaign, ca. September 1967 (quote), Wise Papers.
67. *Southern Patriot,* June 1967; Wise, "A Report on a Strike of Textile Workers at the National Spinning Company," Wise Papers. A similar situation to the Whiteville campaign had unfolded the previous year at the Chatham Manufacturing plant in Elkin. There, the TWUA won a hotly contested union election in August 1965. However, in a harbinger of things to come in Whiteville, workers remained without a contract 18 months after the election. In the meantime, the Elkin plant was in a constant state of turmoil: Chatham had fired pro-union workers and coerced new workers into rejecting the union, while the union had led two walkouts to protest the company's actions and had won the active support of progressive college students in North Carolina. *Charlotte Observer,* 5 July 1967; Swaity, "Address to Southern Conference of Textile Workers, Students, and Church Leaders," 23 April 1967, Greensboro, N.C., "The Chatham Story," 1967, and "Report on Students and Labor," #2, April 1967, Wise Papers.
68. Interview, Guerrero; interview, Orrock.
69. Ibid.; *New South Student,* 4:3 (April 1967); *Cone Workers Union Voice,* February 1967, Wise Papers.
70. Lyn Wells, "Draft Proposal to SSOC," 30 March 1967, Wise Papers; interview, Wells; interview, Frank Goldsmith, 17 March 1995. By the time she had finished her stint as North Carolina campus traveler, Wells had visited more than 45 colleges and universities in the state and had helped to create SSOC chapters at almost half of them.
71. *Labor Analysis and Forecast,* 11:12 (15 June 1967), 982–985, Wise Papers; *Daily Tar Heel,* 14 April 1967; *Charlotte Observer,* 5 July 1967; *Southern Patriot,* June 1967; Untitled Manuscript on Cone Mills Campaign, ca. September 1967, *Cone Workers Union Voice,* 14 February, 19 April and ca. 9 May 1967 (quote), and "Report On: Students and Labor," ca. May 1967, all in Wise Papers; interview, Guerrero; interview, Goldsmith.

72. *Charlotte Observer*, 8 July 1967; *Our Union Speaks*, newsletter of the National Spinning Company Union Members, TWUA, AFL-CIO, 10 August 1967 and "Press Release," 27 July 1967, Wise Papers; interview, Orrock; interview, Guerrero. On the Whiteville strike, see *Whiteville (North Carolina) News Reporter*, 1, 5, 15, 19, 22 June 1967, Untitled Essay on Whiteville Campaign, 1967, "National Spinning, Whiteville, North Carolina," 22 May 1967, Wise, "A Report on a Strike of Textile Workers at the National Spinning Company, Columbus County, North Carolina," July 1967, "There's Glory in Whiteville," and *Textile Labor*, 28:7 (July 1967), 3–5, all in Wise Papers.
73. *National Observer*, 15 May 1967 (quote); *New South Student*, 4:3 (April 1967).
74. "Report on Students and Labor," #2, April 1967, and *Our Union Speaks*, newsletter of the National Spinning Company Union Members, TWUA, AFL-CIO, 20 July 1967, both in Wise Papers.
75. Unidentified unionist quoted in Untitled Manuscript on Cone Mills Campaign, ca. September 1967, Wise Papers; interview, Orrock; interview, Guerrero; *New South Student*, 4:3 (April 1967); *Southern Patriot*, June 1967; Wise, "A Report on a Strike of Textile Workers at the National Spinning Company, Columbus County, North Carolina," July 1967, Wise Papers.
76. Livingstone College student Cathy Slade quoted in "Report On: Students and Labor," ca. May 1967, Wise Papers.
77. Ibid.; interview, Wells; interview, Guerrero; *Cone Workers Union Voice*, ca. 9 May 1967, and "Report On: Students and Labor," ca. May 1967, Wise Papers; *New South Student*, 4:3 (April 1967).
78. Steve Wise to George Brosi, 26 June 1967, Wise Papers; interview, Guerrero.
79. "The Role of the Southern Radical in the American New Left," SSOC Spring Conference, Buckeye Cove, North Carolina, 5–7 May 1967, David and Ronda Kotelchuck Papers (Southern Student Organizing Committee Papers, 1959–1977), Special Collections, University of Virginia Library, Charlottesville, Virginia; *New Left Notes*, 2:20 (22 May 1967); interview, Nelson Blackstock, 8 August 1995. Among the non-SSOC members on the conference program were Greg Calvert, SDS national secretary, Hugh Fowler, Dubois Club executive secretary, and Ed Clark, southern organizer for the Progressive Labor Party, all of whom made presentations to the assembled activists.
80. Tom Gardner and David Nolan, "Toward a Southern Student Organizing Committee," 4 May 1967, and "The New SSOC," Membership Mailing Number Eight, 17 May 1967, in the author's possession.
81. Tom Gardner and David Nolan, "Toward a Southern Student Organizing Committee," 4 May 1967 (quote), and The New SSOC," Membership Mailing Number Eight, 17 May 1967, in the author's possession; Tom Gardner to Bruce Smith, 5 June 1967, Smith Papers; *New Left Notes*, 2:20 (22 May 1967).

82. Interview, Levin; interview, Blackstock.
83. Interview, Levin; interview, Bo Lozoff; interview, Mike Lozoff.
84. Interview, Palmour; Interview, Harlon Joye, 15 March 1995; interview, Harlon and Barbara Joye (interviewed by Ronald Grele), 11 November 1984, Columbia Oral History Project, Columbia University, New York, New York.
85. *Harvest of Shame*, 55 min., CBS Reports, New York, 25 November 1960; *Florida Alligator*, 21 and 22 February 1967; interview, Bo Lozoff; interview, Mike Lozoff. On the migrant organizing project of the summer of 1966, see the *New South Student*, 3:6 (September 1966) and 3:6 (October 1966). On the Belle Glade drive, see the *New South Student*, 4:1 (February 1967) and 4:3 (April 1967); Alan Levin to SDS, 19 January 1967, 3:27, reel 22, SDS Papers; UFWOC Newsletter, 1:3 (ca. November 1966), "Executive Committee Minutes," 25–26 February 1967, and Doug Ireland and Steve Max to Fay Bennett and Linda Lewis, National Sharecroppers Fund; Norman Hill, Industrial Union Dept., AFL-CIO; Emmanuel Muravchik, Jewish Labor Committee, 18 January 1967, all in Wise Papers; Jones, "Berkeley of the South," 112–113.
86. "A Preliminary Prospectus for SLAM: Southern Labor Action Movement; Toward New, Militant Workers Labor Action Movement," in the author's possession; interview, Mike Lozoff.
87. Interview, Bo Lozoff; interview, Nolan; interview, Harlon Joye, 15 March 1995.
88. Interview, Nolan; Mary Stanton, *Freedom Walk: Mississippi or Bust* (Jackson: University Press of Mississippi, 2003), 179–181, 184; "The New SSOC," Membership Mailing Number Eight, 17 May 1967, in the author's possession; Ed Clark, "The South Must Be Won," *PL: Progressive Labor*, 7, no. 3 (November 1969): 19–28. The exact vote on the SLAM proposal is unclear. While the report on the meeting, "The New SSOC," recorded a vote of 79–56 in favor of SLAM, Clark said that it was 83–56. Clark's numbers might be suspect given that they appear in an essay published two-and-a-half years after the meeting. There is no evidence that any of the representatives of other activist groups in attendance voted on the SLAM proposal.
89. "The New SSOC," Membership Mailing Number Eight, 17 May 1967, in the author's possession; "Executive Committee Meeting Minutes, June 10–11, 1967, Buckeye Cove, N.C.," folder 1, box 4, SSOC and Gardner Papers; interview, Gardner, 2 September 1994; interview, Doggett. For the chairmanship, Gardner defeated Jody Palmour, Millsap's David Doggett, and Ed Clark, a member of the Progressive Labor Party and a leader of the New Orleans Movement for a Democratic Society, the PL stronghold in the South.
90. Tom Gardner to All Members, "Participatory Democracy or Death," ca. May 1967, in the author's possession; *New Left Notes*, 2:20 (22 May 1967).
91. *Worklist Mailing*, 1:1 (15 September 1967), folder 29, box 38, Papers of the United States National Student Association, 1955–1969, Archives

and Library, The Martin Luther King, Jr. Center for Nonviolent Social Change, Atlanta, Georgia (document also can be found in Wise Papers). On the Blue Ridge strike see Brenda Mull, "Blue Ridge: The History of Our Struggle Against Levi-Strauss," ca. 1968, in the author's possession; *Southern Patriot*, March 1967; *New South Student*, 5:3 (April 1968); SLAM Newsletter, 1:1 (22 August 1967), Nelson Blackstock Papers (uncatalogued), Special Collections, University of Virginia Library, Charlottesville, Virginia; interview, Harlon Joye 15 March 1995.

92. Interview, Bo Lozoff.

Chapter 7

1. *New South Student*, 6:1 (January 1969).
2. "A Radical Reunion," *Furman Magazine*, 34:1 (1990), 31.
3. On the emergence of women's issues in SNCC and SDS, see, "SNCC Position Paper (Women in the Movement)," 1964, reprinted in Sara Evans, *Personal Politics: The Roots of Women's Liberation in the Civil Rights Movement and the New Left* (New York: Alfred A. Knopf, 1979), 233–235; Casey Hayden and Mary King, "Sex and Caste: A Kind of Memo," 1965, reprinted in *Liberation*, 10 (April 1966); "Liberation of Women," *New Left Notes*, 2:26 (10 July 1967); Clayborne Carson, *In Struggle: SNCC and the Black Awakening of the 1960s* (Cambridge, Mass.: Harvard University Press, 1981), 147–148; Mary King, *Freedom Song: A Personal Story of the 1960s Civil Rights Movement* (New York: Quill, 1987), 437–474; Evans, *Personal Politics*, esp. 83–101, 156–211; Kirkpatrick Sale, *SDS* (New York: Random House, 1973), 252, 356–357, 362–363, 415–416, 508–509, 525–527; Todd Gitlin, *The Sixties: Years of Hope, Days of Rage* (Toronto: Bantam Books, 1987), 362–376; Alice Echols, *Daring to Be Bad: Radical Feminism in America, 1967–1975*, (Minneapolis: University of Minnesota Press, 1989), esp. 23–50; Constance Curry et al., *Deep in Our Hearts: Nine White Women in the Freedom Movement* (Athens: University of Georgia Press, 2000).
4. Interview, Ronda Stilley Kotelchuck, 4 September 1994; author's notes on Sue Thrasher, comments on "Movement Women: Perspectives on Race, Civil Rights and Feminism," *Activism & Transformation: A Symposium on the Civil Rights Act of 1964*, Connecticut College, New London, Connecticut, 4 November 1994.
5. Tom Gardner, "The Southern Student Organizing Committee, 1964–1970" (October 1970), 31, in the author's possession; "Mailing List of Students Attending the Nashville Conference, SSOC" and "Participants at Nashville Conference," Constance Curry Papers (uncatalogued), Special Collections, University of Virginia Library, Charlottesville, Virginia.
6. In 1961, 39 percent of students in the South were women, a figure that grew to 41 percent in 1970. Seymour E. Harris, *A Statistical Portrait of Higher Education* (New York: McGraw-Hill, 1972), 271–272; United States Census Bureau, *Statistical Abstract of the United States, 1962*, 83rd

annual edition (Washington, D.C.: U.S. Government Printing Office, 1962), 132–133; United States Census Bureau, *Statistical Abstract of the United States, 1971,* 92nd annual edition (Washington, D.C.: U.S. Government Printing Office, 1971), 127. The number of women on southern campuses would have been greater had some schools not relied on restrictive admissions policies to keep the number of women students artificially low. Until 1972, for instance, both Vanderbilt University and the University of North Carolina at Chapel Hill strictly limited the number of women admitted in each incoming class, while the University of Virginia's College of Arts and Sciences did not officially open its doors to female undergraduates until 1970, well after it had begun to admit African American students, thus making it the last public university in the nation to coeducate. Pamela Roby, "Institutional Barriers to Women Students in Higher Education," in *Academic Women on the Move,* Alice S. Rossi and Ann Calderwood, eds. (New York: Russell Sage Foundation, 1973), 37–56; Clarence L. Mohr, "World War II and the Transformation of Southern Higher Education," in *Remaking Dixie: The Impact of World War II on the American South,* Neil R. McMillen, ed. (Jackson: University Press of Mississippi, 1997), 45; Alison Jewett, "Women's Agency in Coeducation at the University of Virginia" (unpublished seminar paper: University of Virginia, 1997), 3–6; Tami Lynn Curtis, "Imperfect Progress: Coeducation at the University of Virginia" (Master's Thesis: University of Virginia, 1987), 2–3, 5–9.

7. Mohr, "World War II and the Transformation of Southern Higher Education," 45.
8. Romaine quoted at "Agents of Change: Southern White Students in the Civil Rights Movement," A Public Seminar, 7 April 1994, Charlottesville, Virginia, in the author's possession; Evans, *Personal Politics,* 51–53. On Ruby Doris Smith Robinson, see Cynthia Griggs Fleming, *Soon We Will Not Cry: The Liberation of Ruby Doris Smith Robinson* (Lanham, Md.: Rowman & Littlefield, 1998); King, *Freedom Song,* 316–318; Carson, *In Struggle,* 32–33, 67, 70, 202–204; Evans, *Personal Politics,* 40–41. On Fannie Lou Hamer, see Kay Mills, *This Little Light of Mine: The Life of Fannie Lou Hamer* (New York: Dutton, 1993); Charles M. Payne, *I've Got the Light of Freedom: The Organizing Tradition and the Mississippi Freedom Struggle* (Berkeley: University of California Press, 1995); John Dittmer, *Local People: The Struggle for Civil Rights in Mississippi* (Urbana: University of Illinois Press, 1994); James Forman, *The Making of Black Revolutionaries,* second edition (Seattle, Wa.: Open Hand, 1985 [first edition: New York: Macmillan, 1972]), 290–291, 386–396; Carson, *In Struggle,* 123–129. For an insightful treatment of the role of black women in the Mississippi movement, see Vicki Crawford, "Race, Class, Gender and Culture: Black Women's Activism in the Mississippi Civil Rights Movement," *Journal of Mississippi History,* 58, no. 1 (1996): 1–21. Several of the white women whom SSOC women admired tell the story of their involvement in the movement in Curry et al., *Deep in Our Hearts.*
9. In her study of activist women in the years after World War II, Susan Lynn argues that women who were social movement veterans by the 1960s

played a particularly important role in the lives of progressive women just then coming of age. While relations between young male activists and their adult peers often were strained during these years, Lynn suggests that women with deep experience in progressive movements served as mentors and guides for newly activated women. This was so, she contends, because "Given women's propensity to value connections with others more than many men do, women may have bonded more easily across generational lines than did their male colleagues." Susan Lynn, *Progressive Women in Conservative Times: Racial Justice, Peace, and Feminism, 1945 to the 1960s* (New Brunswick, N.J.: Rutgers University Press, 1992), 176–177.

10. Curry et al., *Deep in Our Hearts*, 106; "Dear Sisters," 20 November 1968, Lyn Wells Papers, Wisconsin Historical Society, Madison, Wisconsin; *Southern Patriot*, March 1969; Evans, *Personal Politics*, 48–51; Catherine Fosl, *Subversive Southerner: Anne Braden and the Struggle for Racial Justice in the Cold War South* (New York: Palgrave, 2002). Author Lillian Smith was another older southern white woman who was an inspiration for SSOC women. The students admired Smith for her bitter denunciations of the "southern way of life" and her refusal to abide by its rules, and until her death in 1966, SSOC women occasionally journeyed to her home on Old Screamer Mountain in the North Georgia town of Clayton to visit her. On Smith, see Anne C. Loveland, *Lillian Smith, a Southerner Confronting the South: A Biography* (Baton Rouge: Louisiana State University Press, 1986),

11. Evans, *Personal Politics*, 50–59.

12. Interview, Sue Thrasher, 1 September 1994. Alice Echols has argued that New Left organizations furthered the cause of women's liberation even when not intending to do so. "White women," she writes, "acquired the skills, confidence, and political savvy necessary to discern the disjuncture between those movements' egalitarian rhetoric and their own subordination." Alice Echols, "We Gotta Get Out of This Place: Notes toward a Remapping of the Sixties," *Socialist Review* 22, no. 2 (1992): 9–33 (quote p. 20).

13. Interview, Nancy Lewis, 30 September 1996.

14. Interview, David Nolan, 13–14 March 1995; interview, Sue Thrasher (interviewed by Ronald Grele), 15 December 1984, Columbia Oral History Project, Columbia University, New York, New York; Evans, *Personal Politics*, 46.

15. Gardner, "The Southern Student Organizing Committee, 1964–1970," 31.

16. Sheldon Vanauken, "Freedom for Movement Girls—Now," 1969, box 17, The Papers of the Boyte Family, Special Collections Library, Duke University, Durham, North Carolina. The Oxford English Dictionary credits Vanauken's essay with the first usage of the term "sexism." *The Oxford English Dictionary*, second edition, vol. XV (Oxford: Clarendon Press, 1989), 112. Ironically, Vanauken later regretted having written the pamphlet because he believed women "not to be inferior but to be *deeply*

different, different through and through. The deep difference was the basis of role differences." Thus, comparing women's place in society to African Americans' "is false because it assumes . . . that a soul-deep difference is the same as a skin-deep one." And unlike the vast majority of feminists, Vanauken insisted that these differences should be celebrated and not denied. Sheldon Vanuauken, *Under the Mercy* (Nashville, Tenn.: Thomas Nelson Publishers, 1985), 88 (quote), 177–180.

17. Anne Koedt, "Women and the Radical Movement," in *Notes from the First Year: Women's Liberation,* Shulasmith Firestone, ed. (New York: New York Radical Women, 1968), quoted in Echols, *Daring to Be Bad,* 60; Evelyn Goldfield, "Toward the Next Step," *The Voice of the Women's Liberation Movement,* August 1968, quoted in Echols, *Daring to Be Bad,* 61. For a thorough discussion of the divisions within the women's movement, see Echols, *Daring to Be Bad,* 51–101.

18. Beverly Jones and Judith Brown, "Toward a Female Liberation," box 17, The Papers of the Boyte Family (quotes from 2–3, 20, 35). Also see Judith Brown, "The Coed Caper," *New South Student,* 6:1 (January 1969); and "The Gainesville Experience," *Southern Patriot,* January 1969. For background on Brown and Jones, see Carol Giardina, "Origins and Impact of Gainesville Women's Liberation, the first Women's Liberation Organization in the South," in *Making Waves: Female Activists in Twentieth-Century Florida,* Jack E. Davis and Kari Frederickson, eds. (Gainesville: University Press of Florida, 2003), 314–319.

19. Interview, Lyn Wells, 16 November 1994 (quotes); Lyn Wells, "American Women: Their Use and Abuse," box 17, The Papers of the Boyte Family, 14–17.

20. Wells, "American Women," 1–10 (quote p. 2).

21. Lyn Wells, "A Movement for Us," *Great Speckled Bird,* 28 February 1969. This essay is a slightly revised version of "American Women."

22. Lyn Wells, Ann Johnson, and Maggie Heggen to "Dear Sisters," 20 November 1968, Wells Papers; "Tentative Agenda: 'Women, Students and the Movement,'" and Randa Holland to "Dear Sisters," ca. February 1969, Steve Wise Papers (Southern Student Organizing Committee Additional Papers, 1963–1979 *and* Southern Student Organizing Committee Papers, 1959–1977), Special Collections, University of Virginia Library, Charlottesville, Virginia; Marilyn Salzman Webb, "Southern Women Get Together," *Guardian,* 22 February 1969; *Southern Patriot,* March 1969; interview, Wells; Christina Greene, " 'We'll Take Our Stand': Race, Class, and Gender in the Southern Student Organizing Committee, 1964–1969," in *Hidden Histories of Women in the New South,* Virginia Bernhard, Betty Brandon, Elizabeth Fox-Genovese, Theda Perdue, and Elizabeth Hayes Turner, eds. (Columbia: University of Missouri Press, 1994), 195–197.

23. *Phoenix,* 1:6 (March 1969) and "NC-SSOC Worklist #2," ca. April 1969 (quote), Wise Papers; Wells, "A Movement For Us," *Great Speckled Bird,* 28 February 1969; interview, Nancy Lewis; Evans, *Personal Politics,*

210–211; Echols, *Daring to Be Bad*, 104–107, 369–377; Giardina, "Origins and Impact of Gainesville Women's Liberation, the first Women's Liberation Organization in the South," 312–321.
24. *Phoenix*, 1:6 (March 1969), Wise Papers; Wells, "A Movement For Us," *Great Speckled Bird*, 28 February 1969; David Littlejohn to all SSOC chapters [in South Carolina], 6 February 1969, box 17, The Papers of the Boyte Family.
25. *Daily Tar Heel*, 1, 4 (quote), and 18 (quote) October 1968; 19 November 1968; 12 December 1968.
26. Interview, David Simpson, 15 March 1995.
27. "Press Release," 14 April 1968, and "A Report on the Demonstrations at the Administration Building," 15 April 1968," 3:28, reel 22, *Students for a Democratic Society Papers, 1958–1970* (Glen Rock, N.J.: Microfilming Corporation of America, 1977); *Worklist Mailing*, April 1968; *Phoenix*, 1:1 (August 1968) and 1:6 (March 1969), Wise Papers; *Great Speckled Bird*, 28 February 1969; *Southern Patriot*, June 1968; *Daily Beacon*, 22 January 1969; *Atlanta Journal and Constitution*, 19 May 1968; interview, Simpson; Thomas G. Dyer, *The University of Georgia: A Bicentennial History, 1785–1985* (Athens: University of Georgia Press, 1985), 347–349.
28. *Daily Beacon*, 23 and 27 November 1968; 3 December 1968; 23 January 1969; 4 (quote), 6, 7, 8, (quote), and 25 February 1969; *Great Speckled Bird*, 28 February 1969; and *Phoenix*, 1:6 (March 1969), Wise Papers.
29. *Phoenix*, 1:5 (ca. February 1969) and "Arkansas SSOC Newsletter," 1 March 1969, in Wise Papers; "Notes from the Field," 15 December 1968, file "Southern Student Organizing Committee," box 2S427, Field Foundation Archives, 1940–1990, The Center for American History, The University of Texas at Austin, Austin, Texas; *Daily Beacon*, 29 March 1968; "Virginia SSOC Mailing, Vol. 1," 5 May 1968, Harlon Joye Papers (uncatalogued), Alderman Library, University of Virginia; SSOC, "Dear Friend" flyer, 29 March 1968, Bruce Smith Papers (Southern Student Organizing Committee Papers, 1959–1977), Alderman Library, University of Virginia.
30. Bruce Smith to SSOC Office, 27 March 1968, and Richmond SSOC "Press Release," 29 March 1968, Smith Papers; "SSOC Conference and Membership Convention," 3–5 May 1968, Wise Papers; Nelson Blackstock, "SSOC Annual Membership Conference Report," 10 May 1968, Nelson Blackstock Papers (uncatalogued), Special Collections, University of Virginia Library, Charlottesville, Virginia; "Notes from the Field," 15 December 1968, file "Southern Student Organizing Committee," box 2S427, Field Foundation Archives, 1940–1990; *Worklist Mailing*, 15 May 1968, 3:172, reel 31, SDS Papers; *Southern Patriot*, May 1968; *Richmond Times-Dispatch*, 30 and 31 March 1968; Bryant Simon, "Southern Student Organizing Committee: A New Rebel Yell in Dixie" (Honors Essay: University of North Carolina, 1983), 82.
31. Tennessee SSOC to "Dear Friend," 16 July 1968, Joye Papers; "Position Paper on SSOC, 1968," *Southern Regional Council Papers, 1944–1968*

(New York: NYT Microfilming Corporation of America, 1983); "Workers Strike—Call for Our Support," ca. December 1967, *Phoenix*, 1:1 (August 1968) and 1:2 (October 1968), *Appalachian Student Press: West Virginia SSOC Newsletter*, January 1969 and ca. February 1969, and "NC-SSOC Worklist no. 1", ca. October 1968, Wise Papers; Lyn Wells, speech at the "Alabama Conference on Radical Social Change," Tuscaloosa, Alabama, 15 March 1969, Wells Papers; *Great Speckled Bird*, 31 March 1969; *Cavalier Daily*, 10 October 1968; *Southern Patriot*, October 1968, November 1968, and May 1969; interview, Tom Gardner, 2 September 1994. On support for the grape boycott among leftists and students, see Peter B. Levy, *The Left and Labor in the 1960s* (Urbana: University of Illinois Press, 1994), 128–134.

32. Jan G. Owen, "Shannon's University: A History of the University of Virginia, 1959–1974" (Ph.D. Dissertation: Columbia University, 1993), 133; *New South Student*, 6:1 (January 1969); "NC-SSOC Worklist #7," April 1968, Wise Papers; Luther J. Carter, "Duke University: Students Demand New Deal for Negro Workers," *Science* 160, no. 3827 (3 May 1968): 513–517; Douglas M. Knight, *Street of Dreams: The Nature and Legacy of the 1960s* (Durham, N.C.: Duke University Press, 1989), 120–124, Erik Ludwig, "Closing in on the 'Plantation': Coalition Building and the Role of Black Women's Grievances in Duke University Labor Disputes, 1965–1968," *Feminist Studies* 25, no. 1 (1999): 79–94; Jeffrey A. Turner, "From the Sit-Ins to Vietnam: The Evolution of Student Activism on Southern College Campuses, 1960–1970," *History of Higher Education Annual*, 21 (2001), 121–122; *Southern Patriot*, April 1969; *Daily Tar Heel*, 22 and 27 February 1969, 5, 7, and 23 March 1969; interview, Scott Bradley (interviewed by Derek Williams), 30 October 1974, Southern Oral History Program, Southern Historical Collection, Wilson Library, University of North Carolina, Chapel Hill, North Carolina; "NC-SSOC Worklist #2," ca. April 1969, Wise Papers; William A. Link, *Power, Purpose, and American Higher Education* (Chapel Hill: University of North Carolina Press, 1995), 144–153; William D. Snider, *Light on the Hill: A History of the University of North Carolina at Chapel Hill* (Chapel Hill: University of North Carolina Press, 1992), 282–283; Arnold K. King, *The Multicampus University of North Carolina Comes of Age, 1956–1986* (Chapel Hill: University of North Carolina Press, 1987), 73–82.

33. Bryan Kay, "The History of Desegregation at the University of Virginia, 1950–1969" (Undergraduate Honors Essay: University of Virginia, 1979), 101–109; Owen, "Shannon's University," 140; Virginius Dabney, *Mr. Jefferson's University: A History* (Charlottesville: University Press of Virginia, 1987), 482–485.

34. "South Carolina Southern Student Organizing Committee Newsletter #1," ca. March 1969, box 17, The Papers of the Boyte Family; *Southern Patriot*, June 1968 and April 1969; *Phoenix*, 1:6 (March 1969), North Carolina SSOC, *Worklist Mailing #1*, 17 February 1969, and "NC-SSOC Worklist #2," ca. April 1969, Wise Papers; Knight, *Street of Dreams*,

134–150; Turner, "From Sit-Ins to Vietnam," 121–122; Kay, "The History of Desegregation at the University of Virginia, 1950–1969," 141–154; Lisa Anne Severson, "A Genteel Revolution: The Birth of Black Studies at the University of Virginia" (unpublished seminar paper: University of Virginia, 1995), in the author's possession; interview, Bill Leary, 2 April 1995; interview, A. V. Huff, 2 October 1996.

35. "Organized Black Students Win," in North Carolina SSOC, *Worklist Mailing #1*, 17 February 1969, Wise Papers.
36. *Florida Alligator*, 16 February 1966; Lucien Cross, "Free Speech at the University of Florida," *New Left Notes*, 1:7 (4 March 1966); *New South Student*, 3:3 (March 1966); Marshall B. Jones, "Berkeley of the South: A History of the Student Movement at the University of Florida, 1963–1968," 61–70, George A. Smathers Libraries, University of Florida, Gainesville, Florida; interview, Alan Levin, 26 December 1994.
37. *Worklist Mailing*, 1:5 (14 February 1968), and *Phoenix*, 1:5 (ca. February 1969), Wise Papers; "Student Tension in East Tennessee," in Dan Hendrickson, "Dear Friend" (letter to Tennessee SSOC), 16 July 1968, Joye Papers; "Notes from the Field," 15 December 1968, file "Southern Student Organizing Committee," box 2S427, Field Foundation Archives, 1940–1990.
38. *Daily Beacon*, 8, 11, and 29 January, 1969; *Phoenix*, 1:6 (March 1969), Wise Papers; *Guardian*, 29 March 1969.
39. Speaker bans and protests against them had occurred previously in the South. The most notorious speaker ban to emerge anywhere in the country, in fact, developed in North Carolina in 1963, even before the Berkeley Free Speech Movement. Unlike later policies, which were crafted by individual school administrations, the North Carolina ban took the form of a state statute. Students at the University of North Carolina at Chapel Hill repeatedly assailed the law, participating in demonstrations, inviting "banned" individuals to campus to speak, and, ultimately, successfully challenging its constitutionality in federal court. William J. Billingsley, *Communists on Campus: Race, Politics and the Public University in Sixties North Carolina* (Athens: University of Georgia Press, 1999); Link, *William Friday*, 109–141.
40. *Daily Beacon*, 17 and 18 April 1968, 19 September 1968 (quote); Ruth Anne Thompson, " 'A Taste of Student Power': Protest at the University of Tennessee, 1964–1970," *Tennessee Historical Quarterly* 57, no. 1 (1998): 84–86; James Riley Montgomery (Stanley J. Folmsbee, and Lee Seifert Greene), *To Foster Knowledge: A History of the University of Tennessee, 1794–1970* (Knoxville: University of Tennessee Press, 1984), 286–288.
41. *Daily Beacon*, 26 and 28 September 1968, 1, 15, and 19 October 1968, 28 and 29 (quote) January 1969, 4, 5, 7, 12, 13, and 21 February 1969, 19 April 1969, 20 June 1969; *Phoenix*, 1:6 (March 1969), Wise Papers; Thompson, " 'A Taste of Student Power,' " 84–86.
42. *Worklist Mailing*, 5 April 1968, Wise Papers; "Big Issue on a Small Campus," *Furman Magazine*, 17:3 (1969), 5; "Trustee Resolution on

the Campus Platform," Furman university, 23 January 1968, "SSOC (Southern Student Organizing Committee)" folder, Gordon Blackwell Presidential Papers, Furman University Archive, James Buchanan Duke Library, Furman University, Greenville, South Carolina; interview, Jack Sullivan, 31 August 1994; interview, John Duggan, 16 March 1995; Alfred Sandlin Reid, *Furman University: Toward a New Identity, 1925-1975* (Durham, N.C.: Duke University Press, 1976), 223-224.
43. Interview, Steve Compton, 12 March 1995; interview, Bill Lavrey, 17 March 1995; interview, O. Vernon Burton, 22 January 1996; interview, Sullivan; interview, Duggan; *El Burro*, no. 2 (14 October 1969), in the author's possession; Free Speech Rally flyer, May 1969, in the author's possession; "Big Issue on a Small Campus," 38-41; "A 'Radical' Way of Thinking," *Furman Magazine*, 17:3 (1969), 9, 26-27; Reid, *Furman University*, 229-230, 247-248.
44. Interview, Sullivan; interview, Greg Wooten, 24 December 1994; interview, Duggan; *Greenville (South Carolina) News*, 13 February 1968; *Spartanburg (South Carolina) Times*, 13 February 1968; Reid, *Furman University*, 224.
45. Interview, Wooten; interview, Duggan; *Bonhomie*, v. 69 (Greenville, S.C.: Furman University, 1969), 280; also see *Bonhomie*, v. 68 (Greenville, S.C.: Furman University,1968), 177.
46. Interview, Bradley; interview, Compton; interview, Wooten.
47. Tom Gardner quoted in *Worklist Mailing*, 20 April 1968, Wise Papers; interview, Mike Welch, 1 September 1994; "Constitution of the Southern Student Organizing Committee, 'Revised,'" July 1968 (adopted August 1968), in Randy Shannon to "Brothers and Sisters," 25 July 1968, Wise Papers; Tom Gardner to Nik [Alan] Levin, 21 January 1968, folder 3, box 3, SSOC and Gardner Papers; "SSOC Handbook: Spring 1969," box 17, The Papers of the Boyte Family, 4-5; Harlon E. Joye, "Dixie's New Left," *Trans-action*, 7, no. 11 (September 1970), 56.
48. "SSOC Handbook: Spring 1969," box 17, The Papers of the Boyte Family.

Chapter 8

1. David Doggett, "*The Kudzu:* Birth and Death in Underground Mississippi," in *Voices from the Underground: Insider Histories of the Vietnam Era Underground Press*, Ken Wachsberger, ed. (Tempe, Ariz.: Mica Press, 1993), 222.
2. Anne Braden to the Southern Student Organizing Committee Thirtieth Reunion and Conference, 5 April 1994, in the author's possession.
3. Ed Hamlett, "Southern Student Organizing Committee 1964-1969[:] R.I.P.," Edwin Hamlett Papers (uncatalogued), Special Collections and University Archives, The Jean and Alexander Heard Library, Vanderbilt University, Nashville, Tennessee.
4. Tom Gardner, "Press Release," 31 July 1967, in the author's possession.

5. C. Vann Woodward, "The Search for Southern Identity," in C. Vann Woodward, *The Burden of Southern History* (Baton Rouge: Louisiana State University Press, 1960), 16 (originally published in *The Virginia Quarterly Review*, XXXII [1956]: 258–267); *New South Student*, 3:7 (December 1966); interview, Tom Gardner, 2 September 1994; Tom Gardner, "Culture, Identity, and Resistance: Reflections on Counterhegemonic Steps by White Southern U.S. Activists in the 1960s," (Unpublished seminar paper: University of Massachusetts, 1994), in the author's possession (quote); interview, Mike Welch, 1 September 1994.
6. "Conference on Radical Southern History," April 1969, Hamlett Papers (quote); Alex Hurder, "Dear Sisters and Brothers," 23 March 1969, Steve Wise Papers (Southern Student Organizing Committee Additional Papers, 1963–1979 *and* Southern Student Organizing Committee Papers, 1959–1977), Special Collections, University of Virginia Library, Charlottesville, Virginia; *Southern Patriot*, March and May 1969.
7. Interview, Lyn Wells, 16 November 1994; Executive Committee Meeting Minutes, 12 April 1967, in "Get the Word Out, Membership Mailing Number Seven," 18 April 1967, folder 2, box 1, Southern Student Organizing Committee and Thomas N. Gardner Papers, 1948–1994 [1958–1975], Special Collections, University of Virginia Library, Charlottesville, Virginia; "The New SSOC, Membership Mailing Number Eight," 17 May 1967, in the author's possession.
8. Interview, Gardner, 2 September 1994; *New South Student*, 4:2 (March 1967).
9. (North Carolina) Student Committee Against the War, *Worklist Mailing #5*, February 1968, 3:55, reel 24, *Students for a Democratic Society Papers, 1958–1970* (Glen Rock, N. J.: Microfilming Corporation of America, 1977); Tom Gardner to Don West, 2 July 1967 and Tom Gardner to Steve Wise, 22 February 1968, folder 2, box 1, SSOC and Gardner Papers; "We Secede," Informational Mailing, April 1968, in the author's possession.
10. Gene Guerrero quoted in "SSOC executive committee meeting, 2/68," in "Material from SOC [*sic*] Files," in the possession of Harlon Joye; "We Secede"; Mike Welch, "Dear Fellow Southerner," ca. March 1968, in the possession of Dorothy Burlage; SSOC, "Dear Friend" flyer, 29 March 1968; *New Left Notes*, 3:11 (25 March 1968). Among the activities planned during the Southern Days of Secession were an anti-draft protest by Davidson College students, draft counseling workshops at the University of Virginia, and an antiwar march by Duke and University of North Carolina students. Lyn Wells and Tom Gardner, "SSOC: The New Rebels Secede," 19 March 1968, folder 1, box 4, SSOC and Gardner Papers (reprinted in *New Left Notes*, 3:11 [25 March 1968]); *New Left Notes*, 3:14 (22 April 1968); "10 Days-UVa," ca. March 1968, Bruce Smith Papers (Southern Student Organizing Committee Papers, 1959–1977), Special Collections, University of Virginia Library, Charlottesville, Virginia.
11. Jody Palmour quoted in *New South Student*, 4:2 (March 1967).

12. Tom Gardner, "Sherman to Nixon—No Change," 18 January 1969 (quotes), folder 2, box 4, SSOC and Gardner Papers (also reproduced in *Phoenix*, 1:5 [ca. February 1969], Wise Papers); "Inauguration Mobilization," Wise Papers; Harlon Joye, "Dixie's New Left," *Transaction*, 7, no. 11 (September 1970), 55.
13. Interview, Gardner, 2 September 1994; Lyn Wells, speech at the "Alabama Conference on Radical Social Change," Tuscaloosa, Alabama, 15 March 1969, Lyn Wells Papers, Wisconsin Historical Society, Madison, Wisconsin; interview, Ed Hamlett, 6 May 1993.
14. "SSOC executive committee meeting, 2/68," in "Material from SOC [*sic*] Files," in the possession of Harlon Joye.
15. Don West to *New South Student*, 13 January 1969, folder 2, box 6, Nolan Papers.
16. Interview, Brett Bursey, 11 March 1995 (quote); *Columbia Record*, 25 April 1969; "South Carolina SSOC Newsletter," 18 February 1969, box 17, The Papers of the Boyte Family, Special Collections Library, Duke University, Durham, North Carolina.
17. Steve Wise, "Southern Consciousness," *Great Speckled Bird*, 17 March 1969; flyer distributed by Young Americans for Freedom chapter at the University of South Carolina, February 1969, reprinted in "SSOC Staff Newsletter," February 1969, and *Phoenix*, 1:6 (March 1969), Wise Papers.
18. Tom Gardner to the author, 30 August 1995, in the author's possession (quote); *Phoenix*, 1:6 (March 1969), Wise Papers; "South Carolina SSOC Newsletter," 18 February 1969, box 17, The Papers of the Boyte Family; interview, Bursey; interview, Bruce Smith, 2 April 1995.
19. Interview, Tom Gardner, 5 May 1993; interview, Gardner, 2 September 1994 (quote); Tom Gardner to the author, 30 August 1995, in the author's possession (quote); interview, Jack Sullivan, 31 August 1994.
20. William J. Billingsley, "The New Left in the New South: SDS and the Organization of Southern Political Communities," delivered at the Organization of American Historians' Eighty-Seventh Annual meeting, 16 April 1994, Atlanta, Georgia; Robb Burlage to Vivian Franklin, 7 September 1962, 2A:21, reel 9, SDS Papers (quote).
21. "A Resolution Concerning SDS's Role in the South and the Relationship Between SDS and the Southern Student's [*sic*] Organizing Committee," submitted by the Executive Committee of the Southern Student's Organizing Committee, May 1964, 2A:115, reel 9, SDS Papers; Nelson Blackstock, "SSOC Annual Membership Conference Report," 10 May 1968, Nelson Blackstock Papers (uncatalogued), Alderman Library, University of Virginia; *Southern Patriot*, May 1969. On SDS at the University of North Carolina, see William J. Billingsley, *Communists on Campus: Race, Politics and the Public University in Sixties North Carolina* (Athens: University of Georgia Press, 1999). On SDS at the University of Texas, see Douglas C. Rossinow, *The Politics of Authenticity: Liberalism, Christianity, and the New Left in America* (New York: Columbia University Press, 1998).

22. *Daily Tar Heel*, 2 May 1969; Ed Clark, "Some Notes on Changing the South (or Towards a 'Port Nashville Statement')," 16 April 1967, David and Ronda Kotelchuck Papers (Southern Student Organizing Committee Papers, 1959–1977), Special Collections, University of Virginia Library, Charlottesville, Virginia; Ed Clark, "The South Must Be Won," *PL: Progressive Labor* 7, no. 3 (November 1969): 24; Kirkpatrick Sale, *SDS* (New York: Random House, 1973), 479. Among the joint SDS–SSOC chapters, the University of Florida one was unique in that it originally was the Gainesville chapter of SDS. In January 1967, the SDS chapter had voted to make itself into a joint SDS–SSOC group because, Alan Levin, Gainesville's leading white student activist explained in a letter to SDS, "of our own feeling of isolation from the Northern campuses and to boost SSOC which is a well deserving organization." Alan Levin to SDS, 19 January 1967, 3:27, reel 22, SDS Papers.
23. Ed Hamlett letter in *New Left Notes*, 1:20 (3 June 1966); "The New SSOC," Membership Mailing Number Eight, 17 May 1967, in the author's possession; Bryant Simon, "Southern Student Organizing Committee: A New Rebel Yell in Dixie" (Honors Essay: University of North Carolina, 1983), 70–71.
24. Interview, Gardner, 5 May 1993; Tom Gardner to Cathy Archibald and Les Coleman, 13 February 1968, folder 3, box 3, SSOC and Gardner Papers; Clark, "The South Must Be Won," 23–24; interview, Wells.
25. Rossinow, *The Politics of Authenticity*, 191.
26. Bernadine Dohrn to SDS membership, ca. February 1969, Wise Papers; Steve Wise to Bernadine Dohrn, 14 March 1969, folder 1, box 6, Nolan Papers; Billingsley, "The New Left in the New South," 30.
27. *New Left Notes*, 3:39 (23 December 1968). Sale, *SDS*, 467–469, 506–510, 515–516.
28. *New Left Notes*, 2:21 (29 May 1967); Sale, *SDS*, 332–334, 471–472, 485–486; Leigh David Benin, *The New Labor Radicalism and New York City's Garment Industry: Progressive Labor Insurgents in the 1960s* (New York: Garland Publishing, 2000), 3–26, 41–43. On PL's criticisms of the Soviet Communist Party and veneration of the Chinese Communists, see "Road to Revolution-I," *PL* (March 1963), reprinted in Progressive Labor Party, *Revolution Today: U.S.A.: A Look at the Progressive Labor Movement and the Progressive Labor Party* (New York: Exposition Press, 1970), 89–147; "Road to Revolution-II," *PL* (December 1966), reprinted in Progressive Labor Party, *Revolution Today*, 148–205; "Build a Base in the Working Class," *PL* (June 1969), reprinted in Progressive Labor Party, *Revolution Today*, 13–78.
29. Sale, *SDS*, 264, 396–397, 456, 476, 485, 509–510; Benin, *The New Labor Radicalism and New York City's Garment Industry*, 65–67; Todd Gitlin, *The Sixties: Years of Hope, Days of Rage* (New York: Bantam Books, 1987), 189–191. PL's anti-nationalism further distinguished it from RYM and the rest of SDS. PL argued that nationalist movements were misguided because they failed to recognize that class interests rather than national identity was the fundamental division in society. This view led PL

to become an outspoken critic of Black Power advocates, radical black student groups, and the Black Panthers. Where RYM and other white student groups championed black nationalism and argued that blacks were at the forefront of the revolutionary movement, PL responded that only black workers had revolutionary potential since they were the most exploited segment of the working class. Socialism, not nationalism, would be their salvation. As PL claimed, "the essence of Black liberation is working class liberation." "Program for Black Liberation," *PL* (February 1969), reprinted in Progressive Labor Party, *Revolution Today*, 244–267; Sale, *SDS*, 508, 533–537.

30. Sale, *SDS*, 332–334, 398–399 (quote), 409–411, 461–467, 470–472, 507–510; Max Elbaum, *Revolution in the Air: Sixties Radicals Turn to Lenin, Mao and Che* (London: Verso, 2002), 69–70.
31. For trenchant critiques of PL as well as late SDS, see Andrew Kopkind, "The Real SDS Stands Up," in Andrew Kopkind, *The Thirty Years' War: Dispatches and Diversions of a Radical Journalist, 1965–1994*, JoAnn Wypijewski, ed. (London: Verso, 1995), 161–168 (originally published in *Hard Times*, 30 June–5 July 1969); and Carl Oglesby, "Notes on a Decade Ready for the Dustbin," in *Toward a History of the New Left: Essays from within the Movement*, R. David Myers, ed. (Brooklyn, N.Y.: Carson Publishing, 1989), 21–48 (originally published in *Liberation* 14 [1969]: 5–19).
32. Interview, David Doggett, 30 August 1994; Nelson Blackstock, "SSOC Annual Membership Conference Report," 10 May 1968, Blackstock Papers; Doggett, *"The Kudzu,"* 217; Sale, *SDS*, 538–539. The Austin meeting was not the first attempt by PL to convince SDS to break with SSOC. According to Ed Hamlett, a PL organizer from the South whom he did not name first proposed such a motion at SDS's spring 1968 National Council meeting. Although the motion failed, it made clear to him that PL had set its sights on SSOC. Hamlett, "Southern Student Organizing Committee 1964–1969," Hamlett Papers.
33. "Build SDS in the South," Blackstock Papers.
34. Ibid.
35. Interview, Doggett.
36. Jim Gwin, "The South Moves," *Great Speckled Bird*, 31 March 1969; Lyn Wells, "SSOC, Southern Consciousness and the Building of a Southern Liberation Movement," ca. March 1969, Wells Papers; "Position Paper of the Southern Student Organizing Committee," ca. March 1969, 3:172, reel 31, SDS Papers.
37. Interview, Doggett; Doggett, *"The Kudzu,"* 222; David Simpson, Lyn Wells, and George Vlasits, "Dear Sisters and Brothers," letter to SSOC membership, 26 May 1969, box 17, The Papers of the Boyte Family.
38. "Alabama Conference on Radical Social Change" flyer, Blackstock Papers; Lyn Wells speech at "Alabama Conference on Radical Social Change," 15 March 1969, Wells Papers; Howard Romaine, "Movement South," *Great Speckled Bird*, 28 April 1969; interview, Wells; Simon,

"Southern Student Organizing Committee," 91–92. The sources conflict on the exact date of the Alabama conference. The conference flyer states that it was called for 21–23 March, while the date on Lyn Wells's speech was 15 March. But given that a SSOC executive meeting took place on 15 March in Atlanta—a meeting Wells almost assuredly attended—it is more likely that the Alabama conference took place on the later dates.

39. Klonsky quoted in Ed Clark, "The South Must be Won," 26–27; Sale, *SDS*, 538.
40. Interview, Wells; Michael Klonsky, "SSOC: the Man gets what he pays for," *New Left Notes*, 4:16 (24 April 1969).
41. David Simpson, Lyn Wells, and George Vlasits, "Dear Sisters and Brothers," letter to SSOC membership, 26 May 1969, box 17, The Papers of the Boyte Family; Clark, "The South Must Be Won," 26–27; *Kudzu*, 1:11 (5 April 1969); Doggett, *"The Kudzu,"* 222; interview, Doggett; interview, Wells.
42. *New Left Notes*, 4:13 (4 April 1969); *Kudzu*, 1:11 (5 April 1969); Doggett, *"The Kudzu,"* 222–223 (quote); interview, Doggett; interview, Wells; interview, Margaret Hortenstine Bradley, 3 June 1997.
43. *New Left Notes*, 4:13 (4 April 1969); Clark, "The South Must Be Won," 26–27; Howard Romaine, "Movement South," *Great Speckled Bird*, 28 April 1969; interview, Wells; Sale, *SDS*, 538.
44. Clark, "The South Must Be Won," 26–27; interview, Wells; Tom Gardner, "An Open Letter to Mike Klonsky (read 'God')—SDS National Secretary—On SSOC-SDS," ca. April 1969, folder 5, box 4, Gardner Papers.
45. Interview, David Nolan, 13–14 March 1995; Alex Hurder to Tom Gardner, 12 April 1969, folder 4, box 3, SSOC and Gardner Papers.
46. Klonsky, "SSOC." On the ties between the CIA and the NSA, see Sol Stern, "A Short Account of International Student Politics & the Cold War with Particular Reference to the NSA, CIA, Etc.," *Ramparts*, March 1967, 29–38; Philip G. Altbach, *Student Politics in America: A Historical Analysis* (New Brunswick, N.J.: Transaction Publishers, 1997), 119–132 (originally published New York: McGraw-Hill, 1973).
47. Klonsky, "SSOC"; Tom Gardner to Staughton and Alice Lynd, 15 April 1969, folder 5, box 4, SSOC and Gardner Papers; interview, Nolan. For other criticisms of and reactions to Klonsky, see Jim Bains, "Unity in the Movement: A Position Paper from Alabama SSOC," May 1969, Wise Papers; Jim Rumley and Margie Updike, "A constructive reply to Klonsky," ca. May 1969, Harlon Joye Papers (uncatalogued), Special Collections, University of Virginia Library, Charlottesville, Virginia; David Doggett, Letter to the Editor, *New Left Notes*, 4:19 (20 May 1969).
48. Interview, Wells.
49. David Simpson, Jim Skillman, and Lyn Wells "Dare to Struggle . . . Dare to Win," *Great Speckled Bird*, 5 May 1969 (also published in *New Left*

Notes, 4:18 [13 May 1969]); Lyn Wells, "South: Where We Go From Here," George Vlasits, "Critique of Our Work," and David Simpson, "Why SSOC: Basic Criticisms," all attached to and following David Simpson, Lyn Wells, and George Vlasits, "Dear Sisters and Brothers," letter to SSOC membership, 26 May 1969, box 17, The Papers of the Boyte Family. Although not distributed until the end of May, the working papers most likely were written in late April, for they served as the basis for the "Dare to Struggle" article.

50. Wells, "South: Where We Go From Here" (quote), and Vlasits, "Critiques of Our Work," box 17, The Papers of the Boyte Family; Simpson, Skillman, and Wells, "Dare to Struggle . . . Dare to Win" (quote).
51. Simpson, Skillman, and Wells, "Dare to Struggle . . . Dare to Win"; Simpson, "Why SSOC," (quote) and Simpson, Wells, and Vlasits, "Dear Sisters and Brothers," letter to SSOC membership, 26 May 1969, box 17, The Papers of the Boyte Family (quote).
52. Interview, Smith.
53. Romaine, "Movement South"; "SSOC Struggles," membership mailing, May 1969, Wise Papers; Hamlett, "Southern Student Organizing Committee 1964–1969," Hamlett Papers; Simon, "Southern Student Organizing Committee," 97–99.
54. Interview, Welch; Simon, "Southern Student Organizing Committee," 99.
55. Romaine, "Movement South," *Great Speckled Bird*, 28 April 1969.
56. Interview, Joe Bogle, 12 November 1994; David Doggett, Letter to the Editor, *New Left Notes*, 4:19 (20 May 1969); *Kudzu*, 1:11 (5 April 1969) and 1:13 (13 May 1969); Doggett, "*The Kudzu*," 223; interview, Doggett; interview, Welch; interview, Gardner, 2 September 1994.
57. "SSOC Struggles," May 1969, Wise Papers (quote); *Southern Patriot*, May 1969; interview, Welch; interview, Gardner, 2 September 1994.
58. A list of the conference attendees has not survived. Contemporary reports on the gathering suggest that somewhere between 80 and 100 people were in attendance. Bob Goodman, "SSOC Dissolves," *Great Speckled Bird*, 16 June 1969; Mike Eisenscher, "Memorandum on the SSOC Conference: June 5–8, 1969 in Edwards, Miss.," folder 19, box 5, Nolan Papers; Clark, "The South Must Be Won," 27–28. On Mt. Beulah, see John Dittmer, *Local People: The Struggle for Civil Rights in Mississippi* (Urbana: University of Illinois Press, 1994), 369, 374.
59. Eisenscher, "Memorandum on the SSOC Conference," folder 19, box 5, Nolan Papers (quote); Hamlett, "Southern Student Organizing Committee 1964–1969," Hamlett Papers; *Southern Patriot*, May 1969.
60. Hamlett, "Southern Student Organizing Committee 1964–1969," Hamlett Papers; Eisenscher, "Memorandum on the SSOC Conference," folder 19, box 5, Nolan Papers; interview, Everett Long, 12 November 1994.
61. Eisenscher, "Memorandum on the SSOC Conference," folder 19, box 5, Nolan Papers (quote); Hamlett, "Southern Student Organizing Committee 1964–1969," Hamlett Papers.
62. Ibid.; "Resolution: To Dissolve SSOC," 7 June 1969, Wise Papers.

63. "Resolution: To Dissolve SSOC," 7 June 1969, Wise Papers.
64. Ibid.
65. Eisenscher, "Memorandum on the SSOC Conference," folder 19, box 5, Nolan Papers; Hamlett, "Southern Student Organizing Committee 1964–1969," Hamlett Papers; interview, Long.
66. "Resolution: To Dissolve SSOC," 7 June 1969, Wise Papers; *Kudzu*, 1:14 (ca. 24 June 1969); interview, Long; Doggett, "*The Kudzu*," 223. In his recounting of the incident in the *Kudzu*, Long wrote that "when a SSOC chick confronted Mark Rudd (from Columbia University SDS), she pulled off her blouse and said, 'All right, Rudd, pull off your clothes and show me that you're human, too, that everything you do doesn't have to contribute to the revolution.' He looked mildly shocked and said, 'No, everything I do *does* have to contribute to the revolution and, besides, I like being repressed.' He said it facetiously, but I can't help thinking it was prophetic." Though he did not identify Johnson by name in the article, he did in a subsequent interview.
67. Hamlett, "Southern Student Organizing Committee 1964–1969," Hamlett Papers; interview, Sue Thrasher (interviewed by Ronald Grele), 15 December 1984, Columbia Oral History Project, Columbia University, New York, New York; Constance Curry et al., *Deep in Our Hearts: Nine White Women in the Freedom Movement* (Athens: University of Georgia Press, 2000), 241.
68. The three extant reports documenting the convention's proceedings offer conflicting testimony on the matter of the number of northern voters. CP member Mike Eisenscher, himself a northern supporter of SSOC, bitterly complained that most of the northern SDSers at the meeting "paraded under the guise of working some-how in the South," and therefore deserved a vote. By his estimation, only 6–8 of the northerners elected not to vote, and three of them were Eisenscher and the two other CP members at the meeting. Ed Hamlett, on the other hand, contended that "Most of the Yankees did not vote. . . . I do not believe that a bunch of Yankees voted SSOC out of existence." In his views, the only northerners to vote were the few who actually were involved in the southern movement. The third report, a partial and confused account of the meeting's proceedings by David Davidson, an informer for the Mississippi State Sovereignty Commission whose only impression of SSOC was that it was a "militant" group whose supporters "expounded basic communistic thought," did not indicate how many northerners voted on SSOC's dissolution. Eisenscher, "Memorandum on the SSOC Conference," folder 19, box 5, Nolan Papers; Hamlett, "Southern Student Organizing Committee 1964–1969," Hamlett Papers; David Davidson to Kenneth W. Fairly, 9 June 1969, Mississippi State Sovereignty Commission Records, SCR ID# 2-158-4-20-2-1-1 to 2-158-4-20-8-1-1 (quotes from 4-1-1 and 6-1-1), Mississippi Department of Archives & History, Jackson, Mississippi.
69. Ibid.; "Resolution: To Dissolve SSOC," 7 June 1969, Wise Papers; *Southern Patriot*, June 1969. Gene Guerrero, David Simpson, Lyn Wells,

Jim Bains, and Robert Berschinski were elected to the liquidation committee.

70. Interview, David Simpson, 15 March 1995; Eisenscher, "Memorandum on the SSOC Conference," folder 19, box 5, Nolan Papers; interview, Welch; Steve Wise, informal notes on conversation with Jim Bains, 18 May 1969, Wise Papers.
71. Eisenscher, "Memorandum on the SSOC Conference," folder 19, box 5, Nolan Papers; "SSOC Struggles," membership mailing, May 1969, Wise Papers.
72. Hamlett, "Southern Student Organizing Committee 1964–1969," Hamlett Papers.
73. Interview, Steve Compton, 12 March 1995; interview, Sullivan. In the spring of 1970, Compton recalls that the SSOC chapter renamed itself the Radical Student Movement, finally concluding that there was no point remaining a SSOC group since the parent organization no longer existed.
74. Interview, Doggett; interview, Smith; interview, Hamlett; Howard Romaine, "Movement South," *Great Speckled Bird*, 28 April 1969; interview, Gardner, 5 May 1993.
75. Sue Thrasher, in collaboration with David Nolan, "SSOC Defended," *Great Speckled Bird*, 30 June 1969.
76. "Southern Student Organizing Committee Year Ending, August 31, 1967 Receipts," file "SSOC November 1966," box 2S427, Field Foundation Archives, 1940–1990, The Center for American History, The University of Texas at Austin, Austin, Texas; interview, Gardner, 5 May 1993 and 2 September 1994.
77. Sue Thrasher, in collaboration with David Nolan, "SSOC Defended," *Great Speckled Bird*, 30 June 1969; Romaine quoted in *New South Student*, 2:6 (November 1965); Guerrero quoted in *New South Student*, 3:1 (January 1966); Jim Williams to SSOC Office, ca. March 1967, folder 2, box 4, SSOC and Gardner Papers.
78. *Phoenix*, October 1968; Hamlett, "Southern Student Organizing Committee 1964–1969," Hamlett Papers (quote); interview, Howard Romaine, 17–18 March 1993.
79. Interview, Gardner, 5 May 1993; Howard Romaine to M. Hayes Mizell, 6 May 1969, M. Hayes Mizell Papers, South Caroliniana Library, University of South Carolina, Columbia, South Carolina; Bob Goodman, "SSOC Dissolves," *Great Speckled Bird*, 16 June 1969; interview, Howard Romaine.
80. Tom Gardner, "A Political Criticism of the 'Political' Criticism," ca. April 1969, folder 7, box 3, SSOC and Gardner Papers.
81. Curry et al., *Deep in Our Hearts*, 241–242; interview, Thrasher, 15 December 1984.
82. Interview, Anne Cooke Romaine, 19 March 1993.

Bibliography

Note: In several cases, the individuals who donated the manuscript collections that comprise the SSOC archive at the University of Virginia made their papers available to me prior to transmitting them to the university's library. I concluded my research in these papers and completed most of this book before the library had organized or catalogued the collection. When working with the papers, I named them for the individual who had donated them. The library, however, named them differently. In this book, I use the names I originally gave each collection. For the current names of these collections, please consult the below list:

Frank Goldsmith Papers	Southern Student Organizing Committee Additional Papers, 1964–1969
David and Ronda Kotelchuck Papers	Southern Student Organizing Committee Papers, 1959–1977
Bruce Smith Papers	Southern Student Organizing Committee Papers, 1959–1977
Bill and Betsy Towe Papers	Southern Student Organizing Committee Papers, 1959–1977
Steve Wise Papers	Southern Student Organizing Committee Additional Papers, 1963–1979 *and* Southern Student Organizing Committee Papers, 1959–1977

Papers not yet catalogued by the library retain the names I have given them:
Nelson Blackstock Papers
Constance Curry Papers
Harlon Joye Papers

Interviews—Conducted by Author

Nelson Blackstock. 8 August 1995. Los Angeles, California.
Gordon Blackwell. 2 October 1996. Greenville, South Carolina.
Joe Bogle. 12 November 1994. Alcoa, Tennessee.
Frank Bonner. 17 March 1995. Greenville, South Carolina (telephone).
Anne Braden. 25 April 1993. Louisville, Kentucky (telephone).
Michael Brannon. 30 October 1994. Hyattsville, Maryland.
George Brosi. 12 November 1994. Berea, Kentucky.
Dorothy Dawson Burlage. 4 September 1994. New York, New York.
Robb Burlage. 22 April 1993. New York, New York (telephone).

Robb Burlage. 4 September 1994. New York, New York.
Brett Bursey. 11 March 1995. Peak, South Carolina.
O. Veron Burton. 22 January 1996. Charlottesville, Virginia.
Reverend Will D. Campbell. 18 March 1993. Mount Juliet, Tennessee.
Steve Compton. 12 March 1995. Charleston, South Carolina.
David Doggett. 30 August 1994. Philadelphia, Pennsylvania.
John Duggan. 16 March 1995. Greenville, South Carolina.
Tom Gardner. 5 May 1993. Cambridge, Massachusetts (telephone).
Tom Gardner. 2 September 1994. Amherst, Massachusetts.
Frank Goldsmith. 17 March 1995. Marion, North Carolina.
Lawrence Goodwyn. 6 March 1995. Durham, North Carolina (telephone).
Gene Guerrero. 26 March 1993. Washington, D.C.
Bob Hall. 16 April 1993. Durham, North Carolina (telephone).
Ed Hamlett. 6 May 1993. Maumelle, Arkansas (telephone).
Dan Harmeling (with Carol Giardina and Kathie Sarachild). 14 March 1995. Gainesville, Florida.
Roger Hickey. 26 March 1993. Washington, D.C.
Nancy Hodes. 3 September 1994. Cambridge, Massachusetts.
Margaret Hortenstine Bradley. 3 June 1997. San Antonio, Texas.
A. V. Huff. 2 October 1996. Greenville, South Carolina.
Harlon Joye. 15 March 1995. Atlanta, Georgia.
David Kotelchuck. 31 August 1994. New York, New York.
Ronda Stilley Kotelchuck. 4 September 1994. New York, New York.
Bill Lavery. 17 March 1995. Greenville, South Carolina.
Bill Leary. 2 April 1995. Takoma Park, Maryland.
Alan Levin. 26 December 1994. Oakland, California.
John Lewis. 30 March 1995. Washington, D.C.
Nancy Lewis. 30 September 1996. Gainesville, Florida.
Everett Long. 12 November 1994. Nashville, Tennessee.
Bo Lozoff. 10 March 1995. Durham, North Carolina.
Mike Lozoff. 26 September 1996. Miami, Florida.
Scott Marshall. 22 April 1995. Chicago, Illinois.
Charles McDew. 24 April 1993, Charlottesville, Virginia.
Robert McMahon. 30 October 1994. Hyattsville, Maryland.
Bob Moses. 18 April 1995. Cambridge, Massachusetts (telephone).
Chuck Myers. 11 December 2002. Nashville, Tennessee (telephone).
David Nolan. 13–14 March 1995. St. Augustine, Florida.
Nan Grogan Orrock. 14 November 1994. Atlanta, Georgia.
Jody Palmour. 2 April 1995. Washington, D.C.
Jody Palmour. 18 April 1995. Washington, D.C. (telephone).
Anne Cooke Romaine. 19 March 1993. Gastonia, North Carolina.
Howard Romaine. 17–18 March 1993. Nashville, Tennessee.
David Simpson. 15 March 1995. Atlanta, Georgia.
Bruce Smith. 2 April 1995. Woodbridge, Virginia.
Marilyn Sokalof. 30 September 1996. Gainesville, Florida.
Jack Sullivan. 31 August 1994. New York, New York.
Joseph Tieger. 26 December 1994. Oakland, California.

Sue Thrasher. 1 September 1994. Amherst, Massachusetts.
Elizabeth Tornquist. 10 March 1995. Durham, North Carolina.
Betty Jean Towe. 16 November 1994. Cary, North Carolina.
Bill Towe. 16 November 1994. Cary, North Carolina.
Michael Welch. 1 September 1994. New York, New York.
Lyn Wells. 16 November 1994. North Carolina.
Earl R. Wilson. 11 November 1994. Louisville, Kentucky.
Stanley Wise. 16 March 1995. East Point, Georgia.
Steve Wise. 18 March 1993. Atlanta, Georgia.
Greg Wooten. 24 December 1994. San Francisco, California.

Interviews—Oral History Collections

Columbia Oral History Project, Columbia University, New York, New York (all interviews conducted by Ronald Grele).
    Cathy Cade. 5 January 1995. Oakland, California.
    Bob Hall. 7 November 1984. Chapel Hill, North Carolina.
    Jacquelyn Hall. 7 November 1984. Chapel Hill, North Carolina.
    Barbara Joye. 11 November 1984. Atlanta, Georgia.
    Harlon Joye. 11 November 1984. Atlanta, Georgia.
    Sue Thrasher. 15 December 1984. New Market, Tennessee.
The Moorland-Spingarn Research Center, Howard University, Washington, D.C.
    David Nolan. ca. 1969. Washington, D.C.
    Mike Welch. ca. 1969. Washington, D.C.
Southern Oral History Program, Southern Historical Collection, Wilson Library, University of North Carolina, Chapel Hill, North Carolina.
    Scott Bradley (interviewed by Derek Williams). 30 October 1974. Chapel Hill, North Carolina.

Interviews—Miscellaneous

Ed Hamlett and Howard Romaine. 5 April 1990. In the author's possession.
Ed Hamlett and Howard Romaine. ca. 23–24 January 1992. In the author's possession.

Manuscript Collections

(alphabetized by repository)
Special Collections Library, Duke University, Durham, North Carolina.
    The Papers of the Boyte Family.
Furman University Archive, James Buchanan Duke Library, Furman University, Greenville, South Carolina.
    Gordon Blackwell Presidential Papers.
Archives and Library, The Martin Luther King, Jr. Center for Nonviolent Social Change, Atlanta, Georgia.
    Papers of the United States National Student Association, 1955–1969.
    Southern Student Organizing Committee Papers.

Millsaps College Archives, Millsaps-Wilson Library, Millsaps College, Jackson, Mississippi.
    Administrative Records, President's office (B. B. Graves, 1965–1970).
Mississippi Department of Archives & History, Jackson, Mississippi.
    Mississippi State Sovereignty Commission Records.
South Caroliniana Library, University of South Carolina, Columbia, South Carolina.
    M. Hayes Mizell Papers.
McCain Library and Archives, University Libraries, The University of Southern Mississippi, Hattiesburg, Mississippi.
    Papers of Will D. Campbell.
    Joseph and Nancy Ellin Freedom Summer Collection.
    Ed Hamlett White Folks Project Collection (uncatalogued).
    Paul B. Johnson Family Papers.
    Douglas Tiberiis White Folks Project Collection (uncatalogued).
    Zoya Zeman Freedom Summer Collection.
The Center for American History, The University of Texas at Austin, Austin, Texas.
    Field Foundation Archives, 1940–1990.
Archives, L. Zenobia Coleman Library, Tougaloo College, Tougaloo, Mississippi.
    Ed King Collection.
Special Collections and University Archives, The Jean and Alexander Heard Library, Vanderbilt University, Nashville, Tennessee.
    Edwin Hamlett Papers (uncatalogued).
Special Collections, University of Virginia Library, Charlottesville, Virginia.
    Nelson Blackstock Papers (uncatalogued).
    Constance Curry Papers (uncatalogued).
    Frank Goldsmith Papers (Southern Student Organizing Committee Additional Papers, 1964–1969).
    Harlon Joye Papers (uncatalogued).
    David and Ronda Kotelchuck Papers (Southern Student Organizing Committee Papers, 1959–1977).
    Bruce Smith Papers (Southern Student Organizing Committee Papers, 1959–1977).
    Southern Student Organizing Committee and Thomas N. Gardner Papers, 1948–1994 [1958–1975].
    Southern Student Organizing Committee Papers, 1964–1968.
    Bill and Betsy Towe Papers (Southern Student Organizing Committee Papers, 1959–1977).
    Steve Wise Papers (Southern Student Organizing Committee Additional Papers, 1963–1979 *and* Southern Student Organizing Committee Papers, 1959–1977).
Wisconsin Historical Society, Madison, Wisconsin.
    Carl and Anne Braden Papers.
    George Brosi Papers.
    Robb Burlage Papers.

David Nolan Papers.
Sam Shirah Papers.
Lyn Wells Papers.

Manuscript Collections—Published

Southern Regional Council Papers, 1944–1968. New York: NYT Microfilming Corporation of America, 1983.
Student Nonviolent Coordinating Committee Papers, 1959–1972. Sanford, N.C.: Microfilm Corporation of America, 1982.
Students for a Democratic Society Papers, 1958–1970. Glen Rock, N.J.: Microfilm Corporation of America, 1977.

Miscellaneous—In the Author's Possession

"1968–1969—Zenith & Dissolution." SSOC Reunion. 6 July 2002. Nashville, Tennessee.
"Agents of Change: Southern White Students in the Civil Rights Movement." A Public Seminar. 7 April 1994. Charlottesville, Virginia.
"Alternative Tour of the University of Virginia." Southern Student Organizing Committee Thirtieth Reunion and Conference. 10 April 1994. Charlottesville, Virginia.
Author's notes on Sue Thrasher, comments on "Movement Women: Perspectives on Race, Civil Rights and Feminism," *Activism & Transformation: A Symposium on the Civil Rights Act of 1964,* Connecticut College, 4 November 1994, New London, Connecticut.
Author's notes on Anne Braden, "Remembering Freedom Summer," presented at the Southern Historical Association 60th annual meeting, 9 November 1994, Louisville, Kentucky.
Janet Dewart Bell to SSOClist, 12 May 2003.
Shirley Newton Bliley. "The Sixties—A Personal Perspective." ca. 1977.
Anne Braden to the Southern Student Organizing Committee Thirtieth Reunion and Conference, 5 April 1994.
Dorothy Burlage to author, 1 December 1997.
Cathy Cade to the author, 18 April 1994.
Cassell Carpenter to author, 27 July 2002.
Charles Cobb to Tom Gardner, 25 January 1967.
David Doggett to SSOClist, 16 April 2003.
Federal Bureau of Investigation Files and Records. Samuel Curtis Shirah, Jr., file, no. 44-22674.
Furman University Free Speech Rally flyer, May 1969.
Tom Gardner to All Members, "Participatory Democracy or Death," ca. May 1967.
Tom Gardner. "Press Release." 31 July 1967.
Tom Gardner. "The Southern Student Organizing Committee, 1964–1970." October 1970.

Thomas N. Gardner. Speech at the University of Virginia. 20 November 1990.
Tom Gardner. "Culture, Identity, and Resistance: Reflections on Counterhegemonic Steps by White Southern U.S. Activists in the 1960s." Unpublished Seminar Paper: University of Massachusetts, 1994.
Tom Gardner to author, 30 August 1995.
Tom Gardner to author, 26 April 1996.
Tom Gardner to SSOClist, 11 May 2003.
Tom Gardner and David Nolan. "Toward a Southern Student Organizing Committee." 4 May 1967.
Tom Gardner, Nancy Hodes, and David Nolan. "Florida Peace Tour, February 23–April 7, 1967." ca. May 1967.
Paul Gaston. "Sitting in in the Sixties." Lecture at the University of Virginia, October 1985.
Paul Gaston to the author, 17 February 1999.
Todd Gitlin to SSOC, 7 May 1964.
Gene Guerrero to author, 17 February 2003.
Ed Hamlett to author, 3 April 2003.
Roger Hickey. Speech at the University of Virginia. 20 November 1990.
Nancy Hodes. "Vietnam: The Choices We Made." Harvard/Radcliffe Class of 1968 Twenty-fifth Reunion Symposium. 7 June 1993.
Martin Luther King, Jr., to Sue Thrasher, 23 February 1965.
Brenda Mull. "Blue Ridge: The History of Our Struggle Against Levi Strauss." ca. 1968.
"The New SSOC." Membership Mailing Number Eight. 17 May 1967.
David Nolan to author, 20 August 2003.
"NC-SSOC Worklist no. 1." ca. October 1968.
N.C. Student Committee Against the War. *Worklist Mailing #2*, 30 October 1967.
N.C. Student Committee Against the War. *Worklist Mailing #4*, 20 January 1968.
"A Preliminary Prospectus for SLAM: Southern Labor Action Movement; Toward New, Militant Workers Labor Action Movement."
"Press Release: 75 Picketers Support Atlanta Draft Refuser Gene Guerrero." 4 December 1967.
"Questionnaires," completed by attendees of the Southern Student Organizing Committee Thirtieth Reunion and Conference, April 1994.
"The Role of Culture in the Movement." Southern Student Organizing Committee Thirtieth Reunion and Conference. 9 April 1994. Charlottesville, Virginia.
Sue Thrasher to SSOC, 25 May 1964.
"The Virginia Project." Southern Student Organizing Committee Thirtieth Reunion and Conference. 9 April 1994. Charlottesville, Virginia.
"We Secede," Informational Mailing. April 1968.

Miscellaneous—Other

Robb Burlage. "For Dixie With Love and Squalor," Prospectus and Introduction for an SDS Pamphlet. 1962. In the possession of Dorothy Burlage.

Marshall B. Jones. "Berkeley of the South: A History of the Student Movement at the University of Florida, 1963–1968." George A. Smithers Libraries. University of Florida. Gainesville, Florida.

"Material from SOC [sic] Files." In the possession of Harlon Joye.

Norman Thomas to Archie Eugene Allen, 3 February 1965. In the possession of Archie Allen.

Mike Welch, "Dear Fellow Southerner," ca. March 1968. In the possession of Dorothy Burlage.

Worklist Mailing: Confront the Warmakers Oct. 21, 1:2 (11 October 1967). In the possession of Dorothy Burlage.

Worklist Mailing: SSOC Stops the Draft, ca. January 1968. In the possession of Dorothy Burlage.

Periodicals

American Mercury
Appalachian Student Press: West Virginia SSOC Newsletter
Atlanta Constitution
Atlanta Journal and Constitution
Austin American
Cavalier Daily (University of Virginia)
Charlotte Observer
Columbia Record
Cone Workers Union Voice (North Carolina)
Daily Beacon (University of Tennessee)
Daily Progress (Charlottesville, Virginia)
Daily Tar Heel (University of North Carolina)
Daily Texan (University of Texas)
Duke Chronicle
El Burro (Greenville, South Carolina)
Emory Wheel
Florida Alligator (University of Florida)
Free Biloxi Herald
Free! South Student (Millsaps College)
Furman Magazine
Gamecock (University of South Carolina)
Great Speckled Bird (Atlanta, Georgia)
Greenville (South Carolina) *News*
Guardian
Highland Echo (Maryville College)
Human Relations: A Newsletter Written and Distributed by the Jefferson Chapter of the Council on Human Relations (University of Virginia)
Jackson Clarion-Ledger
Jackson Daily News
Kudzu (Jackson, Mississippi)
Laurel (Mississippi) *Leader-Call*
Liberation

Life Magazine
Miami Herald
Monthly Review
Nashville Banner
Nashville Tennessean
National Guardian
National Observer
New Left Notes
New Rebel
New Republic
New South Student
New York Times
Newsletter: Southern Student Organizing Committee
Newsweek
North Carolina Anvil
Phoenix
PL: Progressive Labor
The Progressive
Purple & White (Millsaps College)
Ramparts
Reckon: The Magazine of Southern Culture
Reflector (Mississippi State University)
Red and Black (University of Georgia)
Richmond Times-Dispatch
Southern Patriot
Spartanburg (South Carolina) *Times*
St. Petersburg Times
Student Voice
Tallahassee Democrat
Time
Vanderbilt Hustler
Village Voice
Whiteville (North Carolina) *News Reporter*
Woodstock (New York) *Times*
Worklist Mailing

Books, Articles, Theses, Dissertations, and Other Works

Altbach, Philip G. *Student Politics in America: A Historical Analysis.* New Brunswick, N.J.: Transaction Publishers, 1997 (first edition: New York: McGraw-Hill, 1973).

Anderson, Terry H. *The Movement and the Sixties.* New York: Oxford University Press, 1995.

Appy, Christian G. *Working-Class War: American Combat Soldiers and Vietnam.* Chapel Hill: University of North Carolina Press, 1993.

Ball, Ed. "The White Issue," *The Village Voice,* 18 May 1993.

Barnard, Hollinger F., ed. *Outside the Magic Circle: The Autobiography of Virginia Foster Durr.* Tuscaloosa: University of Alabama Press, 1985.
Bass, Patrick Henry. *Like a Mighty Stream: The March on Washington, August 28, 1963.* Philadelphia, Penn.: Running Press, 2002.
Benin, Leigh David. *The New Labor Radicalism and New York City's Garment Industry: Progressive Labor Insurgents in the 1960s.* New York: Garland Publishing, 2000.
Billingsley, William J. "The New Left in the New South: SDS and the Organization of Southern Political Communities." Presented at the Eighty-Seventh Annual Meeting of the Organization of American Historians, 16 April 1994. In the author's possession.
———. *Communists on Campus: Race, Politics and the Public University in Sixties North Carolina.* Athens: University of Georgia Press, 1999.
Bonhomie. Vol. 68. Greenville, S.C.: Furman University, 1968.
Bonhomie. Vol. 69. Greenville, S.C.: Furman University, 1969.
Braden, Anne. *The Wall Between.* New York: Monthly Review Press, 1958.
Branch, Taylor. *Parting the Waters: America in the King Years, 1954–1963.* New York: Simon and Schuster, 1988.
———. *Pillar of Fire: America in the King Years, 1963–1965.* New York: Simon and Schuster, 1998.
Cade, Cathy. *A Lesbian Photo Album: The Lives of Seven Lesbian Feminists.* Oakland, Calif.: Waterwoman Books, 1987.
Campbell, Will D. *Brother to a Dragonfly.* New York: Seabury Press, 1977.
———. *Forty Acres and a Goat: A Memoir.* Atlanta, Ga.: Peachtree Publishers, 1986.
Cantwell, Robert. *When We Were Good: The Folk Revival.* Cambridge, Mass.: Harvard University Press, 1996.
Carmichael, Stokely and Charles V. Hamilton. *Black Power: The Politics of Liberation in America.* New York: Vintage Books, 1967.
Carson, Clayborne. *In Struggle: SNCC and the Black Awakening of the 1960s.* Cambridge, Mass.: Harvard University Press, 1981.
Carter, Luther J. "Duke University: Students Demand New Deal for Negro Workers." *Science* 160, no. 3827 (3 May 1968): 513–517.
Chafe, William H. *Never Stop Running: Allard Lowenstein and the Struggle to Save American Liberalism.* New York: Basic Books, 1993.
———. "The Personal and the Political: Two Case Studies." In *U.S. History as Women's History: New Feminist Essays.* Linda K. Kerber, Alice Kessler-Harris, and Kathryn Kish Sklar, eds. Chapel Hill: University of North Carolina Press, 1995. 189–213.
Chappell, David L. *Inside Agitators: White Southerners in the Civil Rights Movement.* Baltimore, Md.: The Johns Hopkins University Press, 1994.
Clark, Ed. "The South Must Be Won." *PL: Progressive Labor* 7, no. 3 (November 1969): 19–28.
Cohen, Robert. "'Two, Four, Six, Eight, We Don't Want to Integrate': White Student Attitudes toward the University of Georgia's Desegregation." *The Georgia Historical Quarterly* LXXX, no. 3 (Fall 1996): 616–645.
Conkin, Paul, assisted by Henry Lee Swint and Patricia S. Miletich. *Gone with the Ivy: A Biography of Vanderbilt University.* Knoxville: University of Tennessee Press, 1985.

Connelly, Thomas L. *Will Campbell and the Soul of the South.* New York: Continuum Publishing Company, 1982.
Couto, Richard A. *Lifting the Veil: A Political History of Struggles for Emancipation.* Knoxville: University of Tennessee Press, 1993.
Crawford, Vicki. "Race, Class, Gender and Culture: Black Women's Activism in the Mississippi Civil Rights Movement," *Journal of Mississippi History* 58, no. 1 (1996), 1–21.
Cummings, Richard. *The Pied Piper: Allard K. Lowenstein and the Liberal Dream.* New York: Grove Press, 1985.
Cunnigen, Donald. "Standing at the Gates: The Civil Rights Movement and Liberal White Mississippi Students." *Journal of Mississippi History* 62, no. 1 (2000): 1–19.
Curry, Constance, Joan C. Browning, Dorothy Dawson Burlage, Penny Patch, Theresa Del Pozzo, Sue Thrasher, Elaine DeLott Baker, Emmie Schrader Adams, and Casey Hayden. *Deep in Our Hearts: Nine White Women in the Freedom Movement.* Athens: University of Georgia Press, 2000.
Curtis, Tami Lynn. "Imperfect Progress: Coeducation at the University of Virginia." M.A. Thesis: University of Virginia, 1987.
Dabney, Virginius. *Mr. Jefferson's University: A History.* Charlottesville: University Press of Virginia, 1981.
Davenport, Garvin. *The Myth of Southern History: Historical Consciousness in Twentieth-Century Southern Literature.* Nashville, Tenn.: Vanderbilt University Press, 1967.
DeBenedetti, Charles and Charles Chatfield, assisting author. *An American Ordeal: The Antiwar Movement of the Vietnam Era.* Syracuse, N.Y.: Syracuse University Press, 1990.
Dellinger, David T. *From Yale to Jail: The Life Story of a Moral Dissenter.* New York: Pantheon Books, 1993.
Diamond, Nancy. "Catching Up: The Advance of Emory University since World War II." *History of Higher Education Annual* 19 (1999): 149–183.
Dittmer, John. *Local People: The Struggle for Civil Rights in Mississippi.* Urbana: University of Illinois Press, 1994.
Doggett, David. "*The Kudzu* Story: Underground in Mississippi," *Southern Exposure* 2, no. 4 (1975), 86–95.
———. "*The Kudzu:* Birth and Death in Underground Mississippi." In *Voices from the Underground: Insider Histories of the Vietnam Era Underground Press*, volume 1. Ken Wachsberger, ed. Tempe, Ariz.: Mica Press, 1993. 213–232.
Draper, Alan. *Conflict of Interests: Organized Labor and the Civil Rights Movement in the South, 1954–1968.* ILR Press: Ithaca, N.Y., 1994.
Dunbar, Anthony P. *Against the Grain: Southern Radicals and Prophets, 1929–1959.* Charlottesville: University Press of Virginia, 1981.
Dyer, Thomas G. *The University of Georgia: A Bicentennial History, 1785–1985.* Athens: University of Georgia Press, 1985.
Eagles, Charles W. "The Closing of Mississippi Society: Will Campbell, *The $64,000 Question*, and Religious Emphasis Week at the University of Mississippi." *Journal of Southern History* 67, no. 2 (2001): 331–372.

Echols, Alice. "We Gotta Get out of This Place: Notes toward a Remapping of the Sixties." *Socialist Review* 22, no. 2 (1992): 9–33.

Egerton, John. *A Mind to Stay Here: Profiles from the South.* New York: Macmillan, 1970.

———. *Speak Now against the Day: The Generation before the Civil Rights Movement in the South.* Chapel Hill: University of North Carolina Press, 1994.

Elbaum, Max. *Revolution in the Air: Sixties Radicals Turn to Lenin, Mao and Che.* London: Verso, 2002.

Evans, Sara. *Personal Politics: The Roots of Women's Liberation in the Civil Rights Movement & the New Left.* New York: Alfred A. Knopf, 1979.

Eynon, Bret. "Community in Motion: The Free Speech Movement, Civil Rights, and the Roots of the New Left." *Oral History Review* 17, no. 1 (1989): 39–6.

———. "Cast upon the Shore: Oral History and New Scholarship on the Movements of the 1960s." *Journal of American History* 83, no. 2 (1996): 560–570.

Fallows, James. "What Did You Do In The Class War, Daddy?" *Washington Monthly* 7, no. 8 (October 1975), 5–19.

Fayette County Project Volunteers. *Step by Step: Evolution and Operation of the Cornell Students' Civil-Rights Project in Tennessee, Summer, 1964.* New York: W. W. Norton & Company, 1965.

Findlay, James F., Jr. *Church People in the Struggle: The National Council of Churches and the Black Freedom Movement, 1950–1970.* New York: Oxford University Press, 1993.

Finn, Chester E. *Scholars, Dollars, and Bureaucrats.* Washington, D.C.: Brookings Institution, 1978.

Flacks, Richard. "Who Protests: The Social Bases of the Student Movement." In *Protest!: Student Activism in America.* Julian Foster and Durward Long, eds. New York: William Morrow and Co., 1970. 134–157.

Fleming, Cynthia Griggs. *Soon We Will Not Cry: The Liberation of Ruby Doris Smith Robinson.* Lanham, Md.: Rowman & Littlefield, 1998.

Foley, Michael S. *Confronting the War Machine: Draft Resistance during the Vietnam War.* Chapel Hill: University of North Carolina Press, 2003.

Forman, James. *The Making of Black Revolutionaries,* second edition. Seattle, Wa.: Open Hand Publishing, 1985 (first edition: New York: Macmillan, 1972).

Fosl, Catherine. *Subversive Southerner: Anne Braden and the Struggle for Racial Justice in the Cold War South.* New York: Palgrave Macmillan, 2002.

"Freedom on My Mind." Field, Connie and Marilyn Mulford, producers. Berkeley, Calif.: Clarity Educational Productions, 1994.

Freeland, Richard M. *Academia's Golden Age: Universities in Massachusetts, 1945–1970.* New York: Oxford University Press, 1992.

Friedlander, Peter. *The Emergence of a UAW Local, 1936–1939: A Study in Class and Culture.* Pittsburgh, Penn.: University of Pittsburgh Press, 1975.

Frisch, Michael. *A Shared Authority: Essays on the Craft and Meaning of Oral and Public History*. Albany: State University of New York Press, 1990.

Frost, Jennifer. *"An Interracial Movement of the Poor": Community Organizing and the New Left in the 1960s*. New York: New York University Press, 2001.

Frye, Patrick K. "*The Great Speckled Bird:* An Investigation of the Birth, Life, and Death of an Underground Newspaper." Unpublished M.A. Thesis: University of Georgia, 1981.

Gabb, Sally. "A Fowl in the Vorticies of Consciousness: The Birth of the *Great Speckled Bird*." In *Voices from the Underground: Insider Histories of the Vietnam Era Underground Press*, volume 1. Ken Wachsberger, ed. Tempe, Ariz.: Mica Press, 1993. 41–50.

Garrow, David J. *Bearing the Cross: Martin Luther King, Jr., and the Southern Christian Leadership Conference*. New York: Vintage Books, 1988 [first edition: New York: William Morrow and Company, 1986].

Giardina, Carol. "Origins and Impact of Gainesville Women's Liberation, the first Women's Liberation Organization in the South." In *Making Waves: Female Activists in Twentieth-Century Florida*. Jack E. Davis and Kari Frederickson, eds. Gainesville: University Press of Florida, 2003. 312–321.

Gitlin, Todd. *The Sixties: Years of Hope, Days of Rage*. New York: Bantam Books, 1987.

Gottlieb, Annie. *Do You Believe in Magic?: The Second Coming of the Sixties Generation*. New York: Times Books, 1987.

Gould, Lewis L. and Melissa R. Sneed. "Without Pride or Apology: The University of Texas at Austin, Racial Integration, and the Barbara Smith Case." *Southwestern Historical Quarterly* 103, no. 1 (1999): 67–87.

Graham, Hugh Davis and Nancy Diamond. *The Rise of American Research Universities: Elites and Challengers in the Postwar Era*. Baltimore, Md.: The Johns Hopkins University Press, 1997.

Greene, Christina. " 'We'll Take Our Stand': Race, Class, and Gender in the Southern Student Organizing Committee, 1964–1969." In *Hidden Histories of Women in the New South*. Virginia Bernhard, Betty Brandon, Elizabeth Fox-Genovese, Theda Perdue, and Elizabeth Hayes Turner, eds. Columbia: University of Missouri Press, 1994. 173–203.

Grele, Ronald J. "Movement without Aim: Methodological and Theoretical Problems in Oral History." In *Envelopes of Sound: Six Practitioners Discuss the Method, Theory and Practice of Oral History and Oral Testimony*, Ronald J. Grele, ed. Chicago, Ill.: Precedent Publishing, 1985. 127–154.

Hagan, John. *Northern Passage: American Vietnam War Resisters in Canada*. Cambridge, Mass.: Harvard University Press, 2001.

Hamburger, Robert. *Our Portion of Hell. Fayette County, Tennessee: An Oral History of the Struggle for Civil Rights*. New York: Links Books, 1973.

Hampton, Henry and Steve Fayer, with Sarah Flynn. *Voices of Freedom: An Oral History of the Civil Rights Movement from the 1950s through the 1980s*. New York: Bantam Books, 1990.

Harrell, G. L. *History of Millsaps College*. Jackson, Miss.: Millsaps College, 1943.

Harris, David. *Dreams Die Hard*. New York: St. Martin's/Marek, 1982.

Harris, Seymour E. *A Statistical Portrait of Higher Education*. New York: McGraw-Hill, 1972.
Harvest of Shame. 55 min. CBS Reports, New York, 25 November 1960.
Havard, William C. and Walter Sullivan, eds. *A Band of Prophets: The Vanderbilt Agrarians after Fifty Years*. Baton Rouge: Louisiana State University Press, 1982.
Hayden, Tom. *Reunion: A Memoir*. New York: Random House, 1988.
Heineman, Kenneth J. *Campus Wars: The Peace Movement at American State Universities in the Vietnam Era*. New York: New York University Press, 1993.
Henry, David D. *Challenges Past, Challenges Present: An Analysis of American Higher Education Since 1930*. San Francisco, Calif.: Jossey-Bass, 1975.
Hershberger, Mary. *Traveling to Vietnam: American Peace Activists and the War*. Syracuse, N.Y.: Syracuse University Press, 1998.
Hobson, Fred. *But Now I See: The White Southern Racial Conversion Narrative*. Baton Rouge: Louisiana State University Press, 1999.
Holt, Len. *The Summer that Didn't End*. New York: William Morrow and Co., 1965.
Horowitz, Helen Lefkowitz. *Campus Life: Undergraduate Cultures from the End of the Eighteenth Century to the Present*. New York: Alfred A. Knopf, 1987.
Horowitz, Roger. *"Negro and White, Unite and Fight!": A Social History of Industrial Unionism in Meatpacking, 1930–1990*. Urbana: University of Illinois Press, 1997.
Hunter-Gault, Charlayne. *In My Place*. New York: Farrar Straus Giroux, 1992.
Isserman, Maurice. *If I Had a Hammer . . . : The Death of the Old Left and the Birth of the New Left*. New York: Basic Books, 1987.
Jencks, Christopher. "Limits of the New Left." *The New Republic* 157, no. 17 (21 October 1967): 19–21.
Jewett, Alison. "Women's Agency in Coeducation at the University of Virginia." Unpublished Seminar Paper: University of Virginia, 1997. In the author's possession.
Joye, Harlon. "Dixie's New Left," *Trans-action* 7, no. 11 (1970), 50–56, 62.
Joyner, Charles W. "Oral History as Communicative Event: A Folkloristic Perspective. *Oral History Review* 7 (1979): 47–52.
Karenakis, Alexander. *Tillers of a Myth: Southern Agrarians as Social and Literary Critics*. Madison: University of Wisconsin Press, 1966.
Kasinsky, Renée G. *Refugees from Militarism: Draft-Age Americans in Canada*. New Brunswick, N.J.: Transaction Books, 1976.
Kay, Bryan. "The History of Desegregation at the University of Virginia, 1950–1969." Undergraduate Honors Essay: University of Virginia, 1979.
Kelley, Robin D. G. *Hammer and Hoe: Alabama Communists during the Great Depression*. Chapel Hill: University of North Carolina Press, 1990.
Key, V. O. *Southern Politics in State and Nation*. New York: Knopf, 1949.
Killian, Lewis M. *Black and White: Reflections of a White Southern Sociologist*. Dix Hills, N.Y.: General Hall, 1994.
King, Arnold K. *The Multicampus University of North Carolina Comes of Age, 1956–1986*. Chapel Hill: University of North Carolina Press, 1987.

King, Mary. *Freedom Song: A Personal Story of the 1960s Civil Rights Movement.* New York: Quill, 1987.

Klibaner, Irwin. *Conscience of a Troubled South: The Southern Conference Educational Fund, 1946–1966.* Brooklyn, N.Y.: Carlson Publishing, 1989.

Kneebone, John. *Southern Liberal Journalists and the Issue of Race.* Chapel Hill: University of North Carolina Press, 1985.

Knight, Douglas M. *Street of Dreams: The Nature and Legacy of the 1960s.* Durham, N.C.: Duke University Press, 1989.

Knotts, Alice G. "Methodist Women Integrate Schools and Housing, 1952–1959." In *Women in the Civil Rights Movement: Trailblazers and Torchbearers, 1941–1965.* Vicki L. Crawford, Jacqueline Anne Rouse, and Barbara Woods, eds. Brooklyn, N.Y.: Carlson Publishing, 1990. 251–258.

Kopkind, Andrew. *The Thirty Years' Wars: Dispatches and Diversions of a Radical Journalist.* JoAnn Wypijewski, ed. London: Verso, 1995.

Landers, Stuart Howard. "The Gainesville Women for Equal Rights, 1963–1978." Master's Thesis: University of Florida, 1995.

Leamer, Laurence. *The Paper Revolutionaries: The Rise of the Underground Press.* New York: Simon and Schuster, 1972.

Levy, Peter B. *The New Left and Labor in the 1960s.* Urbana: University of Illinois Press, 1994.

Linden-Ward, Blanche and Carol Hurd Green. *American Women in the 1960s: Changing the Future.* New York: Twayne Publishers, 1993.

Link, William A. *William Friday: Power, Purpose, and American Higher Education.* Chapel Hill: University of North Carolina Press, 1995.

Loveland, Anne C. *Lillian Smith, a Southerner Confronting the South: A Biography.* Baton Rouge: Louisiana State University Press, 1986.

Ludwig, Erik. "Closing in on the 'Plantation': Coalition Building and the Role of Black Women's Grievances in Duke University Labor Disputes, 1965–1968." *Feminist Studies* 25, no. 1 (1999): 79–94.

Lukas, J. Anthony. *Don't Shoot—We Are Your Children!*, third edition. New York: Random House, 1971 [first edition: New York: Random House, 1968].

Lynd, Alice, ed. *We Won't Go: Personal Accounts of War Objectors.* Boston, Mass.: Beacon Press, 1968.

Lynn, Susan. *Progressive Women in Conservative Times: Racial Justice, Peace, and Feminism, 1945 to the 1960s.* New Brunswick, N.J.: Rutgers University Press, 1992.

Lyon, Danny. *Memories of the Southern Civil Rights Movement.* Chapel Hill: University of North Carolina Press, 1992.

Marsh, Charles. *God's Long Summer: Stories of Faith and Civil Rights.* Princeton, N.J.: Princeton University Press, 1997.

Matusow, Allen J. "From Civil Rights to Black Power: The Case of SNCC, 1960–1966." In *Twentieth-Century America: Recent Interpretations.* Barton J. Bernstein and Allen J. Matusow, eds. New York: Harcourt Brace Jovanovich, 1972. 494–520.

McAdam, Doug. *Freedom Summer.* New York: Oxford University Press, 1988.

McDonough, Julia Anne. "Men and Women of Goodwill: A History of the Commission on Interracial Cooperation and the Southern Regional Council." Ph.D. Dissertation: University of Virginia, 1993.

McWhorter, Diane. *Carry Me Home: Birmingham, Alabama: The Climactic Battle of the Civil Rights Revolution.* New York: Simon and Schuster, 2001.

Mencken, H. L. "Uprising in the Confederacy," *American Mercury* 22 (March 1931): 381.

———. "The South Astir," *Virginia Quarterly Review* 11 (January 1935): 47–60.

Miller, James. *"Democracy Is in the Streets": From Port Huron to the Siege of Chicago.* Cambridge, Mass.: Harvard University Press, 1994 (first edition: New York: Simon and Schuster, 1987).

Mills, Kay. *This Little Light of Mine: The Life of Fannie Lou Hamer.* New York: Dutton, 1993.

Mills, Nicolaus. *Like a Holy Crusade: Mississippi 1964—The Turning of the Civil Rights Movement in America.* Chicago, Ill.: Ivan R. Dee, 1992.

Minchin, Timothy J. *What Do We Need a Union For?: The TWUA in the South, 1945–1975.* Chapel Hill: University of North Carolina Press, 1997.

Mohr, Clarence L. "World War II and the Transformation of Southern Higher Education." In *Remaking Dixie: The Impact of World War II on the American South.* Neil R. McMillen, ed. Jackson: University Press of Mississippi, 1997. 33–55.

Montgomery, James Riley, Stanley J. Folmsbee, and Lee Seifert Greene. *To Foster Knowledge: A History of the University of Tennessee, 1794–1970.* Knoxville: University of Tennessee Press, 1984.

Moore, Charles, text by Michael S. Durham. *Powerful Days: The Civil Rights Movement Photography of Charles Moore.* New York: Stewart, Tabori & Chang, 1991.

Morris, Aldon. *The Origins of the Civil Rights Movement: Black Communities Organizing for Change.* New York: The Free Press, 1984.

Morris, Willie. *North Toward Home.* Boston, Mass.: Houghton Mifflin, 1967.

Mungo, Raymond. *Famous Long Ago: My Life and Hard Times with Liberation News Service.* Boston: Beacon Press, 1970.

Murphy, John. "The Voice of Memory: History, Autobiography, and Oral History." *Historical Studies* 22 (1986): 157–175.

NET Journal: Homefront 1967. Directed by Robert D. Souier. 45 min. Logos Limited, 1987. Videocassette (originally produced by WETA).

Oglesby, Carl. "Notes on a Decade Ready for the Dustbin." In *Toward a History of the New Left: Essays from within the Movement.* R. David Myers, ed. Brooklyn, N.Y.: Carlson Publishing, 1989. 21–48 (originally published in *Liberation* 14 (1969): 5–19).

Owen, Jan G. *Shannon's University: A History of the University of Virginia, 1959–1974.* Ph.D. Dissertation: Columbia University, 1993.

The Oxford English Dictionary, second edition, vol. XV. Oxford: Clarendon Press, 1989.

Pardun, Robert. *Prairie Radical: A Journey through the Sixties.* Los Gatos, Ca.: Shire Press, 2001.

Parr, Stephen Eugene. "The Forgotten Radicals: The New Left in the Deep South, Florida State University, 1960–1972." Ph.D. Dissertation: The Florida State University, 2000.

Passerini, Luisa. "Work Ideology and Consensus under Italian Fascism." *History Workshop* 8 (1979): 84–92.

Payne, Charles. *I've Got the Light of Freedom: The Organizing Tradition and the Mississippi Freedom Struggle*. Berkeley: University of California Press, 1995.

Peck, Abe. *Uncovering the Sixties: The Life and Times of the Underground Press*. New York: Pantheon Books, 1985.

Peneff, Jean. "Myth in Life Stories." In *The Myths We Live By*. Raphael Samuel and Paul Thompson, eds. London: Routledge, 1990. 36–48.

Portelli, Alessandro. *The Death of Luigi Trastulli and Other Stories: Form and Meaning in Oral History*. Albany: State University of New York Press, 1991.

———. *The Battle of Valle Giulia: Oral History and the Art of Dialogue*. Madison: University of Wisconsin Press, 1997.

Pratt, Robert. *We Shall Not Be Moved: The Desegregation of the University of Georgia*. Athens: University of Georgia Press, 2002.

Powledge, Fred. *Free At Last?: The Civil Rights Movement and the People Who Made It*. New York: Little, Brown and Company, 1991.

Progressive Labor Party. *Revolution Today: U.S.A.: A Look at the Progressive Labor Movement and the Progressive Labor Party*. New York: Exposition Press, 1970.

Raines, Howell. *My Soul Is Rested: Movement Days in the Deep South Remembered*. New York: Penguin Books, 1983 [first edition: New York: G. P. Putnam and Sons, 1977].

Reagon, Bernice Johnson. *If You Don't Go, Don't Hinder Me: The African American Sacred Song Tradition*. Lincoln: University of Nebraska Press, 2002.

Reed, John Shelton. *Kicking Back: Further Dispatches from the South*. Columbia: University of Missouri Press, 1995.

Reed, Linda. *Simple Decency & Common Sense: The Southern Conference Movement, 1938–1963*. Bloomington: Indiana University Press, 1991.

Reid, Alfred Sandlin. *Furman University: Toward a New Identity, 1925–1975*. Durham, N.C.: Duke University Press, 1976.

Ridgeway, James. *The Closed Corporation: American Universities in Crisis*. New York: Random House, 1968.

Rivlin, Alice M. *The Role of the Federal Government in Financing Higher Education*. Washington, D.C.: Brookings Institution, 1961.

Roby, Pamela. "Institutional Barriers to Women Students in Higher Education." In *Academic Women on the Move*. Alice S. Rossi and Ann Calderwood, eds. New York: Russell Sage Foundation, 1973. 37–56.

Rogers, Kim Lacy. "Memory, Struggle, and Power: On Interviewing Political Activists." *Oral History Review* 15 (Spring 1987): 165–184.

———. "Oral History and the History of the Civil Rights Movement." *Journal of American History* 75, no. 2 (1988): 567–576.

Rorabaugh, W. J. *Berkeley at War: The 1960s.* New York: Oxford University Press, 1989.
———. "Challenging Authority, Seeking Community, and Empowerment in the New Left, Black Power, and Feminism." *Journal of Policy History* 8, no. 1 (1996): 106–143.
Rossinow, Doug. "'The Break-through to New Life': Christianity and the Emergence of the New Left in Austin, Texas, 1956–1964." *American Quarterly* 46, no. 3 (1994): 309–340.
———. "Secular and Christian Liberalisms and Student Activism in Early Cold War Texas." Delivered at the American Historical Association One Hundred Eighth Annual Meeting, 5–8 January 1995 (Chicago, Ill.). In the author's possession.
Rossinow, Douglas C. *The Politics of Authenticity: Liberalism, Christianity, and the New Left in America.* New York: Columbia University Press, 1998.
Rothschild, Mary Aickin. *A Case of Black and White: Northern Volunteers and the Southern Freedom Summers, 1964–1965.* Westport, Conn.: Greenwood Press, 1982.
Rubin, Louis D., Jr. *The Wary Fugitives: Four Poets and the South.* Baton Rouge: Louisiana State University Press, 1977.
Sale, Kirkpatrick. *SDS.* New York: Random House, 1973.
Schrager, Samuel. "What Is Social in Oral History?" *International Journal of Oral History* 4, no. 2 (1983): 76–98.
Severson, Lisa Anne. "A Genteel Revolution: The Birth of Black Studies at the University of Virginia." Unpublished Seminar Paper: University of Virginia, 1995.
Simon, Bryant. "Southern Student Organizing Committee: A New Rebel Yell in Dixie," Honors Essay: University of North Carolina, 1983.
Sitkoff, Harvard. *The Struggle for Black Equality, 1954–1980.* New York: Hill and Wang, 1981.
Snider, William D. *Light on the Hill: A History of the University of North Carolina at Chapel Hill.* Chapel Hill: University of North Carolina Press, 1992.
Sosna, Morton. *In Search of the Silent South: Southern Liberals and the Race Issue.* New York: Columbia University Press, 1977.
Stanton, Mary. *Freedom Walk: Mississippi or Bust.* Jackson: University Press of Mississippi, 2003.
Stern, Sol. "A Short Account of International Student Politics & the Cold War with Particular Reference to the NSA, CIA, Etc." *Ramparts* 5, no. 9 (March 1967): 29–38.
Sullivan, Patricia. *Days of Hope: Race and Democracy in the New Deal Era.* Chapel Hill: University of North Carolina Press, 1996.
Sutherland, Elizabeth. *Letters from Mississippi.* New York: McGraw-Hill, 1965.
Thompson, Paul. *The Voice of the Past.* New York: Oxford University Press, 1978.
Thompson, Ruth Anne. "'A Taste of Student Power': Protest at the University of Tennessee, 1964–1970." *Tennessee Historical Quarterly* 57, no. 1 (1998): 80–97.

Trillin, Calvin. *An Education in Georgia: The Integration of Charlayne Hunter and Hamilton Holmes*. New York: Viking Press, 1963.

Turner, Jeffrey A. "From the Sit-Ins to Vietnam: The Evolution of Student Activism on Southern College Campuses, 1960–1970." *History of Higher Education Annual* 21 (2001): 103–135.

Twelve Southerners. *I'll Take My Stand: The South and the Agrarian Tradition*. New York: Harper and Brothers, 1930.

United States Bureau of the Census. *Census of Population: 1960*, volume 1 (Characteristics of the Population), part 26: Mississippi. Washington, D.C.: U.S. Government Printing Office, 1963.

United States Census Bureau. *Statistical Abstract of the United States, 1962*, 83rd annual edition. Washington, D.C.: U.S. Government Printing Office, 1962.

———. *Statistical Abstract of the United States, 1971*, 92nd annual edition. Washington, D.C.: U.S. Government Printing Office, 1971.

Van Deburg, William L. *New Day in Babylon: The Black Power Movement and American Culture, 1965–1975*. Chicago, Ill.: University of Chicago Press, 1992.

Vanauken, Sheldon. *Under the Mercy*. Nashville, Tenn.: Thomas Nelson Publishers, 1985.

Wells, Tom. *The War Within: America's Battle over Vietnam*. Berkeley: University of California Press, 1994.

Wilson, Charles Reagan. "Creating New Symbols from Old Conflicts." *Reckon: The Magazine of Southern Culture* 1, nos 1 & 2 (1995): 17, 20.

Wooden, Kenneth. *The Children of Jonestown*. New York: McGraw-Hill, 1981.

Woodward, C. Vann. *The Strange Career of Jim Crow*. New York: Oxford University Press, 1955.

———. "The Search for Southern Identity." In C. Vann Woodward, *The Burden of Southern History* (Baton Rouge: Louisiana State University Press, 1960), 3–25 (originally published in *The Virginia Quarterly Review* XXXII (1956): 258–267).

Young, Marilyn B. *The Vietnam Wars, 1945–1990*. New York: Harper Collins, 1991.

Young, Stephen Flinn. "The Kudzu: Sixties Generational Revolt—Even in Mississippi." *Southern Quarterly* 34, no. 3 (1996): 122–136.

Young, Thomas Daniel. *Waking Their Neighbors: The Nashville Agrarians Rediscovered*. Athens: University of Georgia Press, 1982.

Yow, Valerie Raleigh. *Recording Oral History: A Practical Guide for Social Scientists*. Thousand Oaks, Calif.: Sage Publications, 1994.

Zaroulis, Nancy and Gerald Sullivan. *Who Spoke Up?: American Protest against the War in Vietnam, 1963–1975*. New York: Holt, Rinehart and Winston, 1984.

Zinn, Howard. *SNCC: The New Abolitionists*. Boston, Mass.: Beacon Press, 1964.

Index

Abernathy, Ralph, 16
Acuff, Roy, 115
AFL-CIO, 76, 154, 162
Agee, James, 191
Allen, Archie, 27, 33, 41, 58, 74, 81–82, 84, 92, 110
   SSOC and Bradens, 78–79
   SSOC's first campus traveler, 75, 90, 100
   SSOC fundraising, 65, 76–77
Altizer, Thomas J. J., 104
Anderson, Terry, 35
Anti-Defamation League, 65, 79
anti-Vietnam War movement, 103–06, 143–44, 180
   See also Southern Student Organizing Committee, Vietnam War
Appalachian State Teacher's College, 151
Aptheker, Herbert, 192
Archibald, Cathy, 199
Armour, Don, 184
Atlanta Project, 125
Auburn Freedom League, 93
Austel, Sam, 177

Baez, Joan, 115, 117, 181
Bailey, Bob, 69
Bains, Jim, 215
Baker, Ella, 15, 30, 48
Barbee, William, 92
Barnett, Gov. Ross, 93
Barrett, Kathy, 75, 170
Barry, Marion, 15, 24, 36–38, 49
   leader of Students for Equal Treatment, University of Tennessee, 34, 55
Belmont Abbey College, 152
Benninger, Chris, 97
Benninger, Judy, 95, 174
   See also Judith Brown

Bevel, James, 25
Black Panther Party, 180
Black Power, 48–49, 111, 119–20, 132, 184, 204
Blackstock, Nelson, 34, 37–38, 57–58, 66, 73, 82, 84, 94, 161
Bode, Gerry, 27
Bogle, Joe, 213
Boner, Don, 126
Bowden, Leo, 101
Boyte, Jr., Harry, 34, 156
Braden, Anne, 29–30, 33, 36–38, 44–45, 67, 77–81, 84–85, 189, 192, 246n
   editor of Southern *Patriot*, 171
   role model for SSOC women, 171
   on white organizing, 47–49, 73–74, 123–26
   White Southern Student Project, 14–16
Braden, Carl, 14, 29, 34, 38, 67, 77–81, 108, 214, 241n
Bradley, Scott, 186
Brandon, Peter, 154
Branscomb, Harvey, 26
Brisbane, Robert, 104
Brown, Benjamin, 137
Brown, Judith, 167, 174–75
   See also Judy Benninger
Brown, Pearly, 118
Browning, Joan, 59
Burlage, Dorothy Dawson, 18, 29, 40–41, 51, 60, 64, 170–71, 247n
Burlage, Robb, 18, 29–30, 36, 51, 64, 82, 122, 198, 235n
   "We'll Take Our Stand," 40–43
Bursey, Brett, 196–97

Cade, Cathy, 34, 38, 46, 56, 170
Campbell, Rev. Will D., 3, 107, 123
Campus Grill demonstrations, 27–30, 58, 92

Carawan, Guy, 117
Carmichael, Stokey, 48, 98, 120, 139
Carpenter, Cassell, 136, 140
Carr, Lester, 30
Carson, Clayborne, 6
Carter Family, 115
Carter, Herman, 98, 114
Carter, Hodding, 69
Cash, Johnny, 115
Cason, Sandra, 15, 236n
  See also Casey Hayden
Cassel, Mike, 136
Caudill, Henry, 191
Chan, Lorine, 27–28
Chandler, Len, 118
Chatagnier, Louis, 104
Chicago Eight, 184
Chittick, Sandy, 89–90
Civil Rights Act (1964), 22, 82, 86, 91, 100, 111
civil rights movement, 6–7, 11–13, 20–22, 25–26, 29–30, 65–66
Clark, Ed, 203, 207–08
Cobb, Alice, 27
Coffin, Stephanie, 141
Coffin, Tom, 141
Coleman, Les, 199
Collingsworth, Arthur, 104
Committee of Southern Churchmen, 123
Communist Party (CP), 44, 140, 201, 207, 213–16, 218
communists, 14, 78–80, 109
Compton, Steve, 185, 220
Connelly, Thomas, 115
"Conversation Vietnam," 106, 127, 136
Cooke, Anne, 54, 56–58, 100–02
  See also Anne Cooke Romaine
Coughlin, Richard, 105
Council of Federated Organizations (COFO), 66–67, 69, 72–73, 86
  See also Freedom Vote
Cox, Courtland, 48
Creel, Austin, 95
Cross, Lucien, 182
Curry, Connie, 33–34, 37, 83, 170, 269n

Davidson, Donald, 42, 243n
  See also Nashville Agrarians
Davis, Rennie, 184
Dawson, Dorothy,
  see Dorothy Dawson Burlage
Democratic National Convention 1964, 119
Dewart, Bob, 105, 111, 147, 276n
Dewart, Janet, 144, 170
Doar, John, 86
Doggett, David, 136–37, 140–42, 189, 272–73n
  SSOC dissolution, 204–05, 207, 213, 221
Dohrn, Bernadine, 185, 200
Dombrowski, James, 15, 81
Donaldson, Ivanhoe, 83, 111, 139
Dorsey, John T., 104
Dubois Club, 160, 214
Dunbar, Leslie, 77, 98, 122
  See also Southern Regional Council
Durr, Clifford, 16
Durr, Virginia, 16
Dylan, Bob, 115

Eastland, Sen. James, 111
Echols, Alice, 6, 173, 283n
Eisenscher, Mike, 215, 217, 219
Evans, Sara, 7
Evers, Medgar, 22
Ezra, Rosemary, 34

Farmer, James, 100
Featherstone, Ralph, 139
Fisher, Bob, 101
Folsom, Gov. "Big Jim", 116, 266n
Forman, Jim, 16, 25, 34, 49, 73, 82
Foster, Rives, 138
Free Speech Movement (1964), 96, 98, 100
Freedom Party, University of Florida, 96–97
Freedom Rides, 59
Freedom Summer, 49, 56–57, 60, 64–65, 69–70, 74–75, 83, 86, 100, 117, 119, 135, 139
Freedom Vote, 66
Fulbright, J. William, 143, 152
Furman University, 184–86

Gainey, Jerry, 37
Gainesville Committee to End the War in Viet Nam, 104
Gainesville NAACP's Youth Council, 94
Galbraith, John Kenneth, 40
Gamecock, 23
Gardner, Tom, 105, 107, 120–21, 137–40, 164, 173, 187, 203, 275n, 276n
  anti-Vietnam War activities, 145–52, 160–61, 193–95
  SLAM, 163–65
  southern distinctiveness, 191, 193–95, 197–98
  SSOC dissolution, 199–200, 208–10, 213, 221, 223–55
Gaston, Paul, 44, 54, 99–100, 105, 261n
Gentry, Lavon, 147
Georgia Students for Human Rights, 22, 94
Gitlin, Nanci, 104
Gitlin, Todd, 30, 51–52, 75, 239n, 248n
Golden, Harry, 58
Goldfield, Evelyn, 174
Goldwater, Barry, 59
Goodman, Bob, 224–25
Goodwyn, Lawrence, 82
Graham, Sen. Frank Porter, 54
Gray, Victoria, 73
Great Speckled Bird, 141–42, 223, 225
Greene, Christina, 7, 243n, 243–44n
Greensboro sit-ins, 25–26
Gregory, Dick, 183
Grogan, Nan, 39, 58–60, 103, 127
  labor organizing, 156–58
  See also Nan Grogan Orrock
Gwin, Jim, 205
Guerrero, Gene, 43–44, 50, 54–56, 60, 80, 84–85, 98–99, 104, 112, 115, 131, 134, 141, 146, 224
  draft resistance, 148
  labor organizing, 154–59
  southern distinctiveness, 194
  SSOC dissolution, 218

SSOC fundraising, 75–77
White Folks Project, 66, 70, 72
Guthrie, Woody, 21

Hamer, Fannie Lou, 170
Hamlett, Ed, 1, 23–24, 53–55, 57–58, 60–61, 83, 86, 110, 115, 119, 121, 131–35, 161
  southern distinctiveness, 84, 191, 193, 195
  SSOC dissolution, 189–90, 215, 217–18, 221, 224
  SSOC founding, 33, 36, 38–39
  SSOC-SDS relations, 52, 199
  White Folks Project, 65–66, 68–69, 72, 74
  White Southern Student Project, 23–24, 46, 48
Hammond, Thomas, 99, 104–05
Harmeling, Dan, 34, 38–39, 84, 95
  University of Florida activism, 96–97
Harmeling, Jim, 95, 97
Harris, Don, 276n
Harris, Wesley, 99
"Harvest of Shame," 163
Hayden, Casey, 170–71
  See also Sandra Cason
Hayden, Tom, 51–52, 113, 224
Heggen, Brian, 134, 144, 147
Heggen, Maggie, 175
Hemphill, John, 135
Henderson, Marjorie, 34
Hershey, Lewis, 113
Hickey, Roger, 100
Highlander Folk School, 16, 30, 67, 72, 116–17, 192
Hill, John, 126
Hillary, Mabel, 118
Hirschkop, Phillip, 76
Hodes, Nancy, 149–50, 152
Horne, Richard, 157
Horton, Myles, 30, 72, 82, 122, 192
House Un-American Activities Committee (HUAC), 14, 78
Huff, A. V., 185
Humphrey, Hubert, 180
Hurder, Alex, 208–09

In loco parentis, 1, 38, 96, 160, 176, 178, 182, 210
  See also university reform
  See also women's liberation movement
  See also Southern Student Organizing Committee, university reform
  See also Southern Student Organizing Committee, women's liberation movement
International Ladies' Garment Workers Union (ILGWU), 161

Jefferson, Thomas, 44
Johnson, Ann, 175, 218, 295n
Johnson, Lyndon, 77, 108, 144–45, 218
Johnston, Jr., Earle, 136
Joint University Council on Human Relations (JUC), 22, 92–93
Jolly, Ralph, 23
Jones, Beverly, 95, 174–75
  See also Southern Student Organizing Committee, women's liberation movement
Jones, Marshall, 89, 98, 104
  advent of SGER, 95
Jones, Thomas F., 183
Jordan, Clarence, 192
Joye, Harlon, 7, 162–63

Kahn, Tom, 78
Kamerman, Sidney, 101
Kennedy, John F., 22, 28, 109, 111, 119, 135–36
Kennedy, Robert, 180
Key, V. O., 44, 191
King, Rev. Ed, 82, 86, 269n
King, Larry, 149
King, Mary, 171, 236n
King, Jr., Rev. Martin Luther, 16, 27, 92, 100, 180–81, 260n
Kissinger, Clark, 87–88
Klonsky, Michael, 201, 205–10, 212, 214–15, 221
Knight, Douglas, 181
Koedt, Anne, 173

Kotelchuck, David, 76, 78–79, 92, 104, 134
  Campus Grill demonstrations, 28–29
  Nashville Movement, 26–27
  SSOC founding, 41
Kotelchuck, Ronda, 3, 169
  See also Ronda Stilley
Kudzu, 142
Ku Klux Klan, 16, 86, 159

Lacey, Fred, 203, 207
Lavery, Bill, 185
Lawson, Rev. James, 24–25, 57, 91, 238n
  See also Nashville Movement
Leary, Bill, 100
Leary, Timothy, 183
Levin, Alan, 119, 143, 146–48, 161–65, 182
Lewis, John, 21, 25, 34, 43, 48–49, 58, 61
Lewis, Nancy, 172
Lingo, Al, 11–12
Link, William, 6
Long, Everett, 136, 142, 217
Looby, Z. Alexander, 25
Lowenstein, Allard, 66
Lozoff, Bo, 162–65
Lozoff, Mike, 162, 163–65
Lynd, Staughton, 104

Mackay, James, 104
Makemson, Lee, 137
March Against Fear, 136–37
  See also James Meredith
March on Washington, 13, 22, 58–59, 112, 139
Marshall, F. Ray, 191
Martin, Tommy, 162–63
Maryville College, 89–90
Maxwell, Bruce, 67, 71, 74
McDermott, John, 104
McDew, Charles, 21
McKinney, Lester, 25, 61
McKinney, Ronald, 185
McNamara, Robert, 145
Meiselman, Michael, 147, 150
Meisner, Maurice, 105

Index

Mencken, H. L., 43
Meredith, James, 57, 136
Miller, Dorothy, 170
MFDP
   See Mississippi Freedom Democratic Party
Millican, Tom, 93, 109, 132
Millspaugh, Frank, 34, 37
Mississippi Christmas Project, 86, 88, 101, 107
   See also Southern Student Organizing Committee, biracialism
Mississippi Council on Human Relations, 69
Mississippi Freedom Democratic Party (MFDP), 69, 72–73, 82–83, 119
Mitchell, H. L., 192
Mizell, Hayes, 79, 85, 87, 122
Mohr, Clarence, 170
Money, Roy, 98
Monroe, Bill, 115
Montgomery, Ben, 102
Moore, William, 12, 21
Morris, Willie, 18
Moses, Bob, 30, 46, 66–67, 69, 116, 153, 245–46n
Movement for Co-Ed Equality, University of Georgia, 178
   See also women's liberation movement
Mt. Beulah Conference Center, 214
   See also SSOC dissolution
Mullins, Phil, 148
Muzorewa, Abel, 27
Myers, Chuck, 28

NAACP
   See National Association for the Advancement of Colored People
Nash, Diane, 25
Nashville Agrarians, 42–43, 190
Nashville Christian Leadership Conference, 27
Nashville Committee for Alternatives to the War in Vietnam, 104
Nashville Leadership Council, 92
Nashville Movement, 25–27, 238n

National Association for the Advancement of Colored People (NAACP), 22, 55, 57, 94–95, 101, 135–37
National Council of Churches, 67, 123
National Guardian, 80
National Mobilization Against the War protests, 145
National Student Association (NSA), 76, 79, 91, 209
   Southern Project, 33, 122, 148
New Left, 4, 6
New Left Notes,
   see Students for A Democratic Society
Newton, Shirley, 135, 140, 144, 170
New York Radical Feminists, 173
Nixon, Richard, 194–95
Nolan, David, 100, 102–3, 105, 121, 137–38, 140–41, 160–61, 164, 172, 210
   anti-Vietnam War activities, 148–52
   SSOC dissolution, 222–23
Norman Fund, 76–77, 116
North Nashville Project, 114, 125–26, 153

Ochs, Phil, 115
Odum, Howard, 40
Ogden, Allan, 101
Orrock, Nan Grogan, 4
   See also Nan Grogan
Owen, Rebecca, 15

Palmour, Jody, 104, 127, 136, 140, 162–63, 165, 194
Parker, Ron, 33, 50, 74, 85, 92
   Campus Grill demonstrations, 28–29
   SSOC and Bradens, 78–80
   SSOC fundraising, 65, 76–77
Patterson, Eugene, 104
Payne, Ray, 114
Peabody College, 13, 22, 28, 33, 91–92
Peace Tour, 143, 148–53, 160, 164, 276n

Pollock, Norman, 192
Populists, 21, 44, 154, 175, 195
Pride, Charley, 116
Progressive Labor Party (PL), 160, 201–02, 291–92n, 292n
 "Build SDS in the South," 203–04
 role in SSOC dissolution, 203–08, 214, 218

Quebec-to-Guantanamo Peace March, 58, 93, 249n

Ransom, John Crowe, 42, 243n
 See also Nashville Agrarians
Reagon, Bernice, 117–18
Reavis, Dick, 207
Reed, John Shelton, 6
Revolutionary Youth Movement (RYM),
 See Students for a Democratic Society, Revolutionary Youth Movement (RYM)
Richards, Dona, 83
Richardson, Bob, 34
Richer, Ed, 95, 98
Ricks, Willie, 122
Robinson, Ruby Doris Smith, 170
Rollins, Rev. J. Metz, 63
Romaine, Anne Cooke, 134, 170, 226
 Southern Folk Festival, 117–18, 267–68n
 See also Anne Cooke
Romaine, Howard, 57, 60, 98, 100–02, 105, 109, 114, 117, 120, 134, 141
 SSOC lack of analysis, 223–25
 SSOC reorganization, 127–28
 On Vietnam War, 112–13
Roosevelt, Franklin, 41, 227, 242n
Rosenbaum, Walter T., 17
Rudd, Mark, 214–15, 218, 221, 295n
Russell, Sen. Richard, 111, 113
Rusk, Dean, 113, 180
Rustin, Bayard, 24, 77

Sale, Kirkpatrick, 6
Sanford, Phil, 206
Savio, Mario, 100

Scarritt College for Christian Workers, 13, 22, 27–28, 33, 36, 58, 61, 91–92
SCEF,
 See Southern Conference Educational Fund
Schiffer, Judy, 67
Schrader, Emmie, 83
Schunoir, Ann, 156–57
Schunoir, Chuck, 156–57
SCHW,
 See Southern Conference for Human Welfare
Schwerner, Mickey, 86
Schwerner, Rita, 86
SDS,
 See Students for a Democratic Society
Seeger, Pete, 115
SGER,
 See Student Group for Equal Rights
Shannon, Randy, 151, 212, 215
Shero, Jeff, 202
Shirah, Sam, 11, 29–31, 56, 236n, 237–38n
 and SLAM, 161–65
 SSOC founding, 33, 36–37
 White Folks Project, 65–66, 68, 71–72, 74
 White Southern Student Project, 12–13, 20–24, 48
Simpson, David, 178–79
 SSOC dissolution, 205, 207–08, 210–12, 219
Sixteenth Street Baptist Church, 23, 58, 139
Skillman, Jim, 210
SLAM
 See Southern Action Labor Movement
SLATE, University of California at Berkeley, 234n
Smith, Barbara, 18
Smith, Bruce, 55, 59, 141
 Peace Tour, 151–52
 SSOC dissolution, 211, 221
 SSOC founding, 34, 38
 on Vietnam War, 110–11
Smith, Charles, 52, 67, 69, 71

Smith, Lillian, 269n, 283n
Smith, Ray, 152
Smith, Willis, 54
SNCC,
    See Student Nonviolent
        Coordinating Committee
Sokalof, Marilyn, 95–96
Sorenson, Soren, 66, 71
Southern Christian Leadership
    Conference, 15, 98, 214
Southern Conference Educational
    Fund (SCEF), 34, 65, 149, 214
    Southern Patriot, 14, 16, 33, 171
    White Southern Student Project,
        14–15, 20, 22
    See also Southern Student
        Organizing Committee,
        relations with SCEF
Southern Conference for Human
    Welfare (SCHW), 14, 19, 192
Southern distinctiveness
    See Southern Student Organizing
        Committee, southern
        distinctiveness
Southern Folk Festival, 114–19, 134
Southern Labor Action Movement
    (SLAM), 160–66
Southern Regional Council (SRC), 77,
    79, 98, 122
Southern Student Organizing
    Committee (SSOC), 1–9, 13
    biracialism, 39, 82–85, 87, 120–21,
        123–25
    campus travelers, 74–75, 171, 197
    Confederate battleflag and SSOC
        symbol, 50, 63, 83–85, 192,
        196–97: *See also* Southern
        Student Organizing
        Committee, southern
        distinctiveness
    Confederate rhetoric, 193–98: *See
        also* Southern Student
        Organizing Committee,
        southern distinctiveness
    dissolution, 202–26, 295n
    factionalism, 189–90, 212–14,
        220–22, 225
    finances and fundraising, 65, 76–77,
        209–10, 222
    founding, 13, 30, 33–43, 50, 61
    founding generation, 27, 30, 33–34,
        36, 52–62, 134, 161, 169
    joint SSOC-SDS chapters, 180,
        199, 291n
    labor organizing: Cone Mills,
        153–60; National Spinning,
        153–60; support of, 132,
        153–54, 168, 180–81; *See also*
        Southern Labor Action
        Movement (SLAM)
    Mt. Beulah Conference, 214–19
    New Rebel, 50, 64–65, 82–83, 85
    New South Student, 125, 144, 147,
        160, 164, 191
    Peace Tour, 143, 148–53, 160, 164
    relations with SCEF, 76–81
    relations with SDS, 50–53, 131,
        202–14, 239n
    relations with SLAM, 163–66
    relations with SNCC, 45–49, 66,
        120–21
    reorganizing, 126–29, 187
    "Resolution: To Dissolve SSOC,"
        216–19
    second generation, 133–43
    and southern history, 43–44,
        191–92: *See also* Southern
        Student Organizing
        Committee, southern
        distinctiveness
    Southwide conferences, 75, 81–85,
        98, 160–65, 199
    southern distinctiveness, 2, 4,
        43–44, 50–51, 115, 190–91,
        197–98, 200, 203–05, 207,
        209–12, 222
    university reform, 98, 181–87,
        211, 217: *See also* university
        reform
    Vietnam War: antiwar activities,
        113–14, 143–52, 193–96;
        opinions of, 108–13, 128, 132;
        See also Vietnam War; *See also*
        anti-Vietnam War movement
    "We'll Take Our Stand," 39–43,
        190, 227–28
    white community organizing,
        71–72, 119–21, 125–26

Southern Student Organizing Committee – *continued*
  women's liberation movement, 168–79
  See also women's liberation movement
Southern Student Organizing Fund, 34
Spencer, Howard, 98
SRC,
  see Southern Regional Council
SSOC,
  see Southern Student Organizing Committee
Stafford, Billie, 37
Stampp, Kenneth, 191
Stembridge, Jane, 15, 30, 119
Stilley, Ronda, 126, 134, 144
  See also Ronda Kotelchuck
Strickland, John A., 66
Stronach, Carey, 102
Student Group for Equal Rights (SGER), University of Florida, 22, 95–96
Student Nonviolent Coordinating Committee (SNCC), 2–3, 6, 26–30, 33–34, 37, 39, 65, 83, 87, 92, 102, 114, 128, 139, 212–13
  Black Power, 48–49, 132
  Freedom Singers, 116–17
  White Southern Student Project, 13–17, 19–25, 235n
  women's liberation movement, 169–71
Student Peace Union, 235n
Student SANE, Harvard University, 235n
Students for a Democratic Society (SDS), 2–6, 31, 53, 85, 87, 128, 131–32, 183, 240–41n
  dissolution, 226
  factional struggle, 200–03
  interest in South, 51–52, 198–200
  joint SSOC-SDS chapters: *See* Southern Student Organizing Committee, joint SSOC-SDS chapters
  March on Washington to End War in Vietnam (first major antiwar demonstration), 112
  New Left Notes, 147, 165, 201
  Port Huron Statement, 40–41, 242n
  Revolutionary Youth Movement (RYM), 200–03, 205–08, 210, 212, 214–15, 225
  role in SSOC dissolution, 189–90, 198–200, 203–08, 210–12, 214–19, 221–26
  Selective Service College Qualification Test protest, 113–14
  SSOC, fraternal ties to, 50–53, 75, 239n
  women's liberation movement, 169, 173–74
Students for Equal Treatment, University of Tennessee, 34, 55
  See also Marion Barry
Students for Integration, Tulane University, 34
Students for Social Action (SSA), University of Virginia, 22, 101, 104–05, 111, 137, 270n
Sullivan, Jack, 167, 185, 197, 220
Sullivan v. New York Times libel case, 16

Tate, Allen, 42, 243n
  See also Nashville Agrarians
Tate, William, 94
Ten Days of Resistance, 193–94
Textile Workers Union of America (TWUA), 155–56, 159, 278n
Tiberiis, Doug, 67
Tieger, Joseph "Buddy," 148
Tillow, Walter, 29, 33
Thomas, Norman, 77, 100, 104, 116, 137
Thrasher, Sue, 1, 45, 52, 54, 58–59, 61, 63–64, 76, 84–85, 88, 92, 116, 125, 127, 133–34, 139
  Campus Grill demonstrations, 27–29
  SSOC and SCEF, 78, 80–81
  SSOC dissolution, 218, 222–23, 225
  SSOC founding, 33, 36–37, 39, 41
  SSOC and SNCC, 48–50

on Vietnam, 106–07, 110, 112
White Folks Project, 66, 69–70, 73–74
women's liberation, 169–73
Towe, Bill, 110
Towe, Betsy Jean, 110
True, Jim, 139
Turner, Gil, 117–18
Turnipseed, Marti, 34

United Farmworkers Organizing Committee, 180
United Packinghouse Workers of America, 154
university reform, 7, 45, 90, 103
   Duke University, 181–82
   University of Florida, 94, 96–97, 182
   University of Georgia, 178
   University of North Carolina, 177, 287n
   University of Tennessee, 179
   University of Virginia, 101
   See also Southern Student Organizing Committee, university reform

Vanauken, Sheldon, 173, 283–84n
Vanderhaag, Ernest, 104
VCHR,
   see Virginia Council on Human Relations
Vietnam War, 108–09, 143, 180
   See also anti-Vietnam War movement
   See also Southern Student Organizing Committee, Vietnam War
Virginia Council on Human Relations (VCHR), 22, 99–101, 104
Virginia Students' Civil Rights Committee (VSCRC), 102, 137–39, 156
Virginia Summer Project, 101–03, 105
Vlasits, George, 148, 210–12
VOICE, University of Michigan, 235n
Voting Rights 1965, 111
VSCRC
   see Virginia Students' Civil Rights Committee

Walden, Eleanor, 118
Wallace, Gov. George, 12, 195
Ware, George, 184
Warren, Robert Penn, 42, 243n
   See also Nashville Agrarians
Watson, Tom, 44
Weaver, Claude, 50
Weissman, Steve, 98
Welch, Mike, 135, 140, 191:
   SSOC dissolution, 212–13, 215, 219–21
Wells; Lyn, 111, 131, 138–39, 145, 152, 170, 193, 195
   "American Women: Their Use and Abuse," 174–75
   labor organizing, support of, 156–60
   SSOC dissolution, 205–08, 210–13, 215, 221–22, 225
   women's liberation, 170–71, 174–75
West, Ben, 25
West, Don, 82, 98, 118, 192, 196
West, Heddy, 118, 198
White Folks Projects, 65–74, 122, 125, 133, 250–51n, 253n
White Southern Student Project
   See Student Nonviolent Coordinating Committee, White Southern Student Project
Whitman, Grenville, 66
Wiley, Bell, 44
William Moore's Freedom Walk, 12, 21
Williams, Jim, 37, 50, 224
Williams, Robert, 73
Wilson, Earl, 127, 138
Wise, Stanley, 48, 122
Wise, Steve, 54, 121, 144, 159, 196–97, 200
Women's International Conspiracy from Hell (WITCH), 176
women's liberation movement, 93, 96–99, 177–79, 184, 282n
   See also Southern Student Organizing Committee, women's liberation movement
   See also university reform

Woodward, C. Vann, 44, 191
Wooten, Greg, 185

Young, Andrew, 98
Young Americans for Freedom (YAF), 118, 197
Young Democrats, 18, 93, 180, 200
Young Republicans, 18

Young Socialists Alliance, 160
YWCA race relations committee, 59

Zellner, Bob, 11, 21–22, 24, 171, 235n
 White Southern Student Project, 15–17, 19–20
Zinn, Howard, 1, 6